Complete Roofing Handbook

Installation
Maintenance
Repair

by JAMES E. BRUMBAUGH

Macmillan Publishing Company
New York

Collier Macmillan Publishers
London

By the same author:

Heating, Ventilating and Air Conditioning Library (3 volumes)
Truck Guide (3 volumes)
Upholstering
Welders' Guide
Wood Furniture

Copyright © 1986 by James E. Brumbaugh

Macmillan Publishing Company
866 Third Avenue, New York, N.Y. 10022
Collier Macmillan Canada, Inc.

Library of Congress Cataloging-in-Publication Data

Brumbaugh, James E.
  Complete roofing handbook.

  "An Audel® book."
  Includes index.
  1. Roofing. 2. Roofs—Maintenance and repair.
  I. Title.
  TH2431.B78   1986   695   86-2952
  ISBN 0-02-517850-4

10 9 8 7 6 5 4 3 2

Printed in the United States of America

While every precaution has been taken in the preparation of this book, the Publisher assumes no responsibility for errors or omissions. Neither is any liability assumed for damages resulting from the use of the information contained herein.

Contents

CHAPTER I

Types of Roofs

• **Roof Design**   • **Roof Construction**

Roofs are constructed in a wide variety of different sizes, shapes, and designs, and there are a number of different ways of classifying them. The three most common methods of classification are based on roof design, roof construction, and the type of roofing material used to cover them. Roofing materials are described in the next chapter.

Roof Design

The easiest way to classify a roof is by its architectural design. The principal types of roof designs are illustrated in figs. 1-1 through 1-3. A *gable roof* is composed of two sloping surfaces that meet and join together along a common ridge line (fig. 1-1A). The triangular end wall formed at each end of the roof is called a gable. The degree of slope or pitch will vary on different roofs.

A roof that slopes away in *four* directions from a common ridge line is called a *hip roof* (fig. 1-1B). The so-called hip is found where a gable would be located on a gable roof. Sometimes a hip roof is used to cover a wing that joins the main portion of the structure covered by a gable roof. This type of roof is frequently called a *gable-and-hip roof*.

Some roofs are designed with a change of slope or double pitch. The *gambrel roof* roughly resembles the gable roof except that the gable formed at each end has a pentagonal shape (fig. 1-1C). The gambrel roof can be modified so that the lower slope or curb occurs on

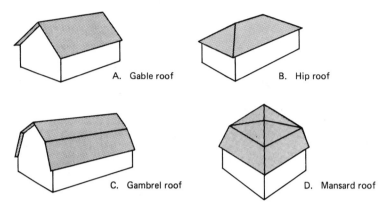

Fig. 1-1. Gable, hip, gambrel, and mansard roofs.

all four sides. A conventional gable is then used above the curb at each end of the roof.

The *mansard roof* design was first introduced in France by the architect François Mansart (1598–1666). It enjoyed initial popularity in the United States in areas of French settlement, particularly in and around New Orleans, but for years it was ignored by American architects and builders. It is now experiencing a revival in popularity, especially in apartment construction.

On a mansard roof, a change of slope or double pitch occurs on all four sides of the roof (fig. 1-1D). The lower slope is longer and steeper than the upper one and has a pitch that is sometimes almost vertical. The sloping surface of a mansard roof is interrupted by a series of evenly spaced dormer windows. A mansard roof differs from a gambrel roof by having a flat top or deck, which is usually covered by built-up roofing materials.

The *French roof* is also designed with a double pitch, and bears a strong resemblance to the mansard roof; however, unlike the mansard roof, the surface is not broken by dormer windows. The two are frequently confused by the layman, and it is common for a French roof to be mistaken for a mansard roof.

In a *pyramidal roof* the four sloping surfaces meet at the same point, thereby eliminating the need for a roof ridge (fig 1-2A). Examples of this type of roof design are found in Victorian architecture where they are used to cover towers. Church steeple roofs provide other examples. The *cone roof* bears a close resemblance to the pyramidal

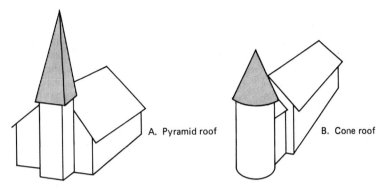

Fig. 1-2. Pyramid and cone roofs.

roof, especially in the ratio of height to base area, but it differs by having no flat surfaces (fig. 1-2B). This roof design is very old, but not uncommon in Europe. It is rarely used in the United States.

The *flat roof* is as common on commercial and industrial structures as the gable and hip roofs are in residential construction (fig. 1-3A). The flat roof is a popular design in some sections of the coun-

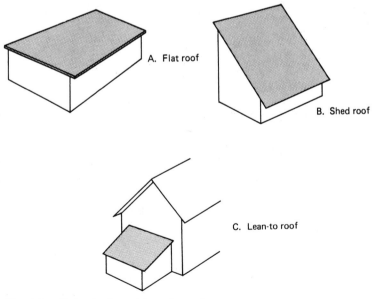

Fig. 1-3. Flat, shed, and lean-to roofs.

try, particularly the Southwest. Every flat roof is built to slope slightly in one direction to provide proper drainage for water. The degree of slope of pitch is almost imperceptible.

A *shed roof* slopes in one direction only (fig. 1-3B). Although it bears some resemblance to a flat roof, its pitch is so much greater that it is regarded as a separate type of roof. This type of roof is generally limited to small frame or industrial buildings, although it is occasionally found on houses and commercial buildings.

A *lean-to roof* is identical to a shed roof except that one end is built against a higher wall of the main structure (fig. 1-3C).

Each of these twelve types of roof designs is subject to some variation. Furthermore, several roofs or roof types will often be used to cover the same structure, particularly when offsets or wings join the main part of the structure. It is not uncommon to find a minor hip roof joined to a major gable roof or vice versa.

The term *major roof* refers to the main roof covering the largest section of the structure. The term *minor roof*, on the other hand, designates any roof that covers a smaller section of the structure and intersects with the major roof to form a roof valley.

Roof Construction

The roof of a structure forms an integral part of an assembly that also includes the ceiling. In some instances, the underside of the roof functions as a ceiling, while the upper side provides a surface for the roofing materials. This type of roofing system is sometimes called an *open timber roof*, because the framing rafters or trusses are exposed to view.

In residential buildings it is common to join the sloping roof rafters with horizontal joists. The underside of these joists support the ceiling, whereas the upper side may be finished to serve as an attic floor.

Several different roof and ceiling assemblies are available. They may be divided into four basic types.

1. Truss roof assembly
2. Wood joist assembly
3. Wood joist and rafter assembly
4. Wood plank and beam assembly

Each of these four types of roof and ceiling assemblies enjoys widespread use in residential construction. Both the truss and wood joist and rafter type assemblies provide a flat ceiling surface for the rooms or spaces below. Because this type of construction seals off an area immediately below the roof, adequate ventilation must be provided to prevent the build-up of condensation and excessive heat in these spaces.

One advantage of the wood-joist and rafter-type construction is that it creates an open area or attic that provides storage space. It can also be finished off as a play area or additional bedroom. This is not possible with truss-type construction because of the bracing between members.

Roof construction details can also be used as a criterion for classifying roofs. For example, different types of roof decks are possible, depending on the choice of materials, and these materials will often determine the type of roofing used to cover the deck. Roofs will also differ according to the type of framing used to support the roof deck.

The flat roof surface on which the roofing materials are laid is called the *roof deck*. The roof deck can be made of either wood or nonwood materials. Most pitched roof decks found on houses are constructed of wood board or plywood panel sheathing.

The roof deck is either flat or slopes downward from both sides of a ridge line. A flat or low-pitched roof is level or almost level, and the frame members supporting the roof deck also serve as ceiling joists in most cases. Flat or low-pitched roofs are said to be of "single roof" construction. A *pitched roof*, which consists of rafters tied together by ceiling joists, has an intermediate or steep slope. The area enclosed by the rafters and joists forms an attic or attic crawl space.

Pitched Roof Framing

The deck of a pitched roof is supported primarily by common rafters and jack rafters (fig. 1-4). The *common rafter* runs square with the wall plate and extends to the *ridge* or *ridge board* The common rafters are the longest and certainly the most numerous rafters on gable, gambrel, and hip roofs.

A *hip rafter* is used on a hip roof, and extends from the outside angle of the wall plate toward the apex of the roof (fig. 1-5). A rafter that runs square with the wall plate and intersects the hip rafter is called a *jack rafter*. Jack rafters never extend the entire distance from

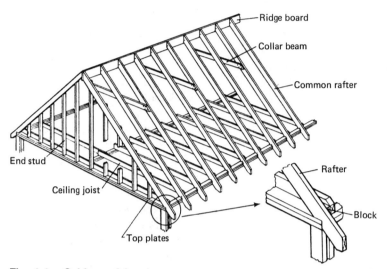

Fig. 1-4. Gable roof framing.

the wall plate to the ridge or ridgeboard. Jack rafters can be divided into hip jacks, valley jacks, and cripple jacks.

A *hip jack* is a jack rafter that runs from the wall plate to a hip rafter. A jack rafter that extends from the ridgeboard to a valley rafter is called a *valley jack*. A *cripple jack* or *cripple rafter* cuts between a valley and a hip rafter.

A structure with a wing added to it will have a roof system that contains a valley where the minor roof (covering the wing) joins the major or main roof. A *valley* is the internal angle formed by the two slopes of the intersecting roofs. A *valley rafter* extends from an inside angle of a wall plate toward the ridge or center line of the structure, and defines the angle of the roof valley (fig. 1-6).

Rafters are nailed to a *wall plate* or *plate*, which generally consists of a 2 × 4 running horizontally across the top of the wall studs. The roof rafters extend a short distance beyond the wall plate on almost all pitched roofs in order to permit water drainage from the roof surface at a suitable distance from the exterior walls and foundation. The portion of the rafter extending beyond the outer edge of the wall plate is called the *eave* or *tail* of the rafter.

A roof rafter is generally cut near its lower end to fit down on

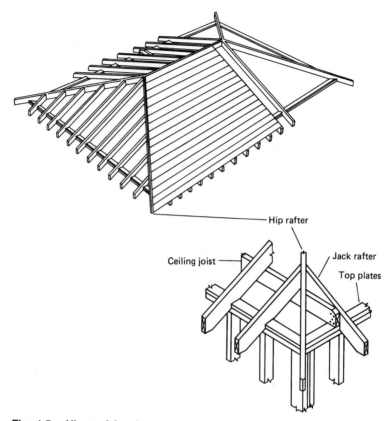

Fig. 1-5. Hip roof framing.

the wall plate. This is referred to as a *seat, bottom,* or *heel cut.* A *ridge, top,* or *plumb cut* is made at the other end of the rafter to allow it to fit against the ridgeboard or, when there is no ridgeboard, against an opposing rafter. Finally, a *side* or *cheek cut* is a bevel cut on the side of a rafter to fit it against another frame member. The various types of rafter cuts are illustrated in figure 1-7.

Some roof framing terms are used specifically for making layout measurements, for example, in determining rafter length or the angle of rafter cut. *Pitch* and *slope* are two of the most commonly used terms belonging to this category of roof-framing terminology. The two terms are often, but incorrectly, used synonymously. Both terms may be used

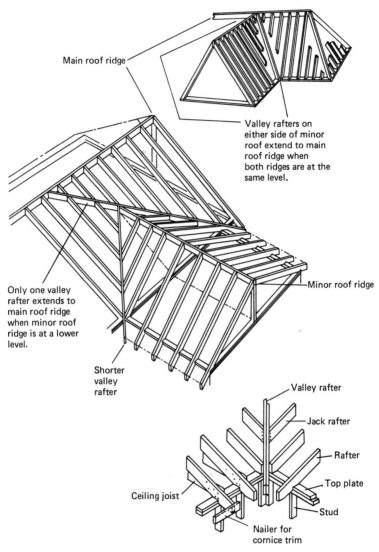

Main roof ridge

Valley rafters on either side of minor roof extend to main roof ridge when both ridges are at the same level.

Only one valley rafter extends to main roof ridge when minor roof ridge is at a lower level.

Minor roof ridge

Shorter valley rafter

Valley rafter

Jack rafter

Rafter

Top plate

Stud

Ceiling joist

Nailer for cornice trim

Fig. 1-6.   Valley framing.

in a general sense to refer to the angle or incline which the roof surface makes with a horizontal plane, but each term more specifically describes a distinctly different mathematical relationship between the total rise and the span of a roof.

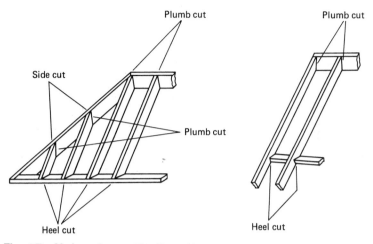

Fig. 1-7. Various types of rafter cuts.

The horizontal (level) distance over which the roof rafter passes is called the *total run* or *horizontal run*. It represents the area covered by the rafter from the outer edge of the wall plate to a vertical line extending down from the exact middle of the ridgeboard, and really has nothing to do with the actual length of the rafter. The total run of a common rafter is one half the width of the structure.

The *span* of the roof is double the total run, and represents the exact width of the structure between the outer edges of opposing wall plates. The *total rise* or *vertical rise*, on the other hand, is the distance extending vertically from the plate line to the bottom of the ridgeboard. The relationships of the terms total run, span, and total rise are il-

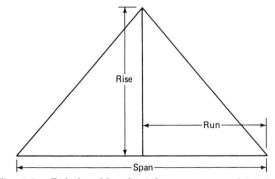

Fig. 1-8. Relationship of total run, span, and total rise.

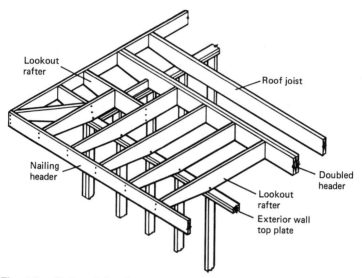

Fig. 1-9. Flat roof framing.

lustrated in figure 1-8. These relationships determine the "pitch" or "slope" of the roof. Both pitch and slope are explained in chapter 2 because they are determining factors in the type of roofing materials used on the roof.

Flat Roof Framing

The flat or low-pitched roof is usually constructed with a slight slope or pitch to improve drainage. The frame members that support the roof deck and also serve as ceiling joists are sometimes called *roof joists*.

Construction details of flat or low-pitched roof framing are illustrated in Fig. 1-9. The look-out rafter provides roof overhang. Each *look-out rafter* is toenailed to the wall plate and nailed at the other end to a double header. The distance from the double header to the wall line is generally twice the overhang measurement. A nailing header for securing the soffit and fascia boards is sometimes nailed to the ends of the look-out rafters.

CHAPTER 2

Roofing and Reroofing

- **Roofing Terminology**
- **Roofing Tools and Equipment**
- **Roofing Materials**
- **Reroofing**

- **Safety Precautions**
- **Estimating Roofing Materials**
- **Hiring a Roofing Contractor**

A properly constructed roof should last from five to over fifty years, depending on weather conditions and the materials used in its construction. As a general rule, the popular asphalt shingle roof has a service life of 15 to 20 years. A roof made of roll roofing lasts about 5 to 10 years. Both slate and tile have a service life of about 50 years or more.

A roof requires a certain amount of annual maintenance and repair. If the roof has deteriorated to a point where simple repairs prove ineffective, it will have to be reroofed or replaced. These two tasks are referred to as *roofing* and *reroofing*.

Roofing is a specialized task in building construction that involves the application of roofing materials to the roof deck. This can occur in new construction or after the existing (but now deteriorated) roofing materials have been stripped from the roof deck. Because of its specialized nature, roofing is usually subcontracted out by the building contractor to specialists in this type of work. As a result, professional roofers and roofing companies can be found in most cities.

Reroofing involves the application of new roofing materials over an existing roof. Depending on the type of roofing material used, reroofing is a task that can often be handled by those without special skills or training. Professional roofers are also available for this type of work.

Roofing Terminology

Certain roofing terms need to be explained at this point because they are frequently encountered in building plans, architectural specifications, and literature from roofing material manufacturers. These terms are also used in the descriptions of the roofing and reroofing methods described in this book.

A knowledge of the pitch or slope of a roof is important in determining the type and application method of the shingle or roofing material that is to be used. Although the terms *pitch* and *slope* both refer to the incline of the roof and are sometimes used synonymously in roofing literature, they are calculated and expressed differently.

Pitch may be defined as the ratio of the rise to total span (or twice total run) and is expressed as a fraction (fig. 2-1). The fraction used to designate pitch becomes larger as the steepness of the roof increases. Thus, a roof with a pitch of ⅓ is steeper than one with a pitch of ⅙.

Roof pitch is calculated by dividing span by rise. For example,

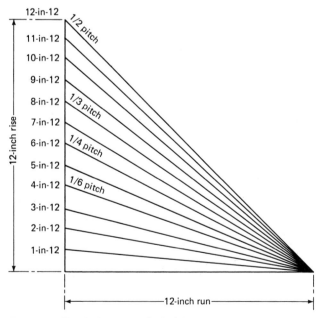

Fig. 2-1. Roof slopes and pitches.

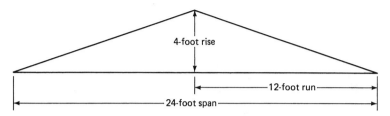

Fig. 2-2. Low-pitched roof.

a roof with a rise of 4 feet and a span of 24 feet has a pitch of $4/24$ or $1/6$ (fig. 2-2). This is a relatively low-pitched roof and limits the roofing material to roll roofing and various types of hot or cold process built-up roofing. Most other types of roofing materials require a roof with a pitch greater than $1/6$. A roof with a relatively steep pitch is shown in figure 2-3. Its rise of 8 feet and its span of 24 feet give it a pitch of $1/3$ ($8/24$ reduced to $1/3$).

In contrast to its pitch, the slope of a roof indicates the ratio of the rise *to the run* (instead of the span) and is expressed as x number of inches per foot of run. For example, the rise of the roof shown in figure 2-3 is 8 feet and the run is 12 feet (one half the span dimension). Slope equals a ratio of 8 feet to 12 feet, or 2 feet of rise for every 3 linear feet. In inches, that equals $24/36$, which reduces to $8/12$. In other words, the roof shown in figure 2-1 rises 8 inches per foot of horizontal run. The slope for this roof is expressed as "8-in-12."

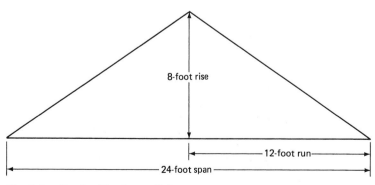

Fig. 2-3. Roof with steep pitch.

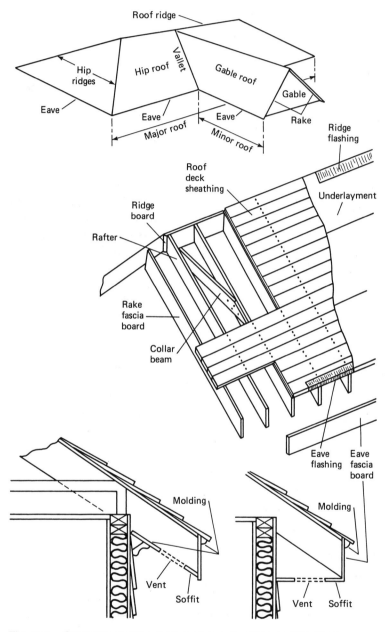

Fig. 2-4. Common roofing and roof construction terminology.

Other commonly used roofing terms are illustrated in figure 2-4. The *ridge* is the peak of the roof or the apex of the angle formed where two roof slopes meet. A *hip ridge*, on the other hand, is the apex of the external angle formed by the meeting of two sloping sides. Both the main roof ridge and hip ridge must be capped with specially cut shingles.

A *gable roof* consists of two sloping roof surfaces joined along a horizontal ridge line. The triangular-shaped endwall formed at each end of the roof is called a *gable*. The roof *rake* is the inclined edge of a gable roof at the endwalls. Each end of a gable roof will have two rakes that incline toward and meet at the ridge. In shingling, it is common to begin at the left rake of the roof and work toward the opposite or right rake.

The roof *eave* is the lower edge of the roof slope that overhangs and extends beyond the face of the exterior wall. Shingle courses are laid to run parallel to the eaves. The ends of the roof rafters at the eaves are covered by a length of trim called the *fascia, fascia board,* or, in this case, an *eave fascia*. The same type of trim attached to a rake is called a *rake fascia*. The underside of the rafters between the eave and the exterior wall may be left open or covered by a *soffit*.

A gable roof rises by inclined planes from two sides of the structure. A *hip roof*, on the other hand, rises by inclined planes from all four sides. As a result, a hip roof has at least four eaves, but no rakes.

The internal angle formed by the junction of two sloping sides of a roof is called a *valley*. Because a roof valley serves as a major channel for water runoff, the valley joint must be made watertight by installing *flashing* material (metal or roll roofing cut to size). Flashing prevents water seepage at roof joints and provides more efficient drainage. Valley and chimney flashing serve both these functions. The drip edge installed along eaves and rakes is an example of flashing used to provide better water runoff.

The *roof deck* is the foundation or nailing base for the roofing materials. Wood or nonwood materials may be used in the construction of a roof deck. A wood roof deck—the type commonly found on houses—consists of wood boards or plywood panels nailed directly to the rafters. These wood boards or plywood panels are called the *sheathing*. The sheathing is covered with an *underlayment* (or *underlay*) of overlapping courses of roofing felt.

Some roofing terms refer to particular aspects of shingling. For example, a shingle *course* is a single row of shingles running the entire length of the roof. Shingling instructions are given in terms of "first course," "second course," "third course," etc.

A *starter course* (or *starter strip*) is a row of roofing material often cut to a smaller size and laid face down along the eaves to provide a solid base for the first course of roofing. The double thickness of roofing material formed by the starter and first courses strengthens the roofing overhang at the eaves and provides greater resistance to strong winds and other severe weather conditions.

Shingle *exposure* is the amount of shingle or roofing material surface exposed to the weather. The amount of exposure is determined by the pitch or slope of the roof. Asphalt shingle roofs with a slope of 4-in-12 or greater will have a 5-inch exposure. When asphalt shingles are used on a roof with a slope of less than 4-in-12, the shingle exposure is only about 3 inches.

Roofing Tools and Equipment

Many of the tools and types of equipment used in roofing and reroofing are illustrated in figures 2-5 to 2-13. One of the most commonly used tools is the 16-ounce claw hammer. In addition to nailing it can also be used for a variety of other purposes ranging from the removal of loose or crooked nails to ripping up old roofing (fig. 2-5). The roofer's or shingler's hatchet, a much more specialized tool, is used primarily for roofing with wood shingles or shakes.

Heavy-duty staplers are frequently used to attach the underlayment courses to the roof deck. They are also used to apply roll roofing and to staple the tabs on certain types of asphalt strip shingles.

Hammers and other tools should be carried in a tool belt secured around the roofer's waist (fig. 2-5). These tool belts are convenient to use, inexpensive, and are available at many hardware stores and local building supply outlets.

Roof and roofing material measurements can be taken with a steel tape or folding rule (fig. 2-6). A chalk line is useful for snapping horizontal and vertical lines across the roof. These chalk lines serve as guides for shingle courses or other roofing materials. The carpenter's level is used to check for horizontally and vertically true surfaces.

Some examples of the different types of saws used in roofing and reroofing are illustrated in figures 2-7 and 2-8. The large carpenter's hand saw is used for heavy-duty work such as cutting plywood or wood

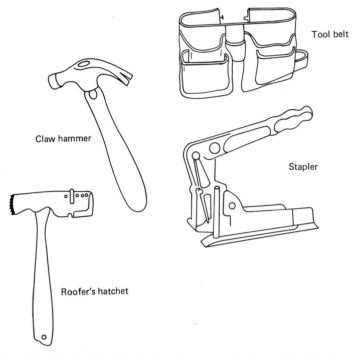

Fig. 2-5. Roofer's hatchet, claw hammer, stapler, and tool belt.

board sheathing. The keyhole saw cuts on the pull-stroke to prevent the buckling of the blade. It is often used to cut small circular holes for vent pipes or stacks because it can enter slots or holes easily. Both the keyhole saw and back saw can be used for making repairs to wood shingle and shake roofs. The hack saw is useful for starting cuts in metal roofing, cutting metal flashing, or cutting off nails when removing roofing. The circular power saw shown in figure 2-8 is recommended for cutting metal roofing panels, concrete and clay tile, and plywood or wood board sheathing.

A roofing knife is useful for cutting shingles, roll roofing, or other asphalt-based roof covering materials (fig. 2-9). A utility knife may also be used for this purpose. Tin snips can be used to cut metal flashing and thinner gauges of metal or vinyl roofing. The shingle ripper, another cutting tool common to roofing, is used to "rip" or cut nails holding down individual shingles, tile, or slate. It is a useful tool for roof repairs.

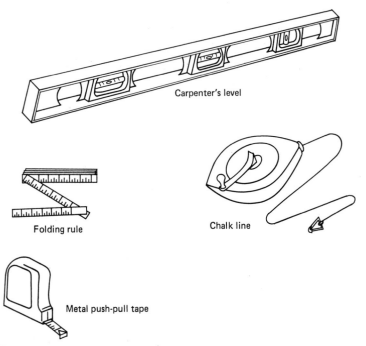

Carpenter's level

Folding rule

Chalk line

Metal push-pull tape

Fig. 2-6. Measuring and marking tools.

A pry bar is essentially a reroofing tool. It is used to remove nails and pry loose shingles, shakes, tiles, or slates when preparing the roof for new roofing materials. A caulking gun is used to apply beads of caulk along flashing joints and other points where different materials meet.

A putty knife can be used for applying mortar or other materials of a consistency too thick and heavy to brush onto the surface. When it is necessary to insert flashing into the wall or chimney joints, the mortar can be removed by a homemade gouging tool.

Roof coatings, asphalt adhesive cements, masonry primers, and other roofing materials are generally of a thin enough consistency to be applied with a brush, mop, broom, or spray gun. Only inexpensive brushes, mops, or brooms should be purchased for this type of work, because they will not be used for anything else and are usually discarded after the roof is completed.

Ladders and scaffolds are the two principal pieces of equipment

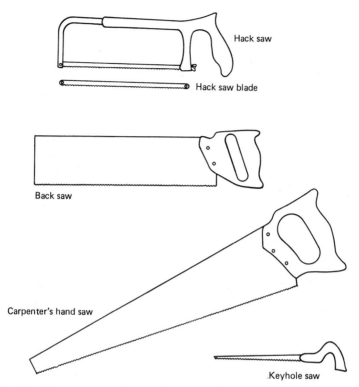

Hack saw

Hack saw blade

Back saw

Carpenter's hand saw

.Keyhole saw

Fig. 2-7. Saws commonly used in roofing.

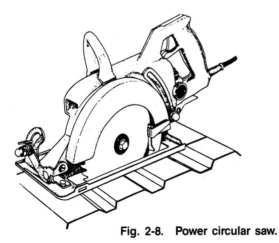

Fig. 2-8. Power circular saw.

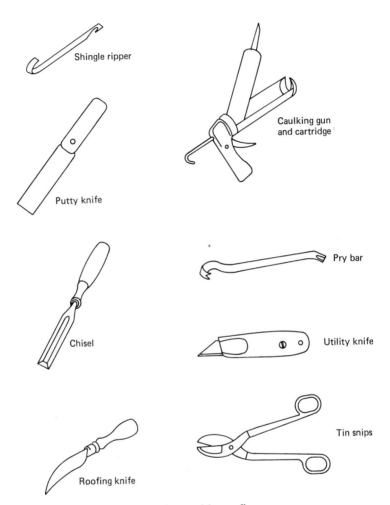

Fig. 2-9. Miscellaneous tools used in roofing.

used in roofing. They can be purchased through local building supply outlets or rented from a tool and equipment rental store.

Ladders must be long enough to reach at least three feet beyond the edge of the roof. The ladder must be placed so that it is on a firm base. Before climbing test the ladder for stability. Some roofers tie the top of the ladder to the structure with a short length of rope as an additional safety precaution.

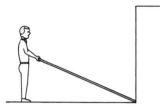

1. Position bottom of ladder against wall base.

2. Push ladder upright.

3. Pull bottom of ladder away from wall.

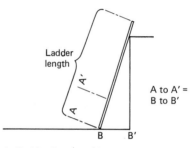

4. Position bottom of ladder about one-fourth its length from wall base.

Fig. 2-10. Positioning a ladder.

A ladder should be raised to the roof by "walking" it (fig. 2-10). Place the bottom end of the ladder against the wall and raise it until it is in a vertical position. Lift the bottom of the ladder and move it away from the wall a distance approximately one fourth its length. The ladder should be positioned so that it is absolutely straight (i.e., not leaning to one side or the other). The bottom of the ladder should be firmly placed so that the bottom of both rails rests solidly and evenly on the ground. If the ladder wobbles, get down immediately and reposition it until it is stable. When raising a ladder, do *not* slam it against a gutter. It is easy to damage a gutter in this way.

Never lean from a ladder, because unevenly placed weight may cause it to fall. Try to keep your hips between the two rails as you climb the ladder. On a pitched roof, a ladder can be used for support by hooking it over the roof ridge (fig. 2-11). Ladder hooks can be homemade or purchased.

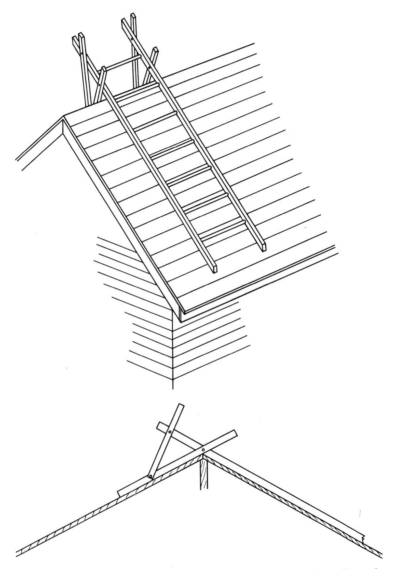

Fig. 2-11. Ladder with braces or supports attached to the rails and set to the angle of the roof.

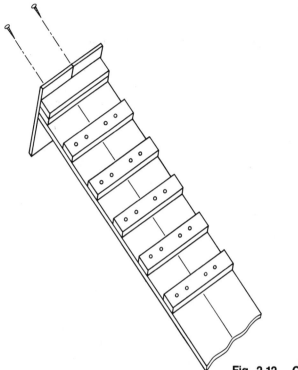

Fig. 2-12. Chicken ladder.

A one-board or so-called chicken ladder (fig. 2-12) is recommended for use on roofs of tile, slate, or other brittle and easily breakable roofing materials, because it spreads your weight more effectively. These are homemade ladders consisting of 1 × 10- or 1 × 12-inch boards to which 1 × 2–inch pieces of wood are nailed. A chicken ladder should also be hooked over the roof ridge.

Metal brackets for supporting a 2 × 4 can be purchased through most local building supply houses. The brackets are nailed to the roof and the 2 × 4 board provides footing on roofs with a steep pitch.

A scaffold can be used not only for getting up and down from a roof, but also for holding the materials nearby while working. This saves time because it is unnecessary to climb down from the roof as often to get more materials. The principal advantage of using a scaffold is that it does not lean against the gutter or eave, and therefore cannot damage these parts of the roof.

A scaffold may be purchased, built, or rented. Most towns have rental stores that will have one or more scaffolds on hand. Moreover, renting is recommended over purchasing because the amount of time spent in roofing a structure is relatively short and any extensive reroofing should not be necessary for a number of years. Do not attempt to build a scaffold unless you are a good carpenter. A poorly built scaffold is worse than no scaffold at all. Scaffold construction details are shown in figure 2-13. The cross bracing is particularly important for safety. All handmade scaffolds should have cross braces.

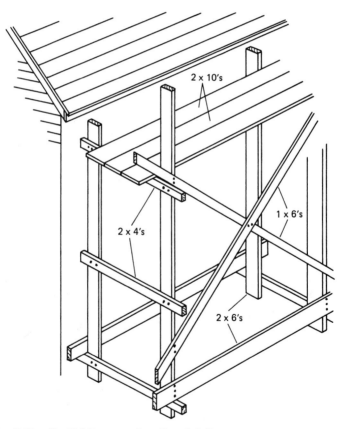

Fig. 2-13. Scaffolding construction details.

Roofing Materials

Roofing materials may be grouped in the following categories: (1) surface materials, (2) underlayment, (3) flashing, (4) fasteners, (5) roofing cements, (6) roof coatings, and (7) roofing tape.

Surface Materials

Roof surface materials form the top layer or external cover of the roof. They are also sometimes called *roofing* or *roof covering materials*. These materials do not include the underlayment or flashing. The principal types of roof covering materials are asphalt shingles, asphalt roll roofing, wood shingles and shakes, tile, slate, metal roofing, and mineral fiber shingles.

Slight shade variations will occur in different production runs of all types of factory-colored roofing materials. When applied to the roof, these shade variations produce a roof surface condition called *color patterning, checkerboarding, shading,* or *stair-stepping*.

Color patterning in asphalt roofing is caused primarily by variations in the depth of granular imbedment into the hot asphalt. Patterning in aluminum or other types of metal roofing panels can be caused by variations between batches of paint or other coloring agents.

A color range spread throughout a slate or clay tile roof is generally considered very attractive. With clay tiles, color patterning may occur when tiles manufactured from clay of varying colors are applied to the same roof. The location of the tiles within the kiln during the baking process may also produce color variations. Concrete tiles are subject to shade variations, but to a much lesser degree than clay tiles.

Slight variations in shades are discernible only after application to a roof and when viewed from a relatively long distance. Because of this, the roofer should scrutinize the work carefully from street level several times during the day. By following this procedure, color patterning can be detected and eliminated by blending the roofing materials over the entire roof deck.

ASPHALT SHINGLES

Asphalt shingles are the most common type of covering material used on pitched roofs. They are made by saturating an organic or fiberglass

Fig. 2-14. Asphalt shingle roof. *(Courtesy Johns-Manville Corp.)*

base with asphalt and then covering the asphalt surface with mineral granules (figs. 2-14 and 2-15). Organic-base asphalt shingles have a class A, B, or C fire rating, depending on how they are made. Fiberglass-base shingles have an A fire rating. Fiberglass-base asphalt shingles are the more durable of the two. Their principal disadvantage is their tendency to become brittle when installed at temperatures of 50° F or lower.

An asphalt shingle roof has a service life of approximately 15 to 20 years. Asphalt shingles are lightweight, comparatively inexpen-

Fig. 2-15. Fiberglass shingle roof. *(Courtesy Johns-Manville Corp.)*

sive, easy to install, and available in many different sizes, shapes, and colors. They are also easily repaired and require little maintenance.

Asphalt shingles are usually manufactured in 12 × 36–inch strips, but individual shingles are also available. The three-tab, self-sealing asphalt shingle is the most popular type in current use. It is usually laid with a 5-inch exposure and a 2-inch horizontal headlap. An 8-foot long strip shingle is also available. This shingle is produced with the upper edge of the shingle material bonded to a plywood nailing strip. The shingle is installed by nailing through the plywood nailer.

ASPHALT ROLL ROOFING

Roll roofing is recommended for flat or low-pitched roofs with a slope of 4-in-12 or less. Roll roofing is easy to install, repair, and maintain. It has a class A or C fire rating, depending on its composition. Roll roofing is less expensive than asphalt shingles, but it results in a less attractive roof and has a shorter service life.

WOOD SHINGLES AND SHAKES

Individual wood shingles are available in lengths of 16, 18, and 24 inches and widths which vary on a random basis (fig. 2-16), while shakes are machine or hand split into 18- and 24-inch lengths. Wood shingles have a smoother surface than shakes because they are sawn instead of split. Most shingles and shakes are made from the highest (No. 1) grade of western red cedar or redwood. Neither are fire resistant unless treated. In many areas, local building codes require that only treated wood shingles and shakes be used on a roof. Chemically treating wood raises the fire resistance to a class C rating.

Wood shingles are also available in the form of roof panels (fig. 2-17). Each panel consists of sixteen 18-inch No. 1 grade cedar shingles bonded to an 8-foot long, 1/2-inch thick plywood base. The use of panels instead of individual wood shingles reduces the roofing cost because of reduced installation time.

The wood-fiber (hardboard) roof panel produced by Masonite Corporation is a comparatively new wood roofing product (figs. 2-18 and 2-19). These heavily textured panels resemble wood shingles and require no finishing. The wood-fiber material weathers to a silver gray like natural wood shingle. Each 12 × 48–inch panel forms a row of 8 to 10 wood shingles of varying width and thicknesses. Shingle thicknesses range from 3/8 to 9/16 inch. Wood-fiber panels have the same material ratings as wood shingles or shakes.

TILE

Roofing tile was introduced to this country by the Spanish, and it is still a popular roofing material in Florida, California, and the Southwest (fig. 2-20). Both clay and concrete tiles are available. Kiln-dried red clay tiles are traditional and are available in mission (barrel-shaped), flat, and other styles in a red, brown, or buff color. Con-

Fig. 2-16. Wood shake roof. *(Courtesy Johns-Manville Corp.)*

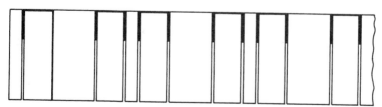

Fig. 2-17. Shingle roof panel.

Fig. 2-18. Masonite Woodruf wood-fiber roof panel. *(Courtesy Masonite Corp.)*

crete extruded tiles are a recent innovation and are available in a wider range of shapes and colors than clay tile. Both clay and concrete tiles are produced integrally colored or with a cured-on glaze. The most widely used tile is integrally nonglazed.

Fig. 2-19. Residence covered with wood-fiber roofing panels. *(Courtesy Masonite Corp.)*

Fig. 2-20. Flat tile roof. *(Courtesy Johns-Manville Corp.)*

The dimensions of a typical roofing tile are 12 × 7 inches. Most are approximately ½ inch thick. Tiles will not deterioriate, wear, or require painting. Although tiles may crack, split, or become brittle, they are easily replaced. A properly constructed tile roof is fireproof and has a service life of 50 years or more.

Tile is a little less expensive than slate, but about twice as expensive as asphalt shingle. Roofing tiles are heavy, expensive to ship, and difficult to install. Because tiles weigh more than most other roofing materials, a strong roof framework is required to support them. In some cases, it may be necessary to strengthen the framework by installing additional rafters and bracing. Roof reinforcement is another factor adding to the cost of a tile roof.

SLATE

Slate is a heavy, durable, nonporous roofing material with a service life of 50 years or more. Because roofing slate resists deterioration, a slate roof requires little maintenance. Cracked or broken slate can be easily replaced.

Although a slate roof is attractive and enjoys a long service life, it is expensive and difficult to install. Like tile, it is heavy and requires a strong roof framework and deck to support it. Its weight also makes it expensive to ship. If the structure being roofed is located far from the few centers in the eastern United States where slate is produced, the shipping costs will add significantly to the total cost. Slating is almost always done by professional contractors who specialize in this type of roofing.

METAL ROOFING

Metal roofing is used on main roof decks or on the flat decks of dormers, porches, and entryways. The roofing metals include aluminum, galvanized steel, copper, copper-coated galvanized steel or aluminum, terne metal, tin, and lead. Metal roofs are used primarily on commercial, industrial, and farm buildings (fig. 2-21).

Metal roofing is available in the form of shingles or shakes, tiles, panels, or sheets. The panels or sheets are produced in several different styles including flat, corrugated or ribbed, and standing seam. Metal roofing is fireproof and has good weather resistance. When properly installed and maintained, a metal roof should have a service life of 20 years or more. Metal roofing usually weighs less than most nonmetal roofing materials per roof square.

Aluminum is the most widely used roofing metal (fig. 2-22). It is available in the form of shingles or shakes, which are produced from lightweight aluminum sheets by a stamp-and-die technique, or roofing panels.

Metal roofing is noisy when it rains. The panels are subject to a certain amount of expansion and contraction as temperatures change,

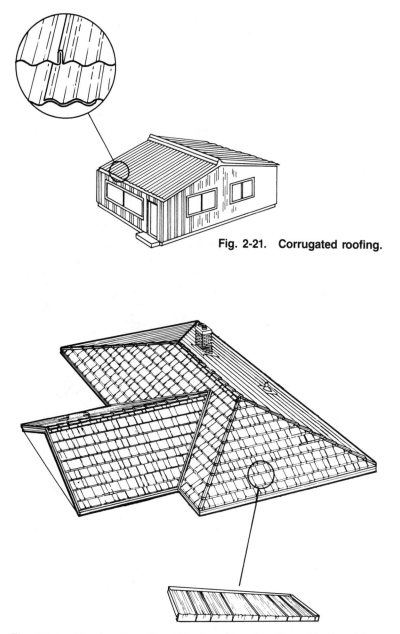

Fig. 2-21. Corrugated roofing.

Fig. 2-22. Aluminum roofing shingles. *(Courtesy Reynolds Metals Co.)*

which may cause the fasteners to pull loose from the nailing base. Aluminum is especially susceptible to denting and scratching. Most metal roofing should be installed by a contractor.

MINERAL FIBER SHINGLES

Mineral fiber (or asbestos-cement) shingles are made of asbestos and cement with a fiberglass reinforcement (fig. 2-23). They are fireproof,

Fig. 2-23. Mineral fiber (asbestos-cement) shingle roof. (Courtesy Johns-Manville Corp.)

very rigid, durable, and strong, and have a service life of 50 years or more. Mineral fiber shingles are thicker and heavier than asphalt shingles, but are not quite as heavy as slate or tile. Like slate or tile, however, they require a strong roof framework to support their weight. Mineral fiber shingles are more expensive than asphalt shingles, but cost considerably less than slate, tile, or wood shingles and shakes. They are easy to install and require little maintenance. Mineral fiber shingles can be damaged by impact, such as by a branch falling on the roof, but the individual shingles are easy to replace.

Mineral fiber shingles are manufactured in a variety of colors, shapes, and textures. They are available in standard shingle sizes or smaller units resembling slate or wood shakes.

Underlayment

The *underlayment* (or *underlay*) is a moisture-resistant layer of roofing felt, which is nailed to the roof deck sheathing before laying the covering material. It is used to provide a dry, flat surface for the roofing, to prevent a chemical reaction from occurring between the wood resins in the sheathing and the roof covering material, and to protect the sheathing, rafters, and interior spaces from water damage.

The underlayment is used under asphalt shingles, mineral fiber (asbestos-cement) shingles, slate, and tile. It can also be used under wood shingles or shakes in certain applications, but this is rarely done and then only under strictly followed conditions.

The roofing felt underlayment is made of dry felt impregnated with an asphalt or coal tar. It is produced in 36-inch wide rolls and in several different weights. The rolls vary in length from 72 feet for a No. 30 weight felt to 144 feet for a No. 15 felt. The most widely used underlayment is No. 15 asphalt-saturated roofing felt; it weighs approximately 15 pounds per roof square (100 square feet of roof surface). No. 15 felt is used as an underlayment under asphalt shingles. Tile and slate roofs use No. 30 asphalt-saturated roofing felt or two layers of No. 15 felt as an underlayment.

Flashing

Flashing is used to protect seams or joints from water seepage. It is installed at the junction formed by the roof and a vertical wall, along roof rakes and eaves, along ridges, in roof valleys, around chimneys, vent pipes and stacks, at intersections of different roof planes, and at other points on the roof where water from rain or melting snow could penetrate the roof and enter the structure. Flashing is made of metal, plastic, or mineral-surfaced roll roofing. Metal and composition materials are used on both residential and nonresidential structures. Plastic flashing is sometimes used on the built-up roofs of commercial and industrial structures.

Galvanized steel is the most commonly used metal flashing, particularly on houses, but metals such as aluminum, copper, and stainless steel are also used. Copper flashing is often used on wood shingle or shake roofs, or as a chimney flashing on other types of roofs with a different material providing the rest of the flashing. Aluminum and stainless steel flashing is frequently used on roofs located near saltwater bodies where corrosion is a problem.

Metal flashing may be cut and formed at the site or preformed valleys, drip edges, gravel stops, and other types of metal flashing may be purchased from a supplier. The valleys, vent pipes, stacks, and eaves of asphalt shingle roofs can also be flashed with 90-pound mineral-surfaced roll roofing, which is available in 36-inch wide rolls. The valley flashing on these roofs consists of an 18-inch wide sheet covered by a 36-inch wide one. A roll of 18-inch wide roll roofing suitable for valley flashing is available from roofing manufacturers.

Fasteners

Roofing nails are usually made of galvanized steel, aluminum, or copper. They are sharp pointed nails with heads ranging from 3/8 to 7/16 inch in diameter, and are available in a variety of different lengths. The length of the nails used for a particular job will depend on the type and thickness of the roofing material, the type of roof-deck sheathing, and whether the roofing material is being applied in new construction or over an existing roof. The nails should penetrate at least 3/4 inch into solid wood-deck sheathing boards. If approved plywood is used in the construction of the deck, the nails should completely penetrate the plywood. Nails with barbed or otherwise deformed shanks provide the best holding power. Recommended nail lengths for several different types of roofing applications are listed in table 2-1.

The number and type of nails required for a specific roofing ap-

Table 2-1. Recommended Nail Lengths for Common Roofing and Reroofing Applications

Application	Nail Length
Roll roofing on new deck	1 inch
Asphalt strip shingles on new deck	1¼ inch
Individual asphalt shingles on new deck	1¼ inch
Asphalt shingles over asphalt shingles or roll roofing	1¼ to 1½ inch
Asphalt shingles over wood shingles	1¾ inch
16-inch and 18-inch wood shingles on new deck	1¼ inch
24-inch wood shingles on new deck	1½ inch
16-inch and 18-inch wood shingles over existing roof	1¾ inch
24-inch wood shingles over existing roof	2 inch

plication will be stated in the instructions provided by the roofing material manufacturer. These instructions will also state the number of nails required per roof square. Additional information about roofing nails and nailing procedures are included in the chapters covering specific types of roofing applications.

Staples may be used instead of nails to fasten the underlayment or roll roofing to the roof deck. They are also sometimes used to hold down asphalt shingle tabs. Before using staples, however, check to determine whether their use is covered by the roofing manufacturer's warranty. The staples used in roofing are 1 inch wide and up to 1¼ inches long. They are applied with a stapling machine.

Special types of fasteners are often used to fasten roofing materials to nonwood decks. The type and number of fasteners and the fastening procedure will usually be specified by the manufacturer of the roof-deck materials.

Roofing Cements and Sealers

Various types of cements and sealers are available for waterproofing seams or bonding overlapping layers of roofing material. The three principal types are plastic asphalt cement, lap cement, and asphalt adhesive cement.

PLASTIC ASPHALT CEMENT

Plastic asphalt cement is an asphalt material used to seal the joint between the roof and a chimney, vent pipe, adjoining wall, or other type of vertical surface. Because plastic asphalt cement forms a part of the flashing assembly, it is sometimes referred to as flashing cement. A good quality plastic asphalt cement will have enough elasticity to compensate for normal expansion and contraction without cracking. Furthermore, it should not flow when the outdoor temperatures are high, or become brittle when they are low.

LAP CEMENT

A lap cement is used to create a waterproof bond between overlapping sections of roll roofing. Lap cement will vary in consistency, depending on the manufacturer, but all lap cements are thinner and easier to work with than plastic asphalt cement. Lap cement is applied with a brush and should cover the entire lapped area.

ASPHALT ADHESIVE CEMENT

Free-tab strip shingles are manufactured without an adhesive on the bottom of each tab. The disadvantage of this is that a strong wind or gust can lift the tab and bend it back. The free tabs on these shingles can be sealed down to the surface by coating the bottom of each tab with an asphalt adhesive cement. These cements are also used for sealing laps of roll roofing when it is applied by the blind nailing method (see chapter 11).

Asphalt adhesive cement is mixed with a solvent that evaporates rapidly when exposed to the air. Consequently, it is a quick setting adhesive and is sometimes difficult to work with. It may be applied with a brush, towel, or gun, depending on consistency.

Roof Coatings

A number of different types of asphalt coatings are used in roofing. Most are used to construct, restore, or resurface the roof membrane of a built-up roof. Some may also be used to resurface roll roofing or to cover a metal roof that shows signs of wear. All are of a thin enough consistency to be applied with a spray gun, broom, or mop.

The membrane of a built-up roof consists of alternate layers of waterproofing bitumen (roofing asphalt or pitch) and roofing felt. Roofing asphalts are available in several grades or classes based on their approximate softening point range. The grade selected will be determined by the roof slope, climatic conditions, type of roof assembly, and other factors. A roof coating called *steep asphalt* is generally recommended for roof slopes of more than 1-in-12.

Asphalt water emulsions are a special type of roof coating consisting of asphalt, or asphalt combined with other ingredients, and emulsified with water. A principal disadvantage of this type of coating is that it must not be exposed to rain for at least 24 hours after application. Another disadvantage is that stored emulsions must not be allowed to freeze during cold weather.

Other Roofing Materials

Masonry primer, roofing tape, and caulk are also used in roofing and reroofing.

MASONRY PRIMER

The joint between the roof and a masonry wall or chimney must be sealed with a suitable asphalt coating or cement to prevent leaks; however, the bond will not be satisfactory unless the masonry surface is first coated with a primer. Masonry primers have been developed for this purpose.

A masonry primer (or asphalt primer) is a fluid substance that can be sprayed or brushed onto a masonry surface. If applied properly, it will be absorbed by the pores of the masonry and will provide a suitable bonding surface for an asphalt coating or cement. Incorrect application will be indicated by the formation of a surface film. If this should occur, the primer should be thinned to the proper consistency and reapplied. Always follow the manufacturer's instructions when thinning a masonry primer.

ROOFING TAPE

Roofing tape is used on built-up roofs to seal and reinforce the joints between rigid insulation boards or panels. The tape also produces a more uniform surface for the application of the roof membrane and

Fig. 2-24. Applying caulking compound along edge of chimney flashing. *(Courtesy Borden Chemical Division/Borden, Inc.)*

prevents the loss of asphalt coating at the insulation joints. Roofing tape is usually made by saturating a porous material, such as cotton or glass fiber, with asphalt. Roofing tape is available in 4- to 36-inch wide rolls. The amount (linear length) of roofing tape contained in a roll will vary depending on the roofing manufacturer.

ACRYLIC LATEX SEALER

Loose flashing or flashing with only minor damage can be sealed with a bead of acrylic latex. The sealer is available in cartridges for application with a caulking gun (fig. 2-24).

Reroofing

Reroofing may involve the complete removal of the existing roof, in which case the method used to apply the new roofing materials is identical to that employed in new construction, or it may involve the application of new roofing materials over an existing roof after it has been prepared to serve as a suitable nailing base.

There are several obvious advantages to reroofing over an existing roof without having to remove the old roofing materials. Two principal advantages are the savings in time and money. A third advantage is the increased protection and insulation provided by the additional layers of new roofing materials. Unfortunately, not all roofs can be reroofed in this manner.

Asphalt shingles are the most commonly used reroofing material. Existing asphalt shingle roofs and wood shingle or shake roofs are frequently reroofed with asphalt shingles. Asphalt shingles can be laid over almost any kind of roofing material *except* tile, slate, or asbestos-cement shingles. These roofing materials are too hard and brittle for nailing and provide a poor base for the new roofing materials. As a result, tile, slate, and asbestos-cement shingle roofs must be completely removed before new roofing materials are applied to the roof deck. The roof should be thoroughly inspected before a decision is made to reroof it. This inspection should not only include the condition of the existing roof covering materials, but also the condition of the sheathing, rafters, and other frame members. Bear in mind that the existing roof must provide a rigid, smooth, and uniform base for the new roof covering materials. If portions of the old roofing materials are missing or too extensively damaged to function as a suitable base, then reroofing over the existing roof is out of the question. The roof will have to be stripped and new roofing materials applied as in new construction.

Safety Precautions

Roofing involves working atop buildings and so requires certain minimal safety precautions. This attention to safety is particularly important for the layman, who may not be accustomed to moving about at these elevations. Although professionals sometimes regard their own safety with a certain disdain, it is not an attitude that should be imitated by the nonprofessional, nor does it weaken the argument for a constant observance of safety standards. More than one professional roofer has been seriously injured or killed because familiarity with the working conditions led to carelessness.

Special attention should be paid to the type of clothing worn on the job. A suitable pair of shoes is especially important, because the roofer is usually working on a sloping surface. Tennis shoes or rubber soled shoes provide the best footing. Shoes with leather soles are *not* recommended, because they will not grip the surface. Working barefoot provides good footing, but the roof surfaces are usually too hot to walk on during the summer months. There is also the possibility of stepping on a loose roofing nail or picking up a splinter from the wood sheathing on an exposed roof deck. Stepping on a nail usually results in stumbling or moving quickly to avoid further pain—an instinctive, but possibly fatal reaction when standing at the edge of a roof deck.

Do not wear clothing that is either too loose or too tight. If the clothing is too loose, it may snag on a shingle or nail and cause you to lose your balance. Shirts should always be buttoned and tucked in. Clothing that is too tight will restrict your movements and will be uncomfortable, especially in warm weather.

On a particularly steep slope, a rope and harness device can be used to provide additional protection against falling. This safety device consists of a leather or rope harness fitted to the worker and attached to a long rope (fig. 2-25). The other end of the rope should be securely tied to a part of the structure or a nearby tree located on the other side of the roof ridge (fig. 2-26). There should be enough play in the rope to allow freedom of movement across the surface of the roof. These rope and harness devices are not difficult to make, and they

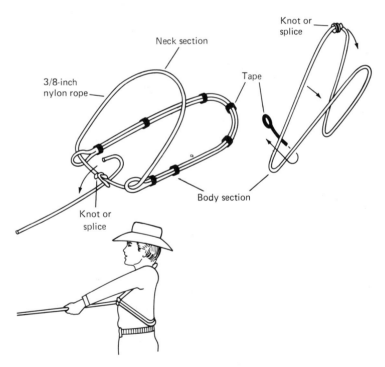

Fig. 2-25. Homemade body harness.

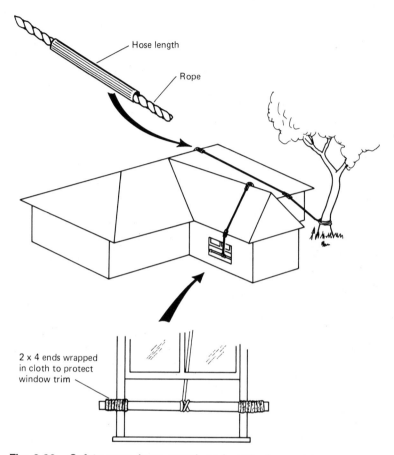

Hose length

Rope

2 x 4 ends wrapped
in cloth to protect
window trim

Fig. 2-26. Safety rope, hose guard, and methods used to secure the rope.

can also be purchased ready-made through a mail order house or from a local building supply dealer. Check with your local dealer first. If he does not stock them, he may know where they can be ordered.

Most professional roofers seem to prefer working without a safety rope or harness, because either one tends to restrict movement. This practice is *not* recommended for the average homeowner.

Roofers have created a number of different devices to provide support or footing when they are working on a roof with a steep slope. These devices are temporarily nailed or otherwise attached to the roof deck, and are moved as the work progresses. They range in complexity from a simple piece of 2 × 4 or strip of wood nailed to the deck to give the roofer extra footing to a shelf-type bracket and board arrangement. Metal brackets may be obtained at a local building supply outlet or hardware store. The chicken ladder shown in figure 2-7 is useful for roofs with steep slopes. A rung ladder with 2 × 4 braces will serve the same purpose.

Weather conditions should be considered before going up on a roof. This may present a problem, especially if time is limited or only a specific time during the day is free for the work. Nevertheless, do not attempt to work on a roof during unfavorable weather conditions. Safety should always come first.

Roofing should never be done when the surface is damp. A damp roof, whether it be from the morning dew or a recent rain, provides poor footing. More than one roofer has been seriously or fatally injured by a fall from a damp and slippery roof.

Working on a roof is not recommended when a rainstorm is approaching. There is always danger from lightning even though the clouds may not be directly overhead. Furthermore, there is a tendency to hurry the work when a storm is approaching, which often results in a careless and sloppy job. When you observe a storm approaching, cover the roofing materials and get down off the roof. Do not go back up until the roof has had time to dry.

Other sources of danger to roofers are chimneys, roof vents, cables, television antennas, and electrical wires. Never lean against an old chimney or use it for support. Loose bricks can be dislodged or the entire chimney above the roof line may collapse. This could be dangerous not only for the roofer, but also for people standing below.

Sometimes gas- or oil-fired heating equipment is vented to the outdoors through a metal chimney or vent pipe that protrudes through the surface of the roof. Grabbing one of these for support can often result in nasty burns. If these pipes are placed near the edge of the roof, sudden reaction to the pain could result in loss of balance and a fall from the roof. Even when cold, these objects are usually too weakly constructed to support the weight of an adult.

Metal cables are often used to support television antennas, metal chimneys, vent pipes, or advertising sign boards mounted on roofs. These sometimes seem to have been deliberately placed to trip the unwary worker. Mentally mark the location of all cables and other potential causes of accidents before doing any work on the roof.

Television antennas, service entrance wires, and other types of electrical wiring are potential sources of electrical shock. Keep hands off of them and avoid brushing up against them. A severe electrical shock can result in a temporary loss of consciousness, and a roof, particularly a sloping one, is the last place for this to happen. If possible, never work on a roof alone. There should be at least one other person on the job, or at least in the house, in case of an accident. Injuries suffered from a fall require immediate attention.

Every type of construction work has certain inherent dangers, some more than others. Awareness of these dangers and observance of safety procedures that specifically apply to the type of work being done will minimize the possibility of injury.

Estimating Roofing Materials

An important initial step in any roofing or reroofing project is to correctly estimate the amount of roofing material required for the job. Overestimating is expensive and wasteful. Underestimating means that additional roofing material must be purchased to complete the job. This often results in slight shade variations between the colors of the new and old materials and results in a condition known as color patterning (see the discussion of color patterning in this chapter).

Roofing is estimated and sold in squares. Each square of roofing is the amount required to cover 100 square feet of roof area. To estimate the amount of roofing required to cover a roof, first compute the total roof area in square feet, add 10 percent for cutting and waste, and divide by 100.

The roof areas of the basic roof types illustrated in figure 2-27 are calculated by multiplying the rake line (or the sum of the rakes)

Fig. 2-27. Basic roof types. *(Courtesy Asphalt Roofing Manufacturers Association)*

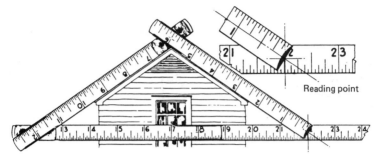

Fig. 2-28. Using a folding rule to determine roof pitch. *(Courtesy Asphalt Roofing Manufacturers Association)*

by the eave line. For example, the area of the rectangular shed roof in figure 2-29 is calculated by multiplying the rake (line A) by the eave (line B). The area of the gable roof in the same illustration is calculated by multiplying the sum of the two rakes (lines A and B) by the eave (line C). Finally, the area of the more complicated gambrel roof is calculated by multiplying the sum of the rakes (lines A, B, C, and D) by the eave (line E). Because these are simple roofs, it is not difficult to climb onto their surfaces and measure them. However, a different method of calculating the total roof surface area is used for roof designs more elaborate than the three basic types illustrated in figure 2-27. The method used does not require climbing onto the roof or resorting to complicated formulas and computations. It requires only that the roof pitch and its horizontal area in square feet be known.

Roof pitch is the relation between the rise and the span of the roof (see the discussion of roofing terminology in this chapter) and is expressed as a fraction. Roof pitch can be determined by forming a triangle with a carpenter's folding rule, as in figure 2-28, and holding it at arm's length so that the roof slopes are aligned with the sides of the rule. The reading is taken on the base section of the rule and the reading point is converted to pitch by using the data in table 2-2. The numerical value used in the line headed *rule reading* will always be the one closest to the reading point on the base section of the rule. In the example shown in figure 2-28, the reading point and rule reading were both 22. However, a reading point of 22⅛ on the rule would also have resulted in a rule reading of 22 because it is closest

Table 2-2. Converting Rule Readings to Pitch and Rise
(Courtesy Asphalt Roofing Manufacturers Association)

RULE READING	20 1/2	20 7/8	21 1/4	21 5/8	22	22 3/8	22 3/4	23 1/16	23 3/8	23 5/8	23 13/16	23 15/16
PITCH FRACTIONS	1/2	11/24	5/12	3/8	1/3	7/24	1/4	5/24	1/6	1/8	1/12	1/24
RISE-INCHES PER FT.	12	11	10	9	8	7	6	5	4	3	2	1

to that figure. Under the numerical value 22 in table 2-2, the roof has a ⅓ pitch and a slope of 8-in-12.

After the roof pitch has been determined, the total ground area (horizontal surface) of the roof should be measured and transferred to a roof plan drawing similar to the one illustrated in figure 2-29. All measurements can be made from the ground or within the attic. Climbing on the roof is not necessary.

The horizontal areas are calculated after all measurements have been made, a roof plan drawn, and the pitches of the various elements of the roof have been determined with a carpenter's folding rule. *Include only those areas in each calculation that come under elements of the roof having the same pitch.* The horizontal area under the main roof of the structure illustrated in figure 2-29 includes

26 ft. × 30 ft. =   780 square feet
19 ft. × 30 ft. =   570 square feet
                 1,350 square feet (total)

The overlapping triangular area of the minor roof (8 ft. × 5 ft. or 40 square feet) and the opening for the chimney (4 ft. × 4 ft. or 16 square feet) must then be subtracted from the 1,350 total square feet to obtain the horizontal area under the main roof. This will give

1,350 sq. ft. – (40 sq. ft. + 16 sq. ft.) = 1,294 square feet

The horizontal area under the minor (ell) roof with the 6-in-12 slope is computed as follows:

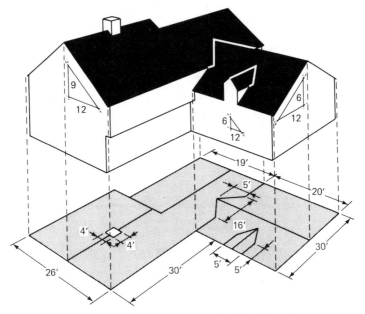

Fig. 2-29. Roof plan. *(Courtesy Asphalt Roofing Manufacturers Association)*

20 ft. × 30 ft. = 600 square feet
8 ft. × 5 ft. = 40 square feet (triangular roof area of
___ minor roof projecting over major roof)
640 square feet (total)

There are cases where one element of a roof will project over another. These duplicated areas must be added to the total horizontal area. For example, a 4-inch projection of the roof eaves on the structure illustrated in figure 2-29 will result in the following duplications (fig. 2-30):

1. A duplication under the dormer eaves of 2(5 ft. × ⅓ ft.) or 3⅓ square feet.
2. A duplication under the eaves of the main roof where they overhang the rake of the minor roof of 2(7 ft. × ⅓ ft.) or 4⅔ square feet.

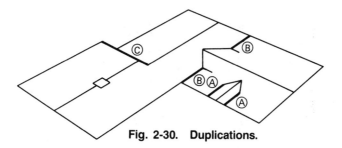

Fig. 2-30. Duplications.

3. A duplication where the rake of the wide section of the main roof overhangs the rake of the small section in the rear or 9½ ft. × ⅓ ft. or 3⅙ square feet.

The total square footage of the duplicated areas is 11⅙ square feet. For these computations, the square footage is rounded off to the next highest number, or 12 square feet, and divided as follows:

$$1{,}294 \text{ sq. ft.} + 8 \text{ sq. ft.} = 1{,}302 \text{ square feet (major roof)}$$
$$640 \text{ sq. ft.} + 4 \text{ sq. ft.} = \underline{644} \text{ square feet (minor roof)}$$
$$1{,}946 \text{ square feet}$$

The horizontal areas of 1,302 square feet (major roof) and 644 square feet (minor roof) can be converted to slope areas by using the data in table 2-3. The horizontal areas are given in column 1 and the corresponding slope areas are given in columns 2–12. To convert horizontal areas to slope areas, begin by finding the slope area in the column under the pitch determined for the roof. The total horizontal area of the main roof of the structure shown in figure 2-29 is 1,302 square feet. Referring to the 9-inch rise column in table 2-3 for the main roof, the following is found:

Horizontal Area	Slope Area under 9-inch Rise
1,000	1,250.0
300	375.0
2	2.5
1,302	1,627.5

Table 2-3. Conversion of Horizontal Distances or Areas to Slope Distances or Areas
(Courtesy Asphalt Roofing Manufacturers Association)

RISE (Inches per ft. of horizontal run)	1″	2″	3″	4″	5″	6″	7″	8″	9″	10″	11″	12″
PITCH (Fractions)	1/24	1/12	1/8	1/6	5/24	1/4	7/24	1/3	3/8	5/12	11/24	1/2
CONVERSION FACTOR	1.004	1.014	1.031	1.054	1.083	1.118	1.157	1.202	1.250	1.302	1.356	1.414
HORIZONTAL (Area in Sq. Ft. or Length in Feet)												
1	1.0	1.0	1.0	1.1	1.1	1.1	1.2	1.2	1.3	1.3	1.4	1.4
2	2.0	2.0	2.1	2.1	2.2	2.2	3.2	2.4	2.5	2.6	2.7	2.8
3	3.0	3.0	3.1	3.2	3.2	3.2	3.5	3.6	3.8	3.9	4.1	4.2
4	4.0	4.1	4.1	4.2	4.3	4.5	4.6	4.8	5.0	5.2	5.4	5.7
5	5.0	5.1	5.2	5.3	5.4	5.6	5.8	6.0	6.3	6.5	6.8	7.1
6	6.0	6.1	6.2	6.3	6.5	6.7	6.9	7.2	7.5	7.8	8.1	8.5
7	7.0	7.1	7.2	7.4	7.6	7.8	8.1	8.4	8.8	9.1	9.5	9.9
8	8.0	8.1	8.3	8.4	8.7	8.9	9.3	9.6	10.0	10.4	10.8	11.3
9	9.0	9.1	9.3	9.5	9.7	10.1	10.4	10.8	11.3	11.7	12.2	12.7
10	10.0	10.1	10.3	10.5	10.8	11.2	11.6	12.0	12.5	13.0	13.6	14.1

20	20.1	20.3	20.6	21.1	21.7	22.4	23.1	24.0	25.0	26.0	27.1	28.3
30	30.1	30.4	31.0	31.6	32.5	33.5	34.7	36.1	37.5	39.1	40.7	42.4
40	40.2	40.6	41.2	42.2	43.3	44.7	46.3	48.1	50.0	52.1	54.2	56.6
50	50.2	50.7	51.6	52.7	54.2	55.9	57.8	60.1	62.5	65.1	67.8	70.7
60	60.2	60.8	61.9	63.2	65.0	67.1	69.4	72.1	75.0	78.1	81.4	84.8
70	70.3	71.0	72.2	73.8	75.8	78.3	81.0	84.1	87.5	91.1	94.9	99.0
80	80.3	81.1	82.5	84.3	86.6	89.4	92.6	96.2	100.0	104.2	108.5	113.1
90	90.4	91.3	92.8	94.9	97.5	100.6	104.1	108.2	112.5	117.2	122.0	127.3
100	100.4	101.4	103.1	105.4	108.3	111.8	115.7	120.2	125.0	130.2	135.6	141.4
200	200.8	202.8	206.2	210.8	216.6	223.6	231.4	240.4	250.0	260.4	271.2	282.8
300	301.2	304.2	309.3	316.2	324.9	335.4	347.1	360.6	375.0	390.6	406.8	424.2
400	401.6	405.6	412.4	421.6	433.2	447.2	462.8	480.8	500.0	520.8	542.4	565.6
500	502.0	507.0	515.5	527.0	541.5	559.0	578.5	601.0	625.0	651.0	678.0	707.0
600	602.4	608.4	618.6	632.4	649.8	670.8	694.2	721.2	750.0	781.2	813.6	848.4
700	702.8	709.8	721.7	737.8	758.1	782.6	809.9	841.4	875.0	911.4	949.2	989.8
800	803.6	811.2	824.8	843.2	864.4	894.4	925.6	961.6	1000.0	1041.6	1084.8	1131.2
900	903.6	912.6	927.9	948.6	974.7	1006.2	1041.3	1081.8	1125.0	1171.8	1220.4	1272.6
1000	1004.0	1014.0	1031.0	1054.0	1083.0	1118.0	1157.0	1202.0	1250.0	1302.0	1356.0	1414.0

Table 2-4. Determination of Valley and Hip Lengths
(Courtesy Asphalt Roofing Manufacturers Association)

RISE (Inches per ft. of horizontal run)	4"	5"	6"	7"	8"	9"	10"	11"	12"	14"	16"	18"
PITCH												
Degrees	18°26'	22°37'	26°34'	30°16'	33°41'	36°52'	39°48'	42°31'	45°	49°24'	53°8'	56°19'
Fractions	1/6	5/24	1/4	7/24	1/3	3/8	5/12	11/24	1/2	7/12	2/3	3/4
CONVERSION FACTOR	1.452	1.474	1.500	1.524	1.564	1.600	1.642	1.684	1.732	1.814	1.944	2.062
HORIZONTAL (Length in Feet)												
1	1.5	1.5	1.5	1.5	1.6	1.6	1.6	1.7	1.7	1.8	1.9	2.1
2	2.9	2.9	3.0	3.0	3.1	3.2	3.3	3.4	3.5	3.6	3.9	4.1
3	4.4	4.4	4.5	4.6	4.7	4.8	4.9	5.1	5.2	5.4	5.8	6.2
4	5.8	5.9	6.0	6.1	6.3	6.4	6.6	6.7	6.9	7.3	7.8	8.2
5	7.3	7.4	7.5	7.6	7.8	8.0	8.2	8.4	8.7	9.1	9.7	10.3

6	8.7	8.8	9.0	9.1	9.4	9.6	9.9	10.1	10.4	10.9	11.7	12.4
7	10.2	10.3	10.5	10.7	10.9	11.2	11.5	11.8	12.1	12.7	13.6	14.4
8	11.6	11.8	12.0	12.2	12.5	12.8	13.1	13.5	13.9	14.5	15.6	16.5
9	13.1	13.3	13.5	13.7	14.1	14.4	14.8	15.2	15.6	16.3	17.5	18.6
10	14.5	14.7	15.0	15.2	15.6	16.0	16.4	16.8	17.3	18.1	19.4	20.6
20	29.0	29.5	30.0	30.5	31.3	32.0	32.8	33.7	34.6	36.3	38.9	41.2
30	43.6	44.2	45.0	45.7	46.9	48.0	49.3	50.5	52.0	54.4	58.3	61.9
40	58.1	59.0	60.0	61.0	62.6	64.0	65.7	67.4	69.3	72.6	77.8	82.5
50	72.6	73.7	75.0	76.2	78.2	80.0	82.1	84.2	86.6	90.7	97.2	103.1
60	87.1	88.4	90.0	91.4	93.8	96.0	98.5	101.0	103.9	108.8	116.6	123.7
70	101.6	103.2	105.0	106.7	109.5	112.0	114.9	117.9	121.2	127.0	136.1	144.3
80	116.2	117.9	120.0	121.9	125.1	128.0	131.4	134.7	138.6	145.1	155.5	165.0
90	130.7	132.7	135.0	137.2	140.8	144.0	147.8	151.6	155.9	163.3	175.0	185.6
100	145.2	147.4	150.0	152.4	156.4	160.0	164.2	168.4	173.2	181.4	194.4	205.2

The total area for the minor roof with a 6-inch rise (6-in-12 slope) or ¼ pitch is 644 square feet.

Horizontal Area	Slope Area under 6-inch Rise
600	670.8
40	44.7
4	4.5
644	720

The total area for both the major and minor roofs combined is 2,347.5 square feet (1,627.5 sq. ft. + 720 sq. ft.). Add 10 percent for wastage to bring the figure to 2,582 square feet or roughly 26 squares of roofing.

The drip edges, valley flashing, and other roof accessories required to complete the job will be determined by linear measurements made along eaves, rakes, ridges, and valleys. Eaves and ridges are horizontal measurements and may be taken off the roof plan. Rakes and valleys run on a slope and their true lengths must be taken from conversion tables (see tables 2-3 and 2-4).

The amount of metal flashing used as drip edges is estimated by calculating the total length of all the roof eaves and rakes. The length of the roof rake is determined by first measuring the horizontal distance over which it extends. For the structure illustrated in figure 2-29, the rakes on the ends of the main roof span distances of 26 feet and 19 feet respectively. Additional rake footage of 13 feet plus 3½ feet is found at the point where the higher roof section joins the lower one. Thus, the total rake footage for the structure illustrated in figure 2-29 is 26 + 19 + 13 + 3½ = 61½ feet. By referring to table 2-3 under the column for a roof with 9-inch rise, the following is found:

Horizontal Run	Length at Rake
60.0	75.0
1.0	1.3
.5	.6
61.5	76.9 feet (actual length of major roof rake)

The minor roof of the structure illustrated in figure 2-29 has a rake with a horizontal distance of 30 feet. The dormer roof rake adds an additional 5 feet for a total of 35 feet. Referring to table 2-3 under the column for a roof with a 6-inch rise, the following is found:

Horizontal Run	Length of Rake
30	33.5
5	5.6
35	39.1 feet (actual length of minor roof rake)

The horizontal length of the eaves for both the major and minor roof should also be measured and added to the total actual length of the rakes. The resulting figure will serve as an estimate of the total amount of metal flashing required as a drip edge along the roof eaves and rakes.

The point at which the minor roof meets the major roof of the structure illustrated in figure 2-29 creates two valleys. To estimate the amount of flashing material required for these valleys, it is first necessary to determine the run of the common rafter and then the length of the valley. The run of the common rafter is always one half the roof span. When used to determine the length of a valley, the run of the common rafter should be taken at the lower end of the valley. For example, the portion of the minor roof that projects over the major roof on the structure in figure 2-29 has a span of 16 feet at the lower end of the valley. The run of the common rafter at this point will be 8 feet (run = ½ span). Because there are two valleys formed at this roof intersection, however, the total run of the common rafter to be used in the calculations is 16 feet.

Horizontal length (16 feet) is converted to true valley length by means of the data provided in table 2-4. Because the intersecting roofs have different rises, however, the length for each rise must be found and the average of the two used. Under the 6-inch rise and 9-inch rise columns in table 2-4, the following is found:

Horizontal Length	6-inch Rise (Minor Roof)	9-inch Rise (Major Roof)
10	15	16.0
6	9	9.6
	24	25.6 = 49.6 feet

49.6 feet ÷ 2 = 24.8 feet (true length of each valley)

The same calculation is used to find the true length of the dormer valleys. As shown in figure 2-29, the run of the common rafter at the dormer is 2.5 feet (½ span of 5 feet). The following is found by using table 2-4:

Horizontal Length	6-inch Rise Dormer Roof
2.0	3.00
.5	.75
2.5	3.75 feet (true length of each dormer valley)
	× 2 = 7.5 feet

Total valley length for the structure in figure 2-29 is 24.8 feet + 7.5 feet = 32.3 feet.

Hiring a Roofing Contractor

A roofing contractor should be hired to do the work if the job requires skills and experience beyond those of the average homeowner. A roofing contractor can be found by looking in the Yellow Pages of the local telephone directory under "Roofing Contractor." Other sources helpful in recommending roofing contractors include the bank willing to finance the work, the local chapter of the National Association of Home Builders or Home Builders Association, local government offices for government-funded or nonprofit-operated home improvement assistance centers, relatives, friends, and neighbors. From these sources, obtain a list of four or five roofing contractors from which to make a final selection.

Ask each contractor for a list of past customers and check whether these individuals were satisfied with the work. Call the local Better Business Bureau to determine if complaints have been lodged against any of the contractors on the list.

Give the same specifications for the job to each of the roofing contractors and ask for an estimate. After the estimates have been reviewed and a contractor selected, draw up a written contract for the job or have one drawn up by a lawyer. Check the contract for work content and warranty before signing it. To make it a valid contract, both the building owner and the contractor must sign it. The contract should contain the same specifications contained in the bid or estimate submitted by the contractor. It should also state the cost of both materials and labor, and the payment method.

Check to determine if the roofing contractor has enough insurance to cover injuries to workers or accidental damage to the structure. If not, the building owner would be wise to purchase the necessary insurance.

CHAPTER 3

Roof Deck Preparation

- **Roof Deck Construction**
- **Roof Deck Preparation for New Construction**
- **Roof Deck Preparation for Reroofing**

The roof deck is the platform or base over which the roofing materials are laid. It is nailed or otherwise fastened to the rafters, the structural framework that supports the roof. The roof deck must provide a smooth, flat, and rigid surface for the roofing materials. If the deck is not rigid, vibration or movement resulting from structural instability may affect the lay of the roofing. An uneven surface will also cause roofing problems. On a flat roof, an uneven roof deck surface may result in drainage problems, wrinkling or buckling of roll roofing, or cracking of built-up roof membranes. These problems may be avoided by careful roof deck preparation before the roofing materials are laid.

Roof Deck Construction

The roof deck may be constructed of either wood or nonwood materials. Most pitched roof decks on houses are made of wood sheathing covered by an underlayment of asphalt-saturated felt. As a general rule, nonwood materials are used in the construction of roof decks on commercial and other nonresidential structures.

Wood Decks

A wood roof deck consists of wood sheathing and, when required, an underlayment or covering of asphalt-saturated felt. The wood sheathing may be boards, plywood panels, or structural panels.

WOOD SHEATHING BOARDS

Wood sheathing boards should be made of clear, sound, well-seasoned lumber of not less than 1 inch nominal thickness and not more than 8 inches nominal width. Wider boards are more likely to swell or shrink in width and cause the roofing material to wrinkle or buckle. Badly warped boards or boards containing excessively resinous areas or loose knots should be rejected. Tongue-and-groove, shiplap, or splined boards are recommended with each board flush nailed to the supporting rafters. Tongue-and-groove boards have the greatest resistance to warping or buckling. All boards and joints should be staggered and provided with adequate bearing or support.

Wood sheathing boards are nailed to each supporting rafter with two eightpenny (8d) common nails or two sevenpenny (7d) threaded nails. Each board must be supported by at least two roof rafters. Vertical joints must be centered over rafters unless end-matched tongue-and-groove boards are used. The joints formed between the ends of these boards may occur between rafters. All board end joints should be staggered.

The roof deck may be of closed or open construction, depending on the type of roofing material used to cover it. On a closed roof deck, the boards are applied without any spacing between them. They are usually nailed parallel to the roof eave. In areas of the country where high winds are a problem, greater racking resistance can be obtained by applying the boards diagonally to the eave. At the rake, the boards may extend beyond the end wall to provide an overhang or they may be cut flush with the wall (fig. 3-1).

Spaced or open sheathing is customarily used as a roof deck for wood shingles or shakes. The spaces between the boards provide enough ventilation to prevent the bottoms of the shingles or shakes from absorbing and retaining moisture. The width of the space between the boards is determined by the shingle or shake weather exposure. Spaced or open sheathing is also used as a roof deck for tile

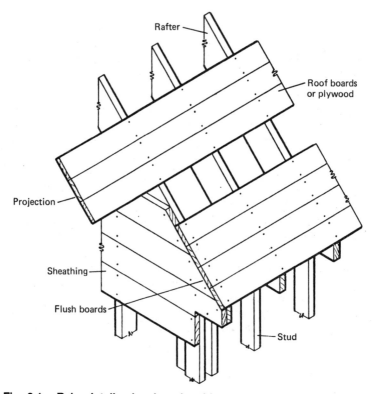

Fig. 3-1. Rake details showing sheathing projection or cut flush with endwall.

roofing and metal or vinyl roofing panels. As shown in figure 3-2, many open roof decks are constructed with closed sheathing along the eaves and sometimes along the eaves as well. Additional information about spaced or open roof decks is provided in chapters 13 and 15.

PLYWOOD SHEATHING

Plywood roof sheathing is available in 4 × 8–foot panels and in different thicknesses. Each panel is nailed to the rafters with its face grain running parallel to the roof eaves, as in figure 3-3. The ends of the panels should be positioned over the center of rafters with vertical seams staggered by at least one rafter spacing to eliminate the possibili-

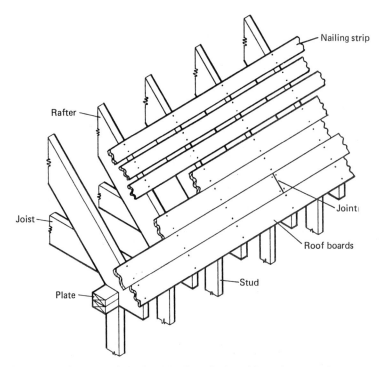

Fig. 3-2. Open roof deck with closed sheathing along roof eave.

ty of a single seam running up the slope of the roof deck. Provide a
1/8-inch edge spacing and a 1/16-inch end spacing between plywood
panels to allow for expansion and contraction.

Only structural plywood that meets the American Plywood
Association's performance rating for roof deck sheathing should be
used. Each panel will be stamped with an APA registered trademark
indicating its performance rating (fig. 3-4).

When roofing with asphalt shingles, roll roofing, or wood shingles
and shakes, plywood panels with a *minimum* thickness of 5/16 are
recommended for use as sheathing when the rafters are spaced the
standard 16 inches apart. A 3/8-inch thick plywood panel is recommended when the rafters are spaced 24 inches apart. Preclips or H
clips should be installed on the horizontal joints of adjoining panels
when the rafters are spaced 24 inches apart to insure a flat, even deck
surface. Better nail penetration and holding power, improved racking

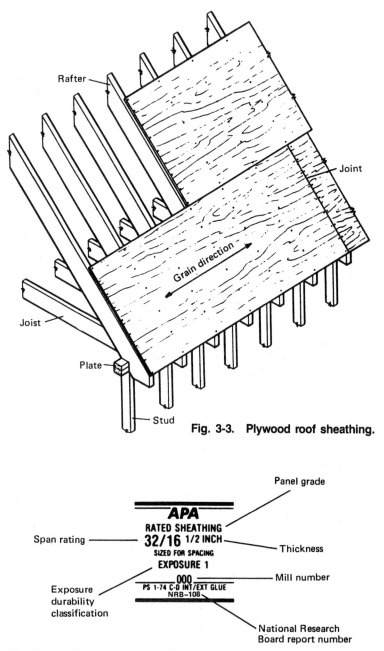

Fig. 3-3. Plywood roof sheathing.

APA

RATED SHEATHING

32/16 1/2 INCH

SIZED FOR SPACING

EXPOSURE 1

000

PS 1-74 C-D INT/EXT GLUE

NRB-108

Panel grade

Span rating

Thickness

Mill number

Exposure durability classification

National Research Board report number

Fig. 3-4. APA plywood sheathing performance rating. *(Courtesy American Plywood Association)*

resistance, and a smoother roof appearance can be obtained by using 3/8-inch thick plywood for 16-inch spacing of rafters and 1/2-inch thick plywood for 24-inch rafter spacings. For 16-inch rafter spacings, 1/2-inch thick plywood is considered minimum when slate and other heavy roofing materials are used.

Nail each plywood sheathing panel 6 inches on center along all bearing edges and 12 inches on center along intermediate members. A sixpenny (6d) common nail or fivepenny (5d) threaded nail should be used for 5/16- and 3/8-inch plywood, and eightpenny (8d) common or sevenpenny (7d) threaded nails for plywood panels of greater thickness. Raw edges of plywood panels should not be exposed to the weather at the gable end of a pitched roof or at the cornice unless the manufacturer has protected the edges with an exterior glue line. Raw edges can be protected by covering them with trim.

Nonwood Roof Decks

Materials such as fiberboard, gypsum products, various forms of concrete, and structural cement-fiber are used in the construction of nonwood roof decks. A deck constructed from one of these materials is sometimes called a *nonnailable roof deck*. Special fasteners and fastening methods are required when applying roofing materials to a nonwood deck. The specifications of the deck material manufacturer should be closely followed to insure that the roofing is properly applied.

Roof Deck Preparation for New Construction

Preparing the roof deck for roofing during new construction is a relatively easy task because the framing and deck materials are new and it is not necessary to strip the deck of old roofing. Roof deck preparation for new construction should begin with a careful inspection for defects in the framing and sheathing before the underlayment (if used) and flashing are applied. Roof ventilation, roof insulation, and the cutting and framing of roof openings should also be considered at this point.

Inspection and Repair

Inspect the sheathing for loose knots or excessively resinous areas. Such defects should be covered with sheet metal patches before the roofing is applied to the deck. The metal patches may be cut from 26-gauge galvanized steel, and should be fastened over the defect with nails of a compatible metal; the edges should be sealed with caulking compound or a suitable asphalt roofing cement.

Make certain that the number and spacing of the rafters is adequate to support the roof deck, the roofing materials, the weight of the workers doing the roofing, and the usual snow and wind loads. Slate, tile, and mineral fiber (asbestos-cement) shingle roofs require greater support than asphalt shingles or roofs covered with relatively lighter roofing materials. Additional rafters and bracing should be installed if the framework is not strong enough to provide adequate support.

Ventilation

Condensation will occur when warm moist air comes in contact with a cold surface. Sources of moisture inside a structure include cooking, bathing, washing clothes, and air conditioning. The moisture rises with warm air and condenses when it reaches the cold roof sheathing. This condensation can damage the sheathing of a wood roof deck, the roof rafters, and the insulation. The problem can be prevented or minimized by permitting air to circulate below the roof deck. This can be accomplished by proper roof ventilation. Roof ventilating methods are described in chapter 5.

Insulation

Insulation can be used to reduce the flow of heat into the attic or attic crawl space or to provide thermal resistance to heat flow in the roof. A vapor barrier installed on the side of the insulation facing toward the interior heated spaces of the structure will reduce moisture penetration. The types of insulating materials and the methods used to install them are described in chapter 4.

Roof Openings

It is usually necessary to cut one or more openings in the roof deck for certain types of construction or for objects that protrude through the roof surface. If the opening is wider than the space between two roof rafters, it will have to be framed. The best time to cut and frame roof openings for chimneys, vent pipes, stacks, skylights, and dormers is during new construction before the underlayment and flashing are applied. An existing roof will require the removal of some of the roofing materials to expose the deck before the opening can be cut and framed. Cutting and framing methods for roof openings are described in chapter 6.

Underlayment

Some roof decks are covered with a protective layer or underlayment of roofing felt. An important function of the underlayment is to protect the interior of the structure from penetration by rain or melting snow, while at the same time allowing moisture from inside the structure to escape. This protection from water penetration is especially necessary when shingles are lifted or removed by high winds or when rain is driven under the roofing.

An underlayment also ensures that shingles will be laid over a dry surface. Moist or wet wood board sheathing may cause buckling and distortion of the shingles. Trapped moisture may also cause the boards to warp. A wet plywood roof deck may result in the delamination of the plywood panels, causing the cross-laminated layers to separate.

Finally, an underlayment prevents any direct contact between the shingles and highly resinous areas on the wood sheathing. Because shingles and resinous spots are chemically incompatible, prolonged contact may result in damage to the shingles.

The method used to apply the underlayment on a roof deck is determined by its pitch or slope. Both pitch and slope, as well as the methods used to calculate them, are described in chapter 2.

Wood decks on pitched roofs with a slope of 2-in-12 or greater over which asphalt shingles are to be laid are usually covered with an underlayment of No. 15 asphalt-saturated roofing felt. Nos. 20 and 30 weight roofing felts are available for other types of roofing applications. All roofing felt is available in 36-inch wide rolls and lengths of 72, 108, and 144 feet per roll. *Never use a coated sheet or heavy felt as an underlayment.* Both types of materials will function as vapor barriers and trap moisture or frost between the roofing material and the roof deck.

If the roof deck has a slope of 4-in-12 or greater (fig. 3-5), nail a metal drip edge along the bottom edge (roof eaves) *before* the felt is laid and proceed as follows:

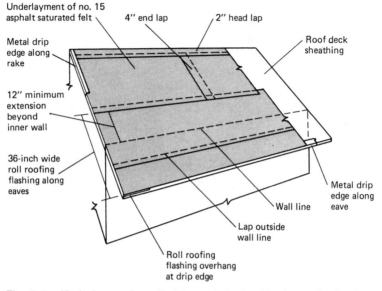

Underlayment of no. 15 asphalt saturated felt

4" end lap

2" head lap

Metal drip edge along rake

Roof deck sheathing

12" minimum extension beyond inner wall

36-inch wide roll roofing flashing along eaves

Metal drip edge along eave

Wall line

Lap outside wall line

Roll roofing flashing overhang at drip edge

Fig. 3-5. Underlayment applied to roof deck with slope of 4-in-12 or more. *(Courtesy Asphalt Roofing Manufacturers Association)*

1. Lay one layer or ply of felt horizontally over the entire roof, lapping each course 2 inches over the underlying one with a 4-inch side lap at end (vertical) joints.
2. Lap the felt 6 inches from both sides over all hips and ridges.
3. Secure the felt to the deck with enough fasteners (roofing nails or staples) to hold it in place until the shingles are applied.
4. Nail a metal drip edge to the rakes of a gable roof *after* the underlayment has been laid.

Where the January average daily temperature is 25° F or less, or where there is a possibility of ice forming along the eaves and causing roof leaks from a backup of water, apply an eaves flashing strip of 36-inch wide heavy roll roofing (50 pound or heavier) along the eaves. The roll roofing should be laid so that it overhangs the metal drip edge ¼ inch and extends up the roof deck to a minimum point of 12 inches inside the interior wall line of the structure. If a second overlapping sheet of roll roofing must be laid to meet this 12-inch minimum requirement, the horizontal lap must occur *outside* the wall line.

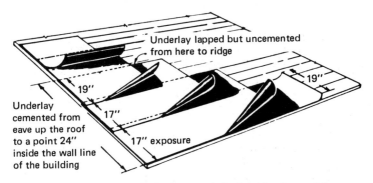

Fig. 3-6. Underlayment applied to roof deck with slope of less than 4-in-12. *(Courtesy Asphalt Roofing Manufacturers Association)*

For roof decks with a slope of 2-in-12 to less than 4-in-12, as in figure 3-6, apply the underlayment as follows:

1. Coat the roof deck with bituminous plastic cement (at two gallons per 100 square feet of roof) from the roof eaves up the roof to a point approximately 24 inches inside the wall line and nail the first course of roofing felt to the deck.
2. Apply two layers of felt parallel with the eaves, laying the first 19-inch wide sheet as a starter course covered by a second layer of 36-inch wide felt.
3. Cover the roof deck with 36-inch wide sheets of felt overlapping the preceding layer by 19 inches to expose 17 inches of the underlying sheet.
4. Secure the underlayment to the roof deck with enough fasteners (roofing nails or staples) to hold the felt in place until the shingles are laid.

Wood decks of pitched roofs with a slope of 2-in-12 or greater over which asphalt shingles are to be laid are usually covered with an underlayment of No. 15 asphalt-saturated roofing felt.

Flashing

Flashing is used to seal the roof against leakage at chimneys, vent pipes, stacks, and other protrusions through the roof surface; at the intersections of different roof planes; and at the abutments of the roof against adjoining vertical walls.

In most cases, the flashing is installed after the underlayment (if used) is laid and before the shingles or other roof covering materials are applied. There are some exceptions. For example, the metal drip edge along the eaves of a gable or hip roof are usually applied before the underlayment is laid. Flashing materials and methods of application are described in chapters 8 and 12.

Gutters

Some gutters are attached to the structure by strap hangers nailed to the roof sheathing (new construction) or to the existing roof surface (reroofing) *before* new shingles or other roof covering materials are applied. When installed, the nailing straps of the strap hangers will be covered by the roofing (see chapter 16).

Roof Deck Preparation for Reroofing

Preparation of the roof deck for reroofing will depend on the condition of the existing roof and the types of materials selected for the new roof. The roof deck requirements for new construction also apply to reroofing. The roof deck must be smooth, flat, and rigid, and must provide a solid nailing base for the new roofing materials.

Inspection and Repairs

Inspect the condition of the old roofing materials to determine whether they should be removed or allowed to remain. Check the roof for loose, cracked, broken, or damaged roofing. Old roofing that shows signs of extensive deterioration must be removed. Tile, slate, and mineral fiber (asbestos-cement) shingles must be removed regardless of their condition because they are too brittle for driving nails. The preparation of the roof deck after removing the existing roofing materials will be the same as described for new construction.

Inspect the rafters and the underside of the sheathing from the attic or attic crawl space for damage or defects and make the necessary repairs (see the discussion of roof deck preparation for new construction). Deteriorated plywood panels and warped, broken, or rotting sheathing boards will have to be replaced. In order to replace damaged sections of the sheathing, it will be necessary first to strip the old roofing from the deck. Only the damaged portions of plywood panel or sheathing board have to be cut away and replaced. Make certain that the replacement piece is large enough to be supported by at least two rafters, nail it flush with eightpenny (8d) nails, and caulk the joints to prevent water from entering the structure along the joints. A rafter can be reinforced without removing the old roofing by nailing a second rafter directly parallel to it.

Roofs are exposed to a variety of weather conditions and eventually develop leaks, which must be repaired before reroofing. The most common types of roof leaks can be traced to one or more of the following causes:

1. Warped, corroded or cracked flashing around chimneys, vent pipes, and other structural interruptions in the roof surface
2. Broken, loose, or missing shingles, tiles, or other types of roofing materials
3. Rusty, loose, or missing nails or other types of fasteners
4. Dried out and cracked roofing compounds or sealers used to seal seams in roof-covering materials
5. Blocked or damaged gutters and downspouts
6. Rotted or cracked plywood sheathing panels or boards
7. Ice dams forming along the roof eaves

Most roof leaks are difficult to trace because the water almost never collects directly under the point at which it enters. It will frequently run down a rafter or beneath the sheathing for a considerable distance before dropping to the attic or attic crawl space floor. To trace a leak, find the point at which the water has been collecting and then look for water stains along the joist above it. These stains will usually leave a path to the point at which the water has been entering. Sometimes a small pinhole of light can be seen. If the leak can be located, shove a straight piece of wire through the hole to indicate its position from the outside. If the point at which the water is entering cannot be determined by this method, calculate the approximate location by measuring the distance from the end of the joist to the point at which the stains end. Add to this measurement the length of the roof overhang, and this will be the approximate distance of the leak from the edge of the roof.

If it is raining, trace the flow of water up along the joist to its approximate entry point. Repairs are impossible until the rain stops, but the point at which the water enters can be marked and a pail

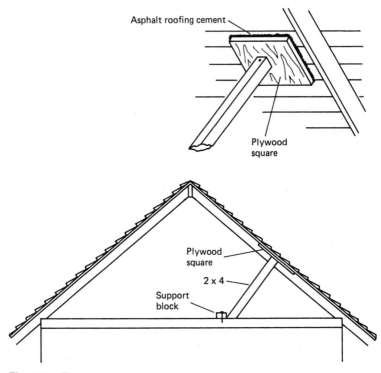

Fig. 3-7. Temporary roof repairs.

placed where the puddle is forming. To ensure that the dripping water does not change its position on the joist, run a string or wire from the joist down to the pail. After the leak has been located from the inside (or at least its approximate location determined), go onto the roof and try to find the cause of the leak.

Repairing a leak is *not* recommended during inclement weather. The leak should be stopped with a temporary repair until permanent repairs can be made (fig. 3-7).

As in new construction, the roof framing must be strong enough to support the deck and the new roofing, especially if heavy roofing material such as slate, tile, or mineral fiber (asbestos-cement) shingles are used. The framework can be strengthened by installing additional rafters and bracing.

Removing Roofing Materials

Begin by removing the ridge shingles or caps with a claw hammer or shovel. If the structure has a hip roof, the hip ridge shingles or caps should be removed next. Removing the covering over the roof ridge will expose the nails along the top edge of the last course of shingles or roofing. Work down the slope of the roof toward the eaves and remove each course of shingles or roofing. Tile, slate, and mineral-fiber shingles are brittle and can be removed by first striking them with a hammer, removing the broken pieces, and then pulling the roofing nails. Built-up roofing membranes can be removed by cutting through the membrane to the roof deck and then prying it up with a shovel. Special care should be taken when removing existing metal flashing, since if kept intact it can be used as a pattern for cutting new flashing.

Deck Preparation

OLD ROOFING REMOVED

If the existing roofing is removed, the roof deck should be prepared and the new roofing materials applied as in new construction. After the roofing materials have been removed, the roof deck should be prepared as follows:

1. Repair the roof framework where necessary to provide a level deck surface.
2. Reinforce the roof framework if roofing materials heavier than the original ones are to be used. This will usually be the case when roofing with slate, tile, or mineral fiber (asbestos-cement) shingles.
3. Remove rotted, warped, split, or broken wood sheathing boards, or delaminated plywood sheathing, and replace them with new sheathing of like kind.
4. Pull all protruding and rusted nails and renail the sheathing at new locations with eightpenny (8d) nails. Fill the old nail holes with asphalt cement.
5. Cover all large cracks, slivers, knot holes, loose knots, pitch spots, and excessively resinous areas with a piece of galvanized sheet metal, securely nail the patch to the sheathing, and

coat the edges of the patch with asphalt cement to prevent leakage.

6. Sweep the deck clean of all loose debris, and, if wet, allow it time to dry thoroughly before applying the new roofing material.

OLD ROOFING RETAINED

The procedures used to prepare the old roof for reroofing are determined by the type of existing roofing materials, their condition, and the type of materials used for reroofing. These procedures are covered in the chapters that describe specific types of roofing materials and their methods of application The procedures for inspecting and repairing the roof framework and roof deck are the same for both new construction and reroofing (see the discussion of roof deck preparation for new construction). If the underlayment is still serviceable, it can be left in place for the new roofing. Its condition can be checked by removing a small section of the old roofing. If the decision is to try to retain the old underlayment, the shingles or roofing materials must be carefully removed to avoid damaging it.

CHAPTER 4

Roof and Attic Insulation

- Insulating Materials
- Vapor Barriers
- Installation Methods

- Roof and Ceiling Applications
- Special Applications
- Hiring an Insulation Contractor

Heat always flows from an area of higher temperature to one of lower temperature. This means that heat flows out of a structure in the winter and into it in the summer. The heat will continue to flow until the temperatures between the inside and outside of the structure are approximately equal. According to studies conducted by the government and others, approximately 27 percent of the heat is lost through the roof and attic floor of a poorly insulated two-story structure during the winter. For a one-story structure, the heat loss can be as high as 70 percent. If there is no insulation in the attic floor or roof, or if the insulation is inadequate, the furnace or boiler will consume huge amounts of fuel or energy to maintain the desired temperature against this constant heat loss. The same holds true for air conditioning equipment during the summer. It is clear then that a properly insulated structure will provide maximum comfort in summer and winter and save thousands of dollars in utility bills during the life of the structure.

In addition to presenting a heating and cooling problem, heat loss from the structure during the cold winter months may also affect the condition of framing and roofing materials. The heat carries water vapor with it as it passes through the attic floor and roof. When this water vapor comes in contact with the cold attic floor or roof deck sheathing, it condenses and the moisture accumulates in the cavities between attic floor joists and roof rafters, in the wood deck sheathing, and in the roofing materials. Accumulations of moisture will damage insulation installed in the framing cavities, cause roofing materials to rot and deteriorate, and rot or warp floor joists, roof rafters, and sheathing boards or panels. Moisture damage can be prevented by installing a vapor barrier with the insulation.

The best time to install insulation and vapor barriers is during new construction when it is easiest to install and adds only about 1 percent to the initial building cost. These materials are more difficult and costly to install in an existing structure, but the protection they provide plus the reduced heating and cooling costs derived from their use more than compensate for the additional effort and expense.

Insulating Materials

Insulating materials are often described in terms of their R-value or thermal resistance, that is to say, their resistance to heat flow. Inches of insulation were at one time used to measure insulation value, but this measurement could not account for the fact that different insulation materials of the same thickness did not always show the same degree of heat resistance. For example, six inches of fiberglass loose-fill insulation has an R-value of R-13, whereas six inches of cellulose loose-fill insulation has an R-value of R-21. Using inches of insulation as a measurement also failed to account for the heat resistance offered by other building materials used with insulation, such as roof deck sheathing, attic floor boards, or drywall. As a result, the insulation industry elected to use R-values to designate the thermal resistance of a material, since they give a more precise index of the resistance to heat flow, regardless of the thickness used.

The type and thickness of insulating material selected for a job will depend on the anticipated heat loss or gain for the structure, local building code requirements, the cost of the material, and other factors. The insulating materials used to insulate the roofs, attics, and attic crawl spaces of houses and small commercial structures against heat transmission are commonly available in the form of batts and blankets, reflective insulation, loose fill insulation, and rigid board insulation.

R-values increase with the thickness of the insulation, but on a

Table 4-1. R-values for Different Types of Insulation

Insulation Type and Material	Approximate R-Value
BATTS AND BLANKETS	
Fiberglass, per inch	3.0
Rock wool, per inch	3.6
Cellulose, per inch	4.0
RIGID BOARD INSULATION	
Fiberglass, 1-inch board	4.0
Polystyrene, 1-inch TG board	3.5
Polyurethane, 1-inch board	6.2
Foil-faced urethane, 1-inch board	8.3
Isocyanurate, 1-inch board	9.0
BLOWN-IN (LOOSE FILL) INSULATION	
Fiberglass, per inch	2.7
Rock wool, per inch	2.73
Cellulose, per inch	3.7
FOAM INSULATION	
Urea-formaldehyde, per inch	5.0
Urethane, per inch	6.2

sliding scale after the first 3 inches. For example, 3 inches of mineral wool blanket insulation will have an insulation value of approximately R-10 or R-11, but 6 inches of the same insulation will not double the R-value. R-values per inch of installed thickness for various types of insulation materials are listed in table 4-1.

Batts and Blankets

Batt-and-blanket-type insulation is probably the most commonly used insulating material in houses and small buildings. This is a flexible insulation made of mineral wool (fiberglass or rock wool) or, less commonly, cellulose fiber (figs. 4-1 and 4-2). Blankets are recommended for long runs between attic floor joists or roof rafters because fewer joints are formed between sections of the material.

Blankets are sold in continuous rolls of up to 64 feet in length, whereas batts are commonly available in 4-foot or 8-foot lengths. Both are generally available in 15-inch widths for installation in structures with standard 16-inch-on-center joist or rafter spacing, and 23-inch

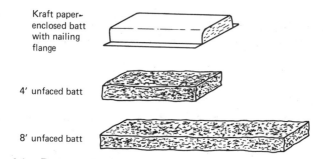

Kraft paper-enclosed batt with nailing flange

4' unfaced batt

8' unfaced batt

Fig. 4-1. Batt-type insulation.

widths for 24-inch spacing. Extra widths are supplied by manufacturers for special applications. The most commonly used batt and blanket thicknesses are 3½ and 6 inches, but other thicknesses are also available. Some insulation is also produced in full widths of 16 and 24 inches. When full width insulation is installed between attic floor joists, it expands over the tops of the framing members to provide a continuous protective layer of insulation.

Batts and blankets are available with a vapor barrier membrane made of a foil/kraft laminate or kraft paper attached to one side. These are called "faced" batts and blankets, and the vapor barrier side must always be installed facing toward the interior of the structure. An "unfaced" batt or blanket has no vapor barrier membrane and should be installed with a separate vapor barrier.

Loose Fill Insulation

Loose fill insulation is recommended for filling in the spaces between the joists of attic floors. It is best suited for irregular or nonstandard

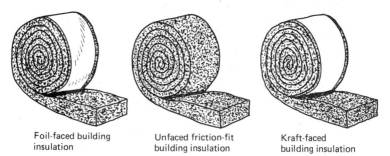

Foil-faced building insulation

Unfaced friction-fit building insulation

Kraft-faced building insulation

Fig. 4-2. Blanket-type insulation.

Table 4-2. R-values per Installed Thickness of Fiberglass, Rock Wool, and Cellulose Fiber Hand-poured Loose Fill Insulation (*Courtesy Mineral Insulation Manufacturers Association, Inc.*)

Insulation value of installed thickness	Fiberglass loose fill	Rock wool loose fill	Cellulose fiber loose fill
R-11	5″	4″	3″
R-19	8–9″	6–7″	5″
R-22	10″	7–8″	6″
R-30	13–14″	10–11″	8″
R-38	17–18″	13–14″	10–11″

joist spacings or where the space between joists has too many obstructions.

Loose fill insulation is made from a fibrous material, such as fiberglass, rock wool, or cellulose, or from a granular material such as vermiculite or perlite. Granular materials are hand poured between framing members. Fibrous materials may be either hand poured or blown in place.

The R-values of fiberglass, rock wool, and cellulose fiber for a number of different installed thicknesses are listed in table 4-2. Cellulose fiber has about 30 percent more insulation value than rock wool. Rock wool, in turn, has a slightly better insulation value than fiberglass for the same installed thickness. Both fiberglass and rock wool are fire and moisture resistant. Cellulose fibers must be chemically treated to be fire and moisture resistant; however, there is evidence that the heat in a hot attic or attic crawl space will cause the treatment to break down and lose its effectiveness. When using cellulose fiber loose fill insulation, check each bag for a label indicating that the material meets federal specifications.

Vermiculite and perlite have about the same insulating value. Both materials are more expensive than fiberglass, rock wool, or cellulose fiber loose fill insulation.

One disadvantage of loose fill insulation, particularly the fibrous type, is its tendency to settle after a period of time and lose some of its effectiveness as an insulating material.

Rigid Board Insulation

Rigid board insulation is commonly available in 2 × 8, 4 × 8, and 4 × 4–foot panels for installation on roof decks under the roofing

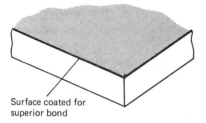

Surface coated for
superior bond

**Fig. 4-3. Perlite rigid board
roof insulation.**

materials or fastened to the bottoms of exposed ceiling beams. They are available in thicknesses ranging from ½ inch to 4 inches and should be installed by qualified, professional insulation contractors.

Rigid insulation boards are made from compressed fiberglass, expanded polystyrene beads, expanded perlite beads, urethane foam, and other materials capable of providing high insulating values (figs. 4-3 to 4-5). The insulating value of rigid insulation boards per inch of installed thickness depends on the material or materials used in its construction. Generally, their R-values are higher than batts and blankets, or loose fill insulation, but slightly lower than foam insulation.

Reflective Insulation

Reflective insulation is made of aluminum foil or specially coated paper designed to reflect radiant heat. It *must* be installed so that there is about a 2¾- to 1-inch air space between it and the ceiling board or sheathing. Reflective insulation will lose almost all of its insulating value if this air space is not provided. Even when properly applied, reflective insulation is neither as effective as other types of insulating materials nor as commonly used.

Aluminum foil reflective insulation is available in single or multiple layers (fig. 4-6). Multiple layer reflective insulation is compressed

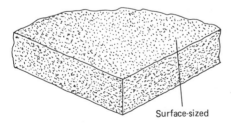

Surface-sized

**Fig. 4-4. Fiberboard roof
insulation.**

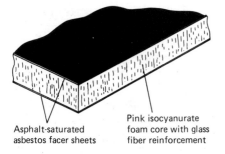

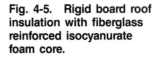

Asphalt-saturated
asbestos facer sheets

Pink isocyanurate
foam core with glass
fiber reinforcement

Fig. 4-5. Rigid board roof insulation with fiberglass reinforced isocyanurate foam core.

when it is packed and shipped from the factory, and it must be pulled apart or expanded prior to installation.

Foam Insulation

Foam insulation requires special equipment for installation. It must be installed by a qualified and professional insulation contractor or the quality of its application may be inconsistent.

Foam insulation has the highest R-value per inch of installed thickness of any of the insulating materials. It may be used to insulate

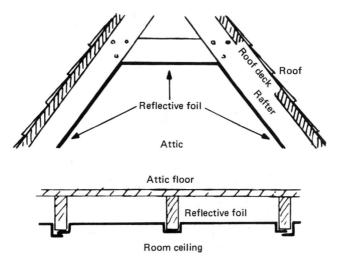

Fig. 4-6. Reflective foil installations. *(Courtesy Howmet Aluminum Corp.)*

unfinished attic floors, finished attic floors, or the top surface of a slate, built-up roof. When used between rafters or joists, it will completely fill the framing cavity if properly installed. Foam insulations are both fire- and moisture-resistant. Because of their moisture resistance, a separate vapor barrier is not required.

Both urea-formaldehyde and urethane are used as foam insulation. Urethane foam produces a toxic gas when burned. Many users have also complained of unpleasant, lingering odors after the application of either type of foam insulation. Some have complained of headaches, nausea, and more serious symptoms. Foam insulations have a tendency to settle after a period of time and lose some of their insulating value.

Vapor Barriers

A vapor barrier is a moisture-resistant material used to block or retard the passage of water vapor. It is always installed on the winter warm side of the insulation, that is, between the insulation and the heated rooms or spaces. Properly applied, the vapor barrier will prevent condensation and the damaging accumulation of moisture in the insulation and other building materials, such as wood decks and wood framing members. A vapor barrier should always be used when the outside January temperature averages 40° F or below, or when the constant relative humidity inside the structure normally reaches 50 percent or higher. Batt or blanket insulation is available faced with either an aluminum foil or kraft paper vapor-resistant membrane. A separate vapor barrier of 4 mil or 6 mil polyethylene sheet should be used when the attic is insulated with unfaced batts or blankets, loose fill insulation, and some types of rigid insulation board (fig. 4-7).

Installation Methods

Insulation can be fastened to the framing, laid in place, poured between framing, blown between framing, or foamed in place, depending on the type used. *Foam insulation should be installed only by professional insulation contractors.*

Wear long-sleeved and loose-fitting clothing, gloves, and goggles when installing insulation. A hard hat should be worn when working

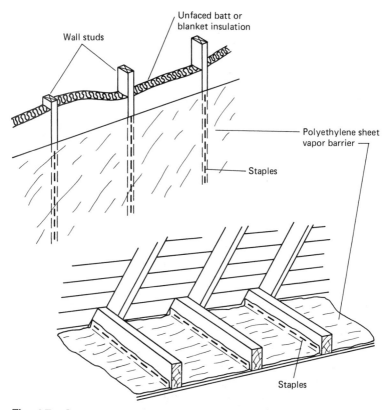

Fig. 4-7. Separate polyethylene sheet vapor barrier.

in an attic or attic crawl space (fig. 4-8). A suitable respirator is recommended when working where there are heavy concentrations of dust or potentially harmful fumes.

Always shower with soap and water after handling insulation. Wash work clothes separately from other clothing and then rinse the washing machine thoroughly before using it again.

A hammer, stapler, measuring tape, and straight edge are useful tools for installing batt and blanket insulation (fig. 4-9). A knife with a serrated-edge blade is recommended for cutting batts and blankets. A rake is recommended for pushing or pulling blankets to the eaves edge when there is not enough headroom. It can also be used to level loose fill insulation.

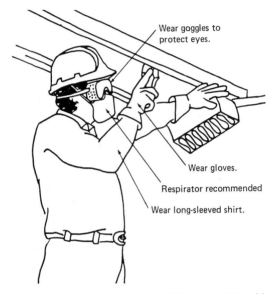

Fig. 4-8. Protective clothing and gear. *(Courtesy Johns-Manville Corp.)*

Several 10- or 12-inch wide walk boards should be used for support when insulating the floor of an attic or attic crawl space. Using walk boards will also eliminate the possibility of stepping through the ceiling. A portable light, such as a mechanic's trouble light, and a long extension cord are recommended for a dark attic or crawl space.

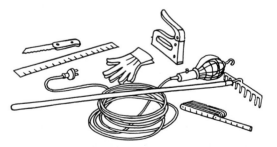

Fig. 4-9. Recommended hand tools for installing insulation. *(Courtesy Mineral Insulation Manufacturers Association, Inc.)*

Installing Batt or Blanket Insulation Between Rafters

Batt or blanket insulation should be stapled to the framing when installed between the roof rafters of an unfinished attic. The fastening methods used are (1) inset stapling and (2) face stapling. Both methods require that the batts or blankets form a tight fit between the rafters.

INSET STAPLING

When inset stapling, press the batts or blankets gently at the sides into the cavity between the rafters until the edge of the stapling flange is flush with the face of the framing (fig. 4-10). Then staple the flange to the side of the rafter. When inset stapling insulation between the inclined framing members of a cathedral ceiling, start stapling at the top and work down. Use enough staples to hold the insulation firmly in place and avoid gaps, loose spaces, and fishmouths between the flanges and framing.

Fig. 4-10. Installing insulation from below (inset stapling). *(Courtesy Johns-Manville Corp.)*

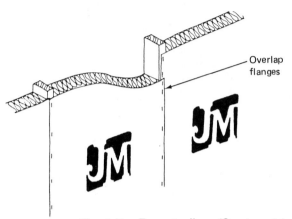

Fig. 4-11. **Face stapling.** *(Courtesy Johns-Manville Corp.)*

FACE STAPLING

Place the insulation between the rafters with the facing material flush with the face of the framing. Both ends should fit tightly and the stapling flanges should overlap the framing. Staple the flanges to the face of the framing every 6 to 8 inches, using enough staples to hold the insulation firmly in place and avoid gaps and fishmouths (fig. 4-11). The flanges of the faced insulation in the next cavity will overlap the previously stapled flange. When 48-inch long batts are used, adjacent batts must be tightly butt-jointed.

Installing Batt or Blanket Insulation Between Attic Floor Joists

Kraft paper faced or aluminum foil faced insulation may be installed from below in new construction before the ceiling is in place by inset stapling as in figure 4-10. Push the insulation between the joists and hold it up with one hand while stapling the flange on both sides every 6 to 8 inches. Faced or unfaced batt or blanket insulation may also be laid between the attic floor joists after the ceiling has been nailed in place. If the insulation is faced, make sure the facings lie toward the heated spaces of the structure. If unfaced batts or blankets are used, a separate vapor barrier will usually be required.

Standard size batt or blanket widths are laid parallel to the attic

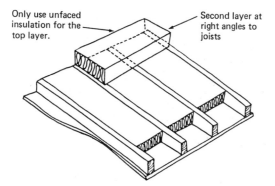

Only use unfaced insulation for the top layer.

Second layer at right angles to joists

Fig. 4-12. Installing second layer of insulation at right angles to joists. *(Courtesy Johns-Manville Corp.)*

floor joists. The insulation may be increased by laying a second layer of *unfaced* insulation at right angles to the joists (fig. 4-12). Full-width R-30 batts or blankets may be used instead of two layers of standard-width insulation. As shown in figure 4-13, full-width insulation butts against the adjoining insulation above the joists.

Cross bracing or bridging between attic floor joists may be in-

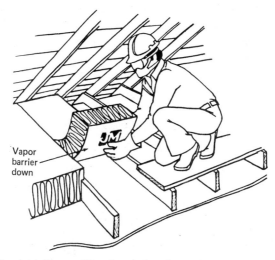

Vapor barrier down

Fig. 4-13. Installing ceiling insulation from above (unfloored attics). *(Courtesy Johns-Manville Corp.)*

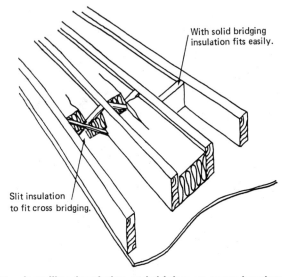

With solid bridging
insulation fits easily.

Slit insulation
to fit cross bridging.

Fig. 4-14. Installing insulation at bridging or cross bracing.
(Courtesy Johns-Manville Corp.)

sulated by slitting the batt or blanket vertically at the center and packing one half into the lower opening and the other half into the upper opening (fig. 4-14). The batts or blankets may also be butted against the cross bracing and the gaps filled with pieces of loose insulation.

Use a board placed across at least three joists for body support while installing the insulation. Avoid stepping on the insulation because the weight of the body may damage the ceiling (see fig. 4-13).

Pouring Insulation Between Attic Floor Joists

Fiberglass, rock wool, cellulose fiber, vermiculite, and perlite loose fill insulation can all be hand poured between joists to the desired level. These loose fill insulating materials are sold in large bags, which are usually labeled for hand pouring. For example, bags containing hand poured rock wool or fiberglass will be labeled *pouring wool*. The bag labels will also contain information indicating the number of bags, the minimum thickness, and the minimum weight per square foot required to reach a specified R-value.

Hand poured loose fill insulation is easy to install, but limited in use to insulating unfinished attic floors. Fiberglass, rock wool, and cellulose fiber insulation become compressed in the bag and must be expanded by shaking or fluffing the material before it is installed, or separating it with a rake as it is being installed in the cavities between the framing. The air pockets in the expanded material will increase the R-value of the insulation to the level estimated for the installed thickness (see table 4-2). Unfortunately, it is difficult to maintain a constant R-value. A leveling board similar to the one shown in figure 4-15 is useful for obtaining a constant insulation thickness.

A vapor barrier must be purchased and installed on the surface next to the heated spaces of the structure to protect the insulation from moisture damage.

Blowing Insulation into Framing Cavities

Fiberglass, rock wool, and cellulose fiber loose fill insulation may be pneumatically installed in both finished and unfinished attic floors. The bags of insulation will be labeled for blowing and the label will contain the same kinds of installation instructions found on the bag labels of pouring insulation.

The insulation may be installed by an insulation contractor or

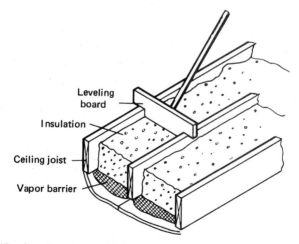

Fig. 4-15. Leveling loose fill insulation.

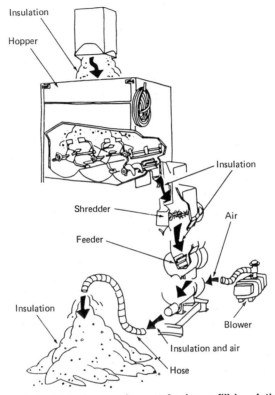

Fig. 4-16. Typical blowing equipment for loose-fill insulation.
(Courtesy Johns-Manville Corp.)

by a homeowner with a blowing machine obtained from a local tool and equipment rental store (fig. 4-16). Some insulation dealers will loan a blowing machine without charge if the insulation is purchased from them.

When loose fill insulation is blown into a closed space, such as between the floor joists of a finished attic, enough must be blown in to fill the entire space. Cellulose fiber will fill small corners and spaces more consistently than either fiberglass or rock wool because it consists of smaller tufts.

Blown-in loose fill insulation has a low moisture resistance. As a result, a vapor barrier must be purchased and installed separately on the surface next to the heated spaces of the structure.

Installing Vapor Barriers

Never add a vapor barrier over existing insulation in an attic floor because it will trap moisture in the ceiling and in older insulating materials. Remove the old insulation, install the vapor barrier, and fill the spaces between the attic floor joists with new insulation. Staple the vapor barrier at wall plates, joists, studs, and rafters (see fig. 4-7). Try to plan joints between sections of the material so that they overlap over a framing member. Make sure the material is smooth and flat as it is stapled to the framing members because gaps will allow air infiltration to carry moisture into the insulation.

Roof and Ceiling Applications

The type of insulation selected for a roof, attic, or attic crawl space, and the thickness used, is largely determined by such factors as the type of construction, insulating values of the materials, estimated heat loss before and after application of the materials, roof slope or pitch, local building codes, insurance ratings, and cost.

Always check the local building code first for restrictions either on the type of insulation that can be used or on its installation method. A building permit may also be required under certain circumstances, particularly if extensive alterations are required.

Joist-and-Rafter Roof Construction

The most common type of residential roof construction consists of pitched rafters joined together at one end to form a ridge and connected by horizontal joists at the other end. The joined rafter and joist framing forms a triangle that encloses an attic or attic crawl space. The methods used to insulate a joist-and-rafter type roof will depend on whether the attic is finished and used as a living space, unfinished with a floor, or unfinished with no floor. The same methods used to insulate an unfinished attic are also used to insulate the smaller attic crawl space.

FINISHED ATTIC

A finished attic should be insulated at the points shown in figures 4-17 to 4-19. No insulation is required in the attic floor between the knee

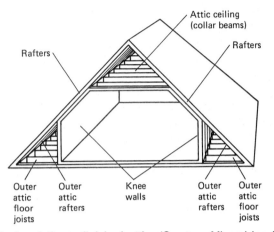

Fig. 4-17. Insulating a finished attic. *(Courtesy Mineral Insulation Manufacturers Association, Inc.)*

walls. If the outer attic rafters are insulated, insulation is not required in the outer attic floor joists or knee walls. The outer attic gables should be insulated if either the outer attic rafters or outer attic floor joists are insulated.

The amount of insulation required for a job will depend on the installed depth and total surface area covered by the material. The

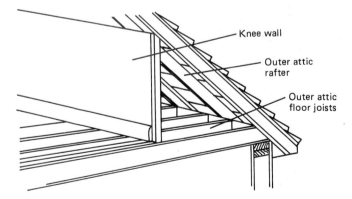

Fig. 4-18. Knee wall, outer attic rafters, and outer attic floor joists. *(Courtesy Mineral Insulation Manufacturers Association, Inc.)*

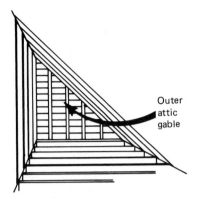

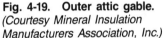

Fig. 4-19. Outer attic gable.
(Courtesy Mineral Insulation Manufacturers Association, Inc.)

surface area is calculated by first determining where the insulation is to be installed and then calculating the square footage of the area or areas. For example, the finished attic of the gable roof shown in figure 4-20 has insulation in the attic ceiling joists (A), the roof rafters (B), the outer attic rafters (C), and the outer attic gables (D). The calculations required to find the total area of insulation are as follows:

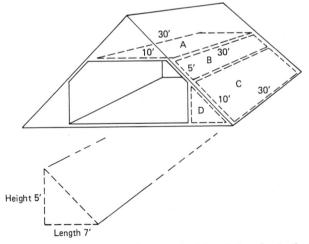

Fig. 4-20. Calculating required amount of insulation for typical finished attic.

1. Attic ceiling (A in fig. 4-22)

    Length (30 ft.) × width (10 ft.) × 1 =   300 sq. ft.

2. Roof rafters (B in fig. 4-22)

    Length (30 ft.) × width (5 ft.) × 2 =   300 sq. ft.

3. Outer roof rafters (C in fig. 4-22)

    Length (30 ft.) × width (10 ft.) × 2 =   600 sq. ft.

4. Outer attic gables (D in fig. 4-22)

    Length (7 ft.) × height (5 ft.) × 2 =   70 sq. ft.

    Total square footage   1,270 sq. ft.

There is only one attic ceiling area for the finished attic illustrated in figure 4-20, but two roof rafter areas (one on each side of the roof). Consequently, the roof rafter area must be doubled in the calculations to find the total square footage for this section of the roof. The same holds true for the outer attic rafters. The outer attic gable area will have to be multiplied by four to find the total square footage because there are two of them at each end of the roof.

A finished attic is more difficult to insulate than an unfinished one unless there is a hatch in the ceiling or openings in the knee walls to provide access to the areas where the insulation will be installed. If there is no access to these areas, a contractor should install the insulation. If access is possible, measure the depth of insulation with a steel rule (fig. 4-21). No additional insulation is required if there is at least six inches or more.

The ceiling of a finished attic may be insulated with batts, blankets, or blown-in insulation. Batts or blankets are installed by laying them between the ceiling joists. Blown-in insulation is installed by holding the hose parallel to the joists and blowing the material

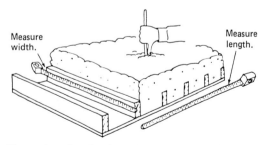

Fig. 4-21. Measuring depth of insulation. *(Courtesy Johns-Manville Corp.)*

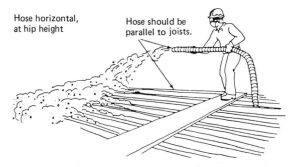

Hose horizontal, at hip height

Hose should be parallel to joists.

Fig. 4-22. Blowing insulation between ceiling joists. *(Courtesy Johns-Manville Corp.)*

into the framing cavities until the minimum recommended depth for the desired R-value is reached (figs. 4-22 and 4-23). This may be predetermined by calculating the number of bags of blown-in insulation for required R-value for 1,000 square feet of ceiling area. The depth of the insulation may be checked with a straight wire and tape measure, as in figure 4-21, and by calculating the R-value for each inch of installed insulation.

Roof Truss Construction

Some roofs are supported by truss assemblies. The lightweight wood truss shown in figure 4-24 is an example of the type used extensively in residential construction. It consists of upper and lower chords (wood framing) reinforced by web members and connected by wood or metal gussets.

A roof supported by trusses is insulated by installing thick mineral wool blankets between the ceiling joists (lower chords), as in figure 25. The mineral wool blankets should have an insulating value of R-19,

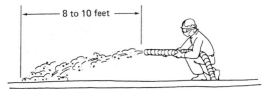

8 to 10 feet

Fig. 4-23. Blowing technique. *(Courtesy Johns-Manville Corp.)*

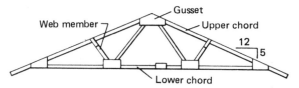

Fig. 4-24.   Lightweight wood truss roof construction.

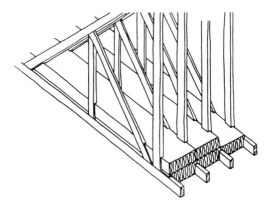

Fig. 4-25.   Installing insulation between joists of truss-type roof.
(Courtesy Johns-Manville Corp.)

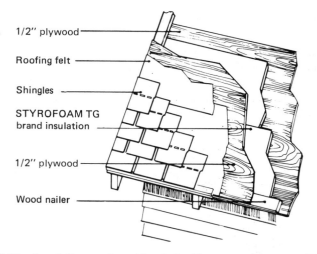

Fig. 4-26.   Insulating roof constructed with exposed beam ceiling.
(Courtesy Dow Chemical Co.)

94

R-22, or R-30, depending on insulation requirements of the structure. The blankets usually expand enough above the lower chord to close the gaps between them except where truss webs meet the chord. Holes or gaps not closed by the expansion of the blankets may be filled with loose mineral wool. If a second layer of blanket insulation is used, the blankets should be laid parallel to the joists (chords) in the same direction as the blankets in the first layer (see fig. 4-25).

Pitched Roofs with Cathedral Ceilings

In structures with pitched or cathedral ceilings, the ceiling joists and roof rafters are combined into the same basic structural element. The method used to install the insulation will depend on whether the rafters are exposed or unexposed in the finished structure.

EXPOSED RAFTERS

If the rafters or beams are exposed, a roof deck composed of materials having the required insulating value and a vapor barrier can be applied above them. If the roof deck is constructed of wood boards or plywood sheathing, rigid insulation of the required R-value can be applied to the top surface of the deck. Construction details of a roof insulated with rigid panels is shown in figure 4-26. In this example, rigid insulation has been installed on top of wood board (plank) sheathing and overlaid with plywood to serve as a nailing base for the shingles. A suitable vapor barrier installed between the insulation and roof deck will protect the insulation and roofing materials from moisture damage.

UNEXPOSED RAFTERS

If the rafters or beams are not exposed, batt or blanket insulation can be stapled or temporarily held in place by friction against the rafters before the ceiling is installed (fig. 4-27). An air space of at least 1 inch should be left between the top of the insulation and the roof sheathing for adequate air movement and ventilation (figs. 4-28 and 4-29). Install a polyethylene vapor barrier of 4 mil thickness or more between the insulation and ceiling for moisture protection.

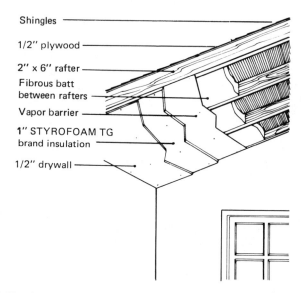

Shingles

1/2" plywood

2" x 6" rafter

Fibrous batt
between rafters

Vapor barrier

1" STYROFOAM TG
brand insulation

1/2" drywall

Fig. 4-27. Insulated sloped ceiling. *(Courtesy Dow Chemical Co.)*

Flat Roof Construction

Flat roofs are commonly covered with built-up roofing systems consisting of several layers of materials. The manufacturer of the roofing materials for a built-up roof will recommend the type of insulation to use, where to install it, and the recommended installation method. Because special equipment, training, and experience are required for

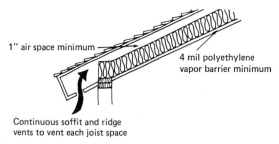

1" air space minimum

4 mil polyethylene
vapor barrier minimum

Continuous soffit and ridge
vents to vent each joist space

Fig. 4-28. Insulating cathedral ceiling of conventional pitched roof. *(Courtesy Johns-Manville Corp.)*

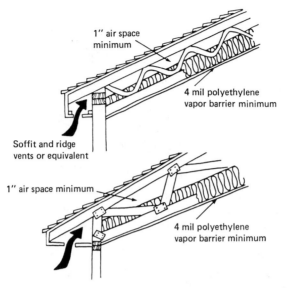

Fig. 4-29. Insulating cathedral ceiling of truss-type roofs. *(Courtesy Johns-Manville Corp.)*

built-up roofing, the insulation and roofing materials should be installed by a qualified, professional roofing contractor.

Special Applications

Attic knee walls, sloping ceilings, soffit vents, water pipes, attic access openings, recessed lights and electrical wiring, and chimneys, stacks, or other small openings must be properly insulated or the roof and attic will lose heat in the winter and gain heat in the summer.

Soffit Vents

Insulation in the attic floor should extend out far enough to cover the top plates of the exterior walls, but should not block the vent openings on roofs constructed with soffit vents (fig. 4-30). Always maintain a minimum 1-inch clearance between the roof sheathing and the insulation to insure the unobstructed movement of air from the soffit vent openings into the attic. If the insulation blocks the air passage,

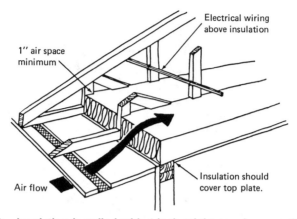

Fig. 4-30. Insulation installed with 1-inch minimum air space for free flow of air from eave vents. *(Courtesy Johns-Manville Corp.)*

compress it by stapling through the blanket or batt to the wall top plate or by installing a baffle made of wood board or heavy corrugated stock (figs. 4-31 to 4-33). Prefabricated baffles are available in either material. Using baffles to shield the soffit vents is especially important when loose fill insulation is blown into the attic or attic crawl space. Sections of insulation batts can also be used as baffles (fig. 4-34).

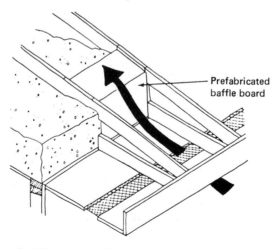

Fig. 4-31. Prefabricated baffle board near soffit vent. *(Courtesy Johns-Manville Corp.)*

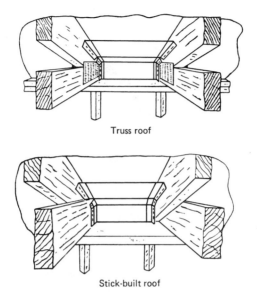

Truss roof

Stick-built roof

Fig. 4-32. Heavy corrugated stock baffle stop applications.
(Courtesy Insulation Products Inc.)

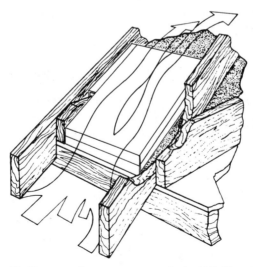

Fig. 4-33. Air flow over heavy corrugated stock baffle stop.
(Courtesy Insulation Products Inc.)

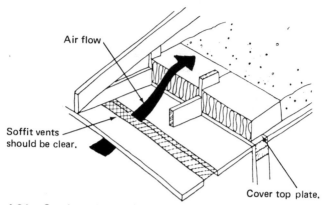

Air flow

Soffit vents
should be clear.

Cover top plate.

Fig. 4-34. Section of batt or blanket insulation used as baffle stop.
(Courtesy Johns-Manville Corp.)

Water Pipes

Water pipes running through an unfinished attic or attic crawl space
should be insulated to protect them from freezing. In moderate
climates, install the same depth of insulation on top of the pipes as
there is between the pipes and a warm-in-winter wall surface. In severe
climates, the pipes should be located as close as possible to the warm-
in-winter wall surface with adequate insulation on top of the pipes.

Chimneys, Stacks, and Other Small Openings

Stuff unfaced fiberglass or rock wool insulation between the framing
and a chimney, stack, vent pipe, duct, or water pipe (fig. 4-35). For
smaller openings, rip hunks of insulation from the blanket or batt and
hand pack the space.

Recessed Lights and Electrical Wiring

Place wood baffles around recessed light fixtures to keep the insulation at least 3 inches away (fig. 4-36). The insulation should never
be allowed to cover a light fixture. Covering a fixture may result in
a fire hazard by causing it to overheat.

Insulation should be installed between the electrical wiring and

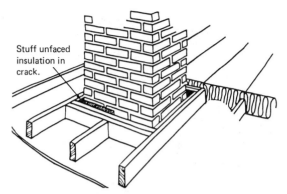

Fig. 4-35. **Insulating around chimneys.** *(Courtesy Johns-Manville Corp.)*

the ceiling of the room below. For best results, make sure all electrical wiring is located above the insulation in the attic or attic crawl space.

Attic Kneewalls and Slopes

Some structures are built with a half story or finished attic. In this type of construction, the top floor ceiling area will usually have both a flat and sloping ceiling extending down to vertical kneewalls. The recommended method of insulating this type of construction is to install batts or blankets in the kneewalls and sloping ceilings (fig. 4-37). Batts, blankets, or blowing wool may be used in the flat surface areas. Blowing wool may also be blown into the kneewalls if a retainer is

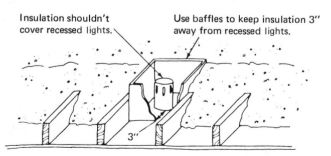

Fig. 4-36. **Insulating around recessed lights.** *(Courtesy Johns-Manville Corp.)*

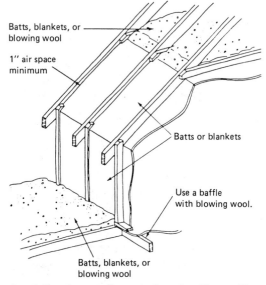

Fig. 4-37. Insulating kneewalls and sloped ceilings. *(Courtesy Johns-Manville Corp.)*

used. Always leave an air passage of at least 1 inch above the insulation in the sloping ceiling. Be sure to insulate any flat areas behind the kneewalls. It may be necessary to cut a hole through a closet wall or attic wall, or make a roof opening to gain access to these areas.

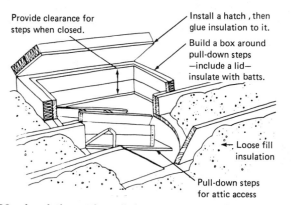

Fig. 4-38. Insulating attic pull-down stairways. *(Courtesy Johns-Manville Corp.)*

These openings should be permanently closed after the insulation is installed.

Hinged or folding pull-down stairs are often used instead of permanent stairs to gain access to attics used primarily for storage. Such stairs operate through an opening in a hall or closet ceiling and swing up into the attic space when not in use. To insulate this type of opening, build a framework around the edge and install a hatch or hinged cover to close it off when the stairs are not in use. Glue batt or blanket insulation to the top of the hatch and lay it up against the framework (fig. 4-38). Scuttle hatches are also insulated by gluing batt or blanket insulation to the top of the hatch (fig. 4-39).

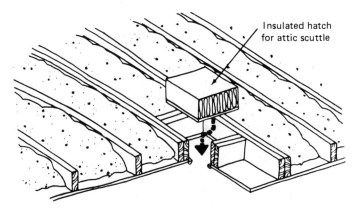

Insulated hatch for attic scuttle

LOOSE FILL INSULATION BETWEEN JOISTS

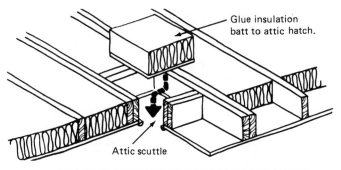

Glue insulation batt to attic hatch.

Attic scuttle

BATT OR BLANKET INSULATION BETWEEN JOISTS

Fig. 4-39. Insulated hatch for attic scuttle and other attic access holes. *(Courtesy Johns-Manville Corp.)*

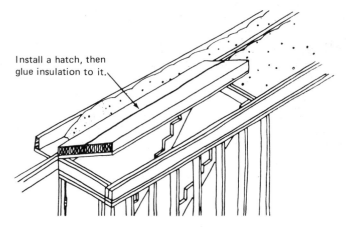

Install a hatch, then glue insulation to it.

Fig. 4-40. Insulated hatch for attic stairway. *(Courtesy Johns-Manville Corp.)*

Attic stairways may be insulated by installing a hatch at the top of the stairs and gluing batt or blanket insulation to the top of the hatch (fig. 4-40). If a door is installed at the bottom of the stairs, insulate the walls enclosing the stairs and the soffit area beneath them (fig. 4-41). The door should be insulated and properly weatherstripped when installed.

Hiring an Insulation Contractor

An insulation contractor should be hired to do the work if the job is too complicated or time consuming. The names of local insulation contractors are listed in the Yellow Pages of the telephone book under "Insulation Contractors" or a similar heading. The local utility company or friends and neighbors may also be able to suggest the names of reliable insulation contractors.

Always ask the insulation contractor for references, including the names of those for whom work has been done, and check them out. Call the Local Better Business Bureau (also listed in the telephone book) and ask if any complaints have been lodged against the contractor, or ask the bank for a report on the contractor's credit rating.

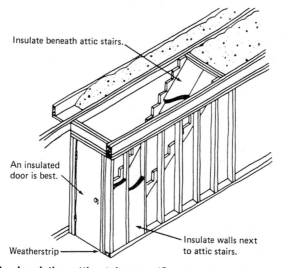

Insulate beneath attic stairs.

An insulated
door is best.

Weatherstrip

Insulate walls next
to attic stairs.

Fig. 4-41. Insulating attic stairways. *(Courtesy Johns-Manville Corp.)*

Ask several contractors to give labor and material estimates based on the same specifications, and require that the estimates be given in terms of R-values and not thicknesses of insulation. Not all insulating materials have the same heat-resistant ability per inch of installed material.

Ask the insulation contractor to explain the information on the insulation bag label before work begins. Do not allow a contractor to install insulation from unlabeled bags because the quality of the material will be unknown. The bag label will indicate the required thicknesses and coverages for various insulating materials. For example, a federal government specification (HH–I–1030A) requires that each bag of loose-fill mineral wool insulation be labeled as shown in figure 4-42. Another government specification (HH–I–515C) requires a similar label for cellulose insulation. *The thicknesses and coverages on the bag label, which apply only to attic floor installations, may be different for different manufacturers.* There is no federal specification for urea formaldehyde insulation.

Make sure the insulation contractor carries adequate insurance

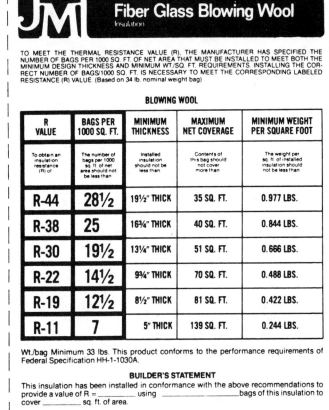

Fiber Glass Blowing Wool Insulation label (Johns-Manville):

Johns-Manville

Fiber Glass Blowing Wool
Insulation

TO MEET THE THERMAL RESISTANCE VALUE (R), THE MANUFACTURER HAS SPECIFIED THE NUMBER OF BAGS PER 1000 SQ. FT. OF NET AREA THAT MUST BE INSTALLED TO MEET BOTH THE MINIMUM DESIGN THICKNESS AND MINIMUM WT./SQ. FT. REQUIREMENTS. INSTALLING THE CORRECT NUMBER OF BAGS/1000 SQ. FT. IS NECESSARY TO MEET THE CORRESPONDING LABELED RESISTANCE (R) VALUE. (Based on 34 lb. nominal weight bag)

BLOWING WOOL

R VALUE	BAGS PER 1000 SQ. FT.	MINIMUM THICKNESS	MAXIMUM NET COVERAGE	MINIMUM WEIGHT PER SQUARE FOOT
To obtain an insulation resistance (R) of	The number of bags per 1000 sq. ft. of net area should not be less than	Installed insulation should not be less than	Contents of this bag should not cover more than	The weight per sq. ft. of installed insulation should not be less than
R-44	28½	19½" THICK	35 SQ. FT.	0.977 LBS.
R-38	25	16¾" THICK	40 SQ. FT.	0.844 LBS.
R-30	19½	13¼" THICK	51 SQ. FT.	0.666 LBS.
R-22	14½	9¾" THICK	70 SQ. FT.	0.488 LBS.
R-19	12½	8½" THICK	81 SQ. FT.	0.422 LBS.
R-11	7	5" THICK	139 SQ. FT.	0.244 LBS.

Wt./bag Minimum 33 lbs. This product conforms to the performance requirements of Federal Specification HH-1-1030A.

BUILDER'S STATEMENT
This insulation has been installed in conformance with the above recommendations to provide a value of R = _____ using _____ bags of this insulation to cover _____ sq. ft. of area.

| Builder's Signature | Company Name (Builder) | Date |
| Applicator's Signature | Company Name (Applicator) | Date |

CAUT Contact with fiber glass may ca~ ~rary irritation. ~this materi~

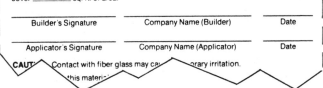

Fig. 4-42. Johns-Manville fiberglass blowing wool bag label.
(Courtesy Johns-Manville Corp.)

to protect the workers if they are injured on the job. The insurance policy for the house or building should also provide protection against work-related injuries or damage.

CHAPTER 5

Roof and Attic Ventilation

• **Ventilation Requirements**   • **Attic Fans**

Daytime summer attic temperatures often reach 150° F and higher. This trapped hot air penetrates downward into the living and sleeping areas of the structure, making them uncomfortably warm both day and night. If the structure is air conditioned, the hot attic air will increase the load on the air conditioner and lower its efficiency. Hot attic air is also potentially dangerous and destructive, sometimes causing spontaneous combustion in overheated attics and attic crawl spaces, and early deterioration of shingles, framing, and insulation.

Most roof decks are covered with an underlayment of waterproof asphalt-saturated felt to prevent rain or melting snow from leaking through the roof deck into the interior of the structure. This waterproof underlayment also tends to block the passage of water vapor from inside the structure through the roof deck to the outdoors. This can be a problem in the winter because moisture from cooking, baths, and a variety of other sources rises with the warm air and condenses when it comes into contact with the cold sheathing boards. The warm air will pass through the roof deck to the outdoors, but the underlayment will block any further movement of the moisture carried by the air. Because the wood in the roof framework and roof deck sheathing is a capillary-type material, it absorbs and traps large amounts of this moisture. Trapped moisture will eventually cause the wood framework and sheathing to warp or decay. The damage may be minimal, resulting in only a few minor leaks, or it may be so extensive that the entire roof structure is seriously weakened. Moisture may also become trapped in the roof, attic, or attic crawl space insulation, causing it to lose its effectiveness.

The most effective method of removing trapped moisture and reducing air temperatures is by adequately ventilating the spaces beneath the roof. Properly ventilating these spaces will cause the air to circulate freely and will remove hot moist air from the structure before the water vapor condenses and causes any damage.

The best time to inspect and improve roof and attic ventilation is during reroofing or before the roofing materials are applied in new construction. *The amount of ventilation should be increased after insulation has been added to the roof or to the floor of the attic or attic crawl space.*

Ventilation Requirements

Always check the local building code or ask a building contractor for the recommended amount of ventilation for a structure. The amount will vary because it depends on such factors as location, wind direction, amount of sun and shade, and roof design.

According to the FHA Minimum Property Standards, attics with a ceiling vapor barrier must be ventilated with at least 1 square foot of free vent area for each 300 square feet of ceiling. Attics without a ceiling vapor barrier must be ventilated with 1 square foot of free vent area for each 150 square feet of ceiling. In either case, the ridge or gable vent must be at least 3 feet above the level of the eave vents.

The three most common kinds of attic vents are eave, gable, and ridge. Proper ventilation is most effectively accomplished with a combination of eave and gable vents or eave and ridge vents. Although any one of the three is better than no ventilation at all, either the eave and gable or eave and ridge combinations allow the best air circulation. Regardless of the type of ventilation system used, it should meet the requirements established by the FHA Minimum Property Standards.

Net Free Vent Area

The net free vent area of a structure is the required minimum total area of the openings through which air can pass unobstructed for adequate ventilation of the spaces immediately below the roof deck. Attics with a ceiling vapor barrier generally must be ventilated with at least 1 square foot of free vent area for each 300 square feet of attic floor or ceiling. For example, a structure with an attic floor of 1,500 square feet requires a net free vent area of 5 square feet equally divided among the various vents. Attics without a ceiling vapor barrier, on the other hand, require 1 square foot of free vent area for each 150 square feet of attic floor space. These two examples show that an attic *without* a ceiling vapor barrier requires twice the net free vent area of one with a barrier because greater amounts of moisture must be removed from the attic or attic crawl space. The size of the net free vent area may also be affected by roof design, prevailing weather conditions, and other factors.

Vents are usually covered with screening or wire cloth to prevent insects or small birds from entering through the opening. Gable vents are covered with both louvers and screening. Covering the vent protects the opening, but it also reduces the net free area and restricts air flow. The amount of reduction will depend on the type of cover used over the vent. As shown in figure 5-1, the small openings in an insect screen reduce the net free area by approximately 50 percent. The ½-inch mesh screen, on the other hand, reduces the net free area by only 10 percent. To maintain adequate ventilation, the area of the vent openings must be increased to offset the obstructing cover. The amount of the increase is determined by multiplying the required net free vent area by a conversion factor for each type of vent cover (table 5-1). For example, if the required net free vent area for a structure is 5 square feet and it is covered with 1/16-inch mesh screen, the gross area of the vent openings should be increased from 5 square feet to 10 square feet (5 times a conversion factor of 2) to provide the required ventilation for the roof. As a general rule of thumb, many building contractors simply double the net free vent area.

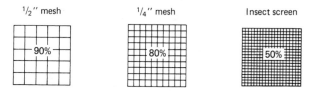

Fig. 5-1. Net free area of vents when covered by different types of screen.

Table 5-1. Opening Size Adjustments for Different Types of Vent Covers (*Courtesy Mineral Insulation Manufacturers Association, Inc.*)

Type of Covering	Size of Opening
¼″ hardware cloth	1 × net vent area
¼″ hardware cloth and rain louvers	2 × net vent area
⅛″ screen	1¼ × net vent area
⅛″ screen and rain louvers	2¼ × net vent area
1/16″ screen	2 × net vent area
1/16″ screen and rain louvers	3 × net vent area

Ventilating Gable Roofs

The gable roof is the most popular type of roof design used in residential construction. The *gable* is the triangular end wall that extends from the level of the eaves to the roof ridge (fig. 5-2A). There is one gable at each end of the roof. The triangular-shaped space enclosed by the gables, roof, and uppermost ceiling of the structure is the attic or attic crawl space.

As shown in figure 5-2B, some gable roofs have outlet vents or louvers installed in each gable end wall, but no inlet vents along the eaves or soffits. This is generally the case when there is little or no roof overhang. Installing one or more roof ventilators along the roof plane close to the ridge is an effective way to improve ventilation for this type of roof. Roof ventilators are especially recommended for low-pitched gable roofs where the endwalls have insufficient area for the installation of outlet vents or louvers.

Positive air movement in an attic or attic crawl space under a gable roof is usually obtained by installing inlet air vents in the soffit area along the eaves and exhaust vents in the gable end walls (fig. 5-2C). These gable vents are covered by horizontal slats or louvers and screens or covers that allow air movement through the opening while excluding the entry of sunlight or rain. The gable vent and louver must be placed as close to the ridge as possible for maximum efficiency or at least 3 feet above the level of the soffit or eave vents.

Vent locations	End elevations	Cross section	Side elevations	Ratio of total minimum net ventilator area to ceiling area	
				Inlet	Outlet
A. Louvered inlet-outlet vents in gable end walls (no overhang and soffit)				0	$\frac{1}{300}$ (Combined)
B. Louvered outlet vents in gable end walls and inlet vents in soffit				$\frac{1}{900}$	$\frac{1}{900}$
C. Louvered outlet vents in gable end walls and dormers with inlet vents in soffit				$\frac{1}{900}$	$\frac{1}{900}$

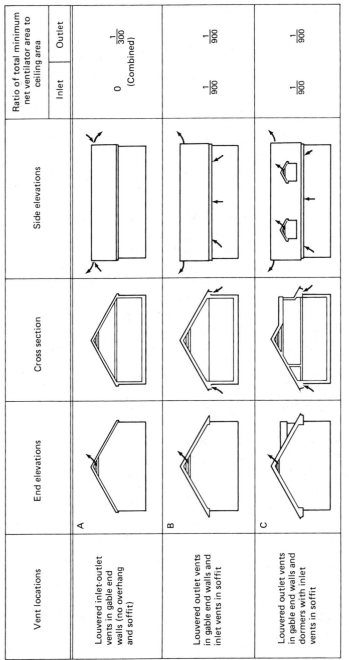

Fig. 5-2. Ventilating areas of gable roofs.

111

After the total net free vent area has been calculated, it must be divided by the number of vents to determine the approximate dimensions for each vent. For example, an attic with a ceiling area of 1,200 square feet requires a total net free vent area of 4 square feet (1/300 of 1,200 square feet). If the roof has no soffit or eave vents, this total net free vent area is divided equally between the two gable vents (one in each gable endwall) to obtain a dimension of 2 square feet for each. If soffit or eave vents are used, use 1 square foot of vent area in each gable (or 2 square feet at the ridge if a ridge vent is used instead of gable vents) and 1 square foot of vent area at each of the two roof eaves.

Several different types of gable louvers are illustrated in figure 5-3. Most are available from manufacturers ready for installation. Some are constructed with metal louvers and frames, whereas others are made entirely of wood. The latter type is enclosed in a wood frame that can be installed in a rough opening in the same way that a window frame is installed. Metal vents or louvers are often made so that they can be adjusted to fit the wall opening. The triangular-shaped gable vent or louver is most efficient. Its shape allows it to be installed close to the ridge of a pitched roof where the hot air collects inside the structure.

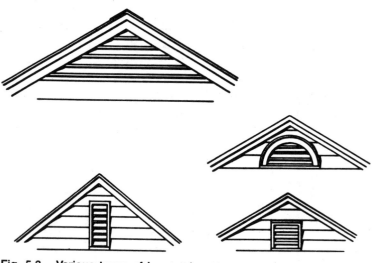

Fig. 5-3. Various types of louvered vents.

Ventilating Hip Roofs

The most satisfactory method of ventilating a hip roof is by installing inlet vents in the soffit area along the eaves and outlet vents at or near the ridge (fig. 5-4A). The difference in temperature between the attic air and the outside air creates an air movement independent of the wind. This arrangement also produces a more positive air movement when there is a wind.

The net area for the inlet vent openings in the soffit should be $\frac{1}{900}$ of the total attic floor or uppermost ceiling area. This is the ratio of the total minimum net vent area to the attic floor or ceiling area. For example, an attic floor or ceiling area of 1,800 square feet requires a minimum total net area of 2 square feet for the inlet vent openings. The minimum total net area for the outlet vents or ventilators is $\frac{1}{1,600}$ of the attic floor or ceiling area.

One or more roof ventilators may be used as outlet openings on this type of roof. On houses with pitched roofs, roof ventilators should be mounted on the rear slope or incline of the roof so as not to be easily visible from the front of the structure.

Some hip roofs are constructed with gable extensions at each end of the ridge (fig. 5-4B). Outlet vents may be installed in the gables on this type of roof. The net area for both inlet vents and the outlet vents or ventilators is $\frac{1}{900}$ of the attic floor or ceiling area.

Ventilating Flat Roofs

A flat or low-pitched roof generally requires a greater ratio of net free vent area to ceiling area than a gable roof because the air movement is less positive and usually dependent on wind force. Furthermore, there should be a clear, unobstructed open space above the ceiling insulation and below the roof sheathing for free air movement from the inlet to the outlet vent openings.

Many flat or low-pitched roofs are constructed with rafters extending beyond the walls to form an overhang. In this type of construction, combined inlet and outlet vent openings can be installed in the soffits either in the form of a continuous slot or as individual vent openings evenly spaced along the overhang (fig. 5-5A). The minimum total net free vent area for both the inlet and outlet vents is $\frac{1}{250}$ of the ceiling area. If the roof is constructed without an overhang, one or more roof and wall vents will provide sufficient ven-

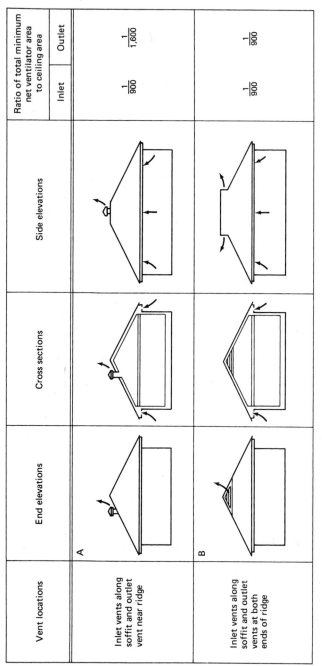

Vent locations	End elevations	Cross sections	Side elevations	Ratio of total minimum net ventilator area to ceiling area	
				Inlet	Outlet
A. Inlet vents along soffit and outlet vent near ridge				$\frac{1}{900}$	$\frac{1}{1,600}$
B. Inlet vents along soffit and outlet vents at both ends of ridge				$\frac{1}{900}$	$\frac{1}{900}$

Fig. 5-4. Ventilating areas of hip roofs.

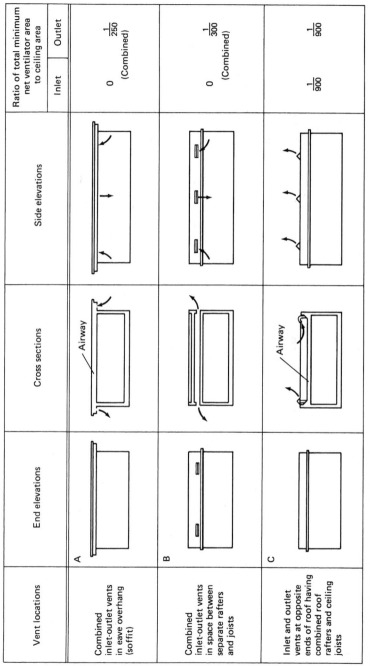

Fig. 5-5. Ventilating areas of hip roofs.

tilation. The minimum total net free area for the combined roof and wall vents is ⅟₉₀₀ of the ceiling area.

Some flat roofs are surrounded with parapet walls. If the roof rafters and ceiling joists of this type of roof are separate frame members, the space between them can serve as a passage for ventilation air (fig. 5-5B). All vent openings function as combined inlet and outlet vents with a minimum total net free area of ⅟₃₀₀ to the ceiling area.

If the ceiling joists and roof rafters in parapet construction are not separate frame members, stack-type inlet ventilators should be mounted on one side of the roof and outlet ventilators on the opposite side (fig. 5-5C). The minimum total net free vent area in this type of roof construction is ⅟₉₀₀ of the ceiling area.

Ventilating Other Types of Roofs

The insulation of roofs with exposed beam ceilings and no attic or attic crawl space should follow the roof slope and be installed so that there is an unobstructed passage of at least 1½ inches between the roof sheathing and insulation for air movement (fig. 5-6).

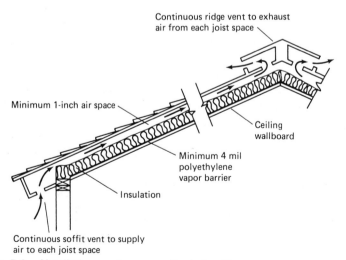

Continuous ridge vent to exhaust air from each joist space

Minimum 1-inch air space

Ceiling wallboard

Minimum 4 mil polyethylene vapor barrier

Insulation

Continuous soffit vent to supply air to each joist space

Fig. 5-6. Ventilating roofs over cathedral ceilings.

Roofs over cathedral ceilings generally require a minimum total net area for the inlet and outlet vents of ⅟₃₀₀ of the total ceiling area. When a vapor barrier is used, cross ventilation can be obtained by locating half the required vent area at each eave. If a vapor barrier is not used, the vent area should be doubled. The same holds true for flat roofs when a vapor barrier is not used.

Roof Vents

The various types of vents and ventilators used on a roof are shown in figure 5-7. The terms *vent* and *ventilator* are frequently used synonymously. Strictly speaking, however, vents are simply openings for the passage of air, whereas ventilators are mechanical devices that force the air to move. Ventilators, which may be either powered or nonpowered, are described in the next two sections.

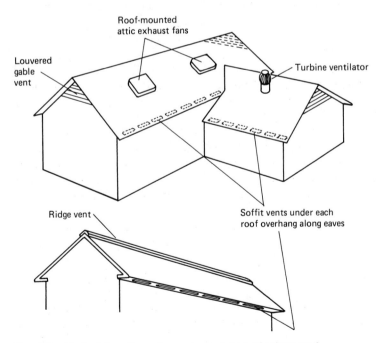

Fig. 5-7. Typical locations for various types of vents and ventilators.

Gable Vents

For structures that do not have vent openings in the gables, a vent opening should be installed in the upper portion of each gable as close to the ridge as possible to provide a sufficient cross draft for ventilation. The size of the vent opening will depend on the amount of attic space that must be ventilated. Both sufficient cross draft and correct vent sizing are required if the attic or attic crawl space is to be properly ventilated.

Read the louver manufacturer's installation instructions carefully before beginning work. Mark the area to be covered by the louvers on the outside wall surface of each gable and remove the siding, brick, or other finishing materials, and the insulation so that the wall studs are exposed. Try not to remove or damage the wall materials below the base level of the louver.

After the outer wall covering materials and the insulation have been removed, go to the inside of the attic or attic crawl space and cut away the upper sections of the wall studs to provide an opening large enough for the louver. The wall studs must be cut back far enough to provide space for both the louver and header (fig. 5-8).

As shown in figure 5-9, the 2 × 4 header provides a base for the louver and a cross sectional nailing support for the wall studs. The header dimensions will depend on the bottom width of the louver. Note also that the header is cut on the oblique at each end to form

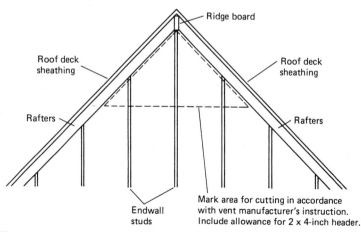

Fig. 5-8. Marking endwall for vent location.

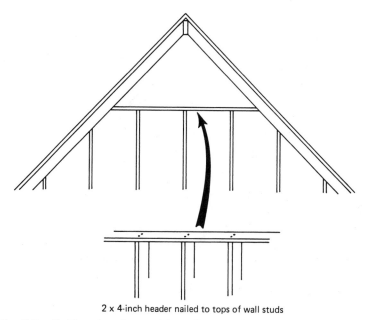

2 x 4-inch header nailed to tops of wall studs

Fig. 5-9. Cutting opening and installing header.

the desired angle against the face of each end rafter. The angle of the cut will depend on the angle formed by the rafters.

Nail the header to the wall studs with 10d or 16d nails. It may be necessary to toenail the oblique butt joint (formed where the ends of the header join the roof rafters). Avoid using too many nails because this may weaken the joint. Fasten the louver in position according to the manufacturer's instructions (fig. 5-10).

Cover the inside of the louver with a screen to prevent insects and debris from blowing into the attic or attic crawl space. The louver manufacturer will usually supply screening material with the louver, or suggest a suitable type and size to use. Ordinary copper or aluminum window screens are generally preferred.

If you have decided to make your own wood louver, instead of purchasing one ready-made at your local building supply outlet, the following suggestions are offered as a guide:

1. The size and number of vents is determined by the size of the area to be ventilated.

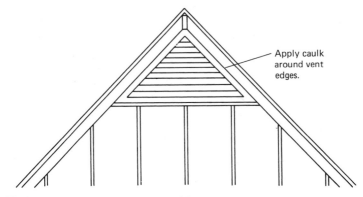

Fig. 5-10.   Installing louvered gable vent.

2. The minimum net open area should be ¼ square inch per square foot of ceiling area.
3. Most louver frames are usually 5 inches wide.
4. The back edge of the louver frame should be rabbeted out for the screening material.
5. Three-quarter-inch slats should be used and spaced about 1³/₄ inch apart.
6. Sufficient slant or slope to the slats should be provided to prevent rain from entering.
7. The louvers should be placed as near the top of the gable as possible.

Soffits or Eave Vents

The overhang of a pitched roof at the eave line is called the *cornice*. The cornice consists of a fascia board nailed to the ends of the rafters, a soffit for a closed cornice, and appropriate moldings. In open-type cornice construction, the bottoms of the rafters in the roof overhang are exposed to view. The underside of the overhang is closed off by a soffit in closed-type cornice construction (fig. 5-11). The soffit forms a connection between the roof and the sidewall, and is made of plywood, metal, hardboard, medium-density fiberboard, or other suitable sheet materials.

The inlet vents of a gable roof are usually installed along the soffit either in the form of individual and equally spaced slots or one con-

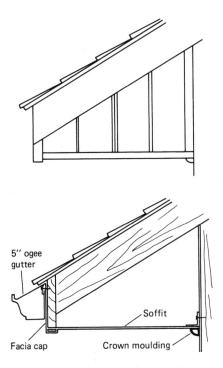

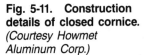

5" ogee gutter

Soffit

Facia cap

Crown moulding

Fig. 5-11. Construction details of closed cornice. *(Courtesy Howmet Aluminum Corp.)*

tinuous slot (figs. 5-12 and 5-13). As shown in figure 5-13, a continuous screened slot should be located near the outer edge of the soffit near the fascia to minimize the possibility of snow entering. This type of inlet vent may also be used on the extension of flat roofs. If the soffit is made of wood or a wood-type material, these small louvered and screened vents can be obtained in most lumberyards or hardware stores and are easy to install. Only small sections need to be cut out of the soffit, and these can be sawed out before the soffit is installed.

Preformed aluminum soffits are also available for installation under the roof overhang. The soffit material has a baked-on enamel or paint surface and is available in coils or panels. Both perforated and solid panels are used to form the soffit system. Air enters the structure through the perforated panels. Soffit coil material is perforated along its entire length. Each metal soffit system is provided with the necessary fascia caps or runners, frieze runners, end caps, and the various hardware and fasteners required to complete the job. Use only aluminum nails to fasten components of the soffit system to wood

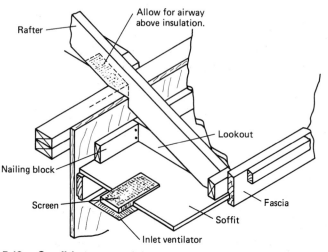

Fig. 5-12. Small insert-type inlet vent in wood soffit.

or other surfaces. When metals of different alloys are used together, an electrolytic action is established that weakens the strength of both metals.

Construction details of a typical soffit installation in which coil material is used is shown in figure 5-14. Soffit panels are produced in 12-inch wide sheets and a variety of different lengths. A soffit panel installation is illustrated in figure 5-15. Installation methods for both types of preformed aluminum soffit systems are described in chapter 20.

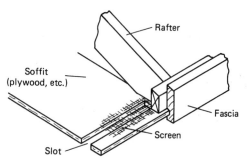

Fig. 5-13. Slot-type inlet vent in wood soffit.

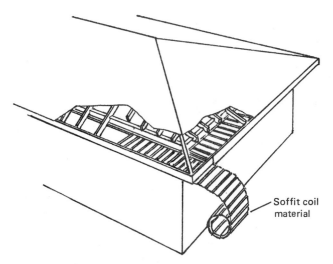

Fig. 5-14. Coil-type soffit installation. *(Courtesy Reynolds Metals Co.)*

Ridge Vents

A cross-sectional view of a typical ridge vent used on agricultural, industrial, and commercial structures is shown in figure 5-16. A smaller version is sometimes installed on house roofs (fig. 5-17). A ridge vent is installed along the entire length of the roof ridge above the space or spaces to be ventilated. It is designed to protect the roof opening from rainfall while at the same time providing a path for the rising, moisture-laden warm air to escape from the structure.

Fig. 5-15. Installing aluminum soffit panels. *(Courtesy Howmet Aluminum Corp.)*

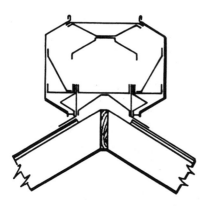

Fig. 5-16. Cross section of continuous ridge vent used on many agricultural and industrial structures.

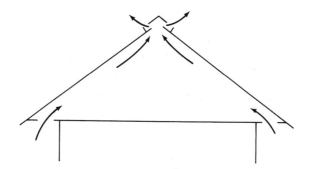

Fig. 5-17. Air flow in attic with continuous ridge vent.

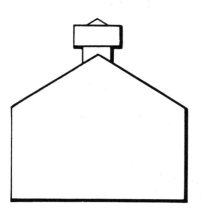

Fig. 5-18. Ridge-mounted turbine ventilator.

Locating the ridge vent at the highest point in the structure results in efficient ventilation because it is at this point that hot air collects after rising from the spaces below. Its location along the roof ridge also results in an even and uniform distribution of the air as it leaves the vent.

A principal disadvantage of using a ridge vent is its failure to take full advantage of wind action unless the wind is blowing directly across it. Another factor to be considered is that a ridge vent must be installed during the construction of the roof deck and before the roofing materials are laid. Because it requires opening the roof ridge, it should not be installed during reroofing.

Turbine Ventilators

A *turbine ventilator* is a nonpowered roof ventilator that depends on air movement for its operation. This type of roof ventilator is commonly used on agricultural, industrial, and commercial structures where it is usually installed along the ridge line (fig. 5-18). It is also sometimes used on houses where it is mounted on the roof plane of a rear slope or incline of a pitched roof (fig. 5-19). When it is mounted on the roof plane of a pitched roof, it is also sometimes called a *roof plane ventilator.*

The principal components of a turbine ventilator are the base, mounting flange, rotating head or top, and blades. The vane-type blades may be mounted externally on top of the unit or designed as an integral part of a globe-shaped head (fig. 5-20). Globe-shaped turbine ventilators are available with either internal or external metal

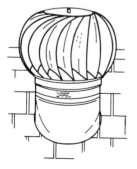

Fig. 5-19. Turbine ventilator mounted on roof slope. *(Courtesy Sears, Roebuck and Co.)*

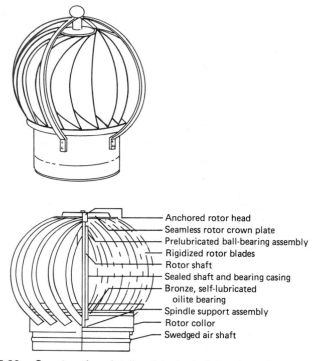

Anchored rotor head
Seamless rotor crown plate
Prelubricated ball-bearing assembly
Rigidized rotor blades
Rotor shaft
Sealed shaft and bearing casing
Bronze, self-lubricated
   oilite bearing
Spindle support assembly
Rotor collor
Swedged air shaft

Fig. 5-20. Construction details of typical globe-shaped turbine ventilator. *(Courtesy Penn Ventilator Co., Inc.)*

bracing. External metal bracing provides extra strength for installations in areas with high winds.

Most turbine ventilators are available in 12- and 14-inch diameter sizes. The turbine ventilator size indicates the diameter of the blades, the height of the ventilator above the roof deck, and the size of the cutout in the roof deck. For example, a 12-inch diameter turbine ventilator has blades with a 12-inch diameter, stands 12 inches above the roof deck, and requires a 12-inch diameter cutout in the roof deck for installation.

One 12-inch turbine ventilator with at least 2 square feet of net free inlet opening can ventilate about 600 square feet of attic. A 14-inch diameter turbine ventilator, on the other hand, can move up to 50 percent more air than the 12-inch model.

Operating Principle

Wind-operated turbine ventilator blades are designed to move with the slightest breeze. When the blades rotate, they create a centrifugal action that reduces the air pressure in the attic. This slight difference in pressure between the attic and the outside creates an air flow that draws fresh air into the attic through inlet vents and exhausts hot, moist air through the "throat" in the ventilator base.

The principal limitation of the turbine ventilator is that its blades draw the air out of the structure only as long as its blades rotate. It may also become noisy, like any heavy rotating body that cannot be kept accurately balanced; has a low capacity when the air is at or near a calm state; and has moving parts that must be occasionally serviced or repaired. Turbine ventilators with globe-shaped heads have a number of disadvantages specific to their design. For example, wind impact on the vanes allows outside air to enter the ventilator on the windward side while air is being exhausted from the structure on the leeward side. This decreases the efficiency of the turbine ventilator because its head must handle both volumes of air simultaneously. Furthermore, the globe-shaped turbine ventilator functions as an open vent space when the blades are not turning. Unless the opening can be closed by an automatic or chain-operated damper, rain water may enter through the opening.

Installing Turbine Ventilators

The number of turbine ventilators installed on a roof will depend on the size of the ventilator and the attic square foot area. The square footage of the attic is determined by multiplying attic length by width in feet. One 12-inch turbine ventilator should be installed for every 600 square feet of attic area and one 14-inch model for every 900 square feet. If two turbine ventilators are installed on the roof, locate one ventilator one fourth of the distance from each end of the structure. Allow a minimum of one square inch of intake vent area for every two square feet of attic floor space. When turbine ventilators are installed in the center of the roof, inlet vents should be spaced an equal distance apart around the overhang (fig. 5-21). Install vents under the eaves if the roof has an overhang and soffit. If the roof has no overhang, install louvers in the gable end walls (fig. 5-22).

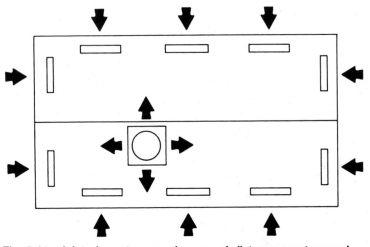

Fig. 5-21. Inlet air vents spaced an equal distance apart around roof overhang.

The procedure for installing a turbine ventilator on the roof plane of a pitched roof is illustrated in figure 5-23 and may be outlined as follows:

1. Locate the projected opening for the ventilator base and make certain that it falls between roof rafters.
2. Measure a hole the size of the ventilator base and cut through the roofing materials and sheathing.
3. Apply a caulking compound or a suitable asphalt roofing cement to the bottom of the mounting flange on the ventilator base.
4. Slide the mounting flange (with the connected base facing upward) under the shingle course above the hole and center the base "throat" or opening directly over the cutout hole in the roof.
5. Nail the flange to the roof and seal the edges of the flange with a caulking compound. Cover the nail heads with caulking compound to protect them from corrosion.
6. Set the turbine ventilator head in the base opening, level it, and fasten it to the base according to the manufacturer's instructions.

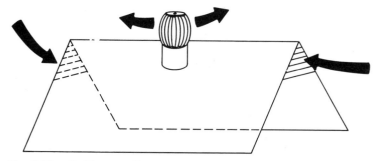

Fig. 5-22. Turbine ventilator mounted on roof slope and inlet louvers in gables.

Attic Fans

Attic fans are used to augment natural air convection by pulling fresh air into the attic or attic crawl space, and by pushing hot air out. Attic fans are of two types: attic exhaust fans and whole-house ventilating fans.

Fan Capacity

The size or capacity of an attic fan is determined by the volume of the room or space that must be ventilated. The volume of a room or space is calculated by multiplying its area (width × length) by its height. This figure will also indicate the amount of air (expressed in cubic feet) in the room or space that must be changed each minute to provide proper ventilation.

The number of air changes per minute is an important consideration when selecting a fan for a whole-house ventilation system. Generally, one air change per minute is sufficient in most sections of the country. In areas with a colder than average climate, one air change for every minute and a half is recommended. Most one-story houses with normal occupancy require an attic fan capable of exhausting air at a minimum rate of 5,000 cubic feet per minute (CFM).

Fan manufacturers provide data that indicate the fan blade diameter (in inches), fan speed (in RPM's), and fan capacity (CFM) for each make and model. Once the volume of air to be changed has

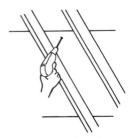

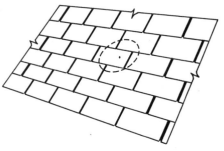

A. Position ventilator base on roof as close to ridge as possible. Measure down from ridge inside attic and drill a hole between two rafters.

B. Draw a circle around the drilled hole equal to the diameter of the turbine ventilator base.

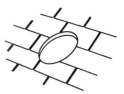

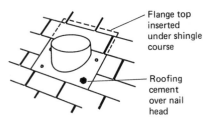

Flange top inserted under shingle course

Roofing cement over nail head

C. Cut out the hole and remove the roofing and roof deck sheathing

D. Apply roofing cement to the bottom of the base flange and center the base opening over the hole. Nail the base flange to the roof deck and cover the nail heads with roofing cement or caulk.

E. Run a bead of roofing cement or caulk along the edge of the ventilator base flange. Install the turbine ventilator top on the base and level it.

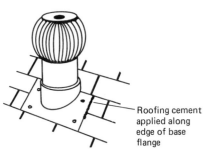

Roofing cement applied along edge of base flange

Fig. 5-23. Installing turbine ventilator.

been determined, choose a fan with a blade diameter and speed capable of meeting these requirements. Quiet fan operation generally requires that the fan tip speed not exceed 3,800 feet per minute.

Attic Fan Exhaust Opening

An attic fan must have an outside opening of adequate size to exhaust the air to the outdoors. The location of the outside opening is usually determined by the design of the structure, but the most common locations are in gables and dormers. For flat roofs, it is necessary to construct a small rectangular or box-shaped structure to house the exhaust vent and louvers.

The size (area) of an exhaust opening is determined by a number of different factors, including (1) fan size, (2) type of screening (screen or cloth), and (3) type of louvers (wood or metal). An unscreened opening will obviously be the smallest for all fan diameters because there is no obstruction to the flow of air. An opening without any protective screening is impractical, however, because it allows insects, rodents, leaves, and debris to enter.

Screening made of ½-inch hardware cloth requires a smaller opening than one screened with 16-mesh wire screen, and the thinner metal louvers on factory produced units require smaller openings than wood louvers. These factors should be taken into consideration when planning the size of the attic fan exhaust opening. The most suitable opening must be large enough to accommodate the volume of air flow and should provide the maximum amount of protection against insects, debris, and weather conditions.

One disadvantage of using an attic fan is that it provides an opening for cold air to enter during the winter months. This can result in the attic air becoming so chilled that temperatures in the rooms below drop to uncomfortable levels. The fan opening can be covered with a piece of plywood or heavy plastic sheeting when the fan is not in use, but it is probably worth the extra expense to invest in an attic fan equipped with thermostatically controlled metal louvers. The thermostat element responds to temperature changes and automatically closes the louvers when the outside temperature drops below a certain level. Although the metal louvers close sufficiently to reduce significant heat loss in cold weather, there is still enough air infiltration through the closed louvers to equalize humidity conditions and prevent the buildup of condensation in the attic. Some thermostatically controlled louvers are also designed to close automatically when there is a fire in the attic, thereby reducing dangerous drafts. These thermostats are heat sensitive, so the fire must be close enough and hot enough to activate them, but they still provide an additional safety factor in preventing the rapid spread of flames.

Attic Exhaust Fans

Attic exhaust fans are usually mounted externally on a roof slope or internally in a gable endwall. They are used to reduce the temperature and moisture content of the air in a closed attic or attic crawl space. They are not used for cooling the house (see the discussion of whole-house ventilating fans).

The size of the roof-mounted or gable-mounted attic exhaust fan and the size of the inlet air vent openings required for the attic or attic crawl space may be calculated as follows:

1. Calculate the square foot area of the attic or attic crawl space by multiplying its length by its width.
2. Allow one CFM per square foot of attic or attic crawl space area and select an attic fan with a CFM capacity nearest that figure.
3. Allow a minimum of one square foot of intake air vent area for every 400 CFM if metal louvers are used and for every 300 CFM if wood louvers are used.

Install louvers in the gables if the roof has no overhang. If the roof has sufficient overhang and soffit area, install the air intake vents along the soffit under the eaves. When the attic exhaust fan is installed externally in the center of the roof, air inlet vents should be spaced an equal distance apart around the roof overhang.

GABLE-MOUNTED ATTIC EXHAUST FANS

A gable-mounted attic exhaust fan is sometimes called an *attic fan, attic exhaust fan, attic ventilator, gable-mounted power ventilator,* or simply a *power ventilator.* Each fan consists of a housing or venturi, fan blade, belt-drive or direct drive AC electric motor, mounting brackets, fan controls (thermostat, limiting fuse, and manual on-off switch), and connecting wiring (figs. 5-24 to 5-27). Some fans are equipped with automatic thermostatically controlled metal shutters (fig. 5-26).

The attic exhaust fan is mounted in or behind one of the gable endwall vent openings where it draws air through the opposite gable vent and the inlet vents along the soffit (fig. 5-27). If there are no

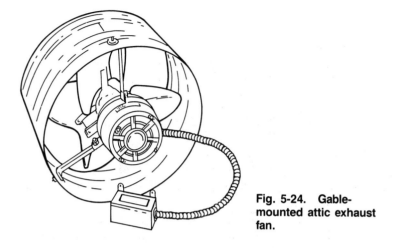

Fig. 5-24. Gable-mounted attic exhaust fan.

existing louvers, a fan with an automatic metal shutter is available for mounting in the gable vent opening (see fig. 5-26). If there are existing louvers, the fan can be mounted directly to the frame. The fastening method will depend in large part on the shape of the louver. A gable-mounted attic exhaust fan can be mounted to a rectangular-type louver or a square louver by fastening through the mounting brackets to the louver frame (fig. 5-28). On a wide louver, the brackets are fastened to the back side of two boards of suitable length (fig. 5-29). These boards can then be fastened to the louver frame. Figure 5-30 illustrates the method of mounting the fan off center in an irregular-shaped louver. On large triangular louvers, the fan brackets may be fastened directly to the louver frame or to boards fastened to the frame to provide a mounting base (fig. 5-31).

If an opening must be cut in the gable endwall to install the fan, square- and rectangular-shaped openings will range in size from 8¼ × 17 inches to 11⅜ × 25¼ inches (table 5-2). Fan blade diameters will not exceed 17¼ inches.

Attic exhaust fans operate on 110–120 volt AC 60 Hz (cycle) current, and all wiring must conform to local codes. Use the National Electric Code if a local code does not exist. When in doubt about local codes or wiring practices, hire an electrician to do the work. *Shut off the electric power at the fuse box or circuit breakers before beginning any work on the wiring.* All wiring should be permanent. Use No. 14 gauge copper wire or its equivalent. Do not use an extension cord. Install a single pole toggle switch as a shut-off switch for the fan.

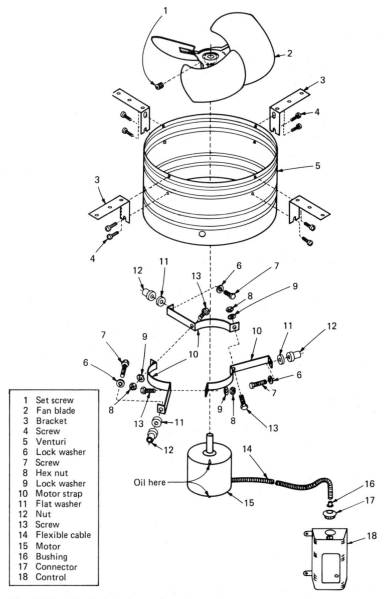

1	Set screw
2	Fan blade
3	Bracket
4	Screw
5	Venturi
6	Lock washer
7	Screw
8	Hex nut
9	Lock washer
10	Motor strap
11	Flat washer
12	Nut
13	Screw
14	Flexible cable
15	Motor
16	Bushing
17	Connector
18	Control

Oil here

Fig. 5-25. Exploded view of typical gable-mounted attic exhaust fan. *(Courtesy Sears, Roebuck and Co.)*

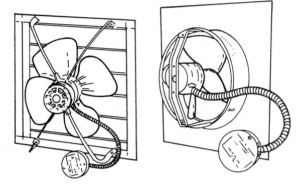

Fig. 5-26. Gable-mounted attic exhaust fans with and without automatically operated metal shutters.

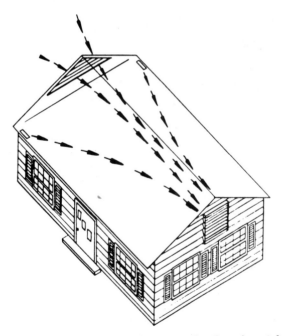

Fig. 5-27. Air movement for gable-mounted attic exhaust fan. *(Courtesy Sears, Roebuck and Co.)*

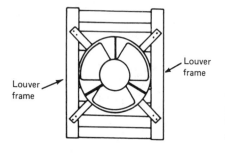

Louver frame

Louver frame

Louver frame

Fig. 5-28. Attic exhaust fan mounted in rectangular louver. *(Courtesy Sears, Roebuck and Co.)*

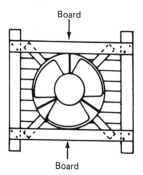

Board

Board

Fig. 5-29. Attic exhaust fan mounted in wide louver. *(Courtesy Sears, Roebuck and Co.)*

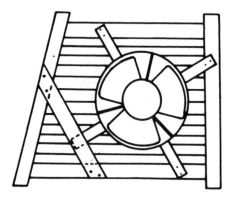

Fig. 5-30. Attic exhaust fan mounted in irregular-shaped louver. *(Courtesy Sears, Roebuck and Co.)*

Fig. 5-31. Attic exhaust fan mounted in triangular-shaped louver. *(Courtesy Sears, Roebuck and Co.)*

An attic fan with an unguarded fan blade should be installed in a location not readily accessible. If the attic fan is accessible to persons or animals, a proper guard should be constructed around the unit. *Do not operate an attic exhaust fan in attics or attic crawl spaces where gas or oil fired equipment such as furnaces or water heaters are operating.* Doing so may impair flue gas ventilation of the equipment in the attic or attic crawl space.

Table 5-2. Fan Blade Sizes, CFM Capacities, and Cutout Opening Dimensions for Different Roof-mounted Attic Exhaust Fans *(Courtesy Sears, Roebuck and Co.)*

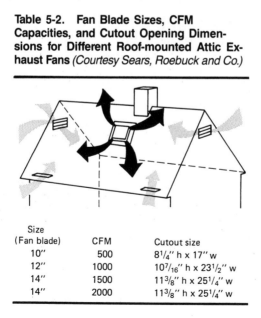

Size (Fan blade)	CFM	Cutout size
10″	500	$8^{1}/_{4}$″ h x 17″ w
12″	1000	$10^{7}/_{16}$″ h x $23^{1}/_{2}$″ w
14″	1500	$11^{3}/_{8}$″ h x $25^{1}/_{4}$″ w
14″	2000	$11^{3}/_{8}$″ h x $25^{1}/_{4}$″ w

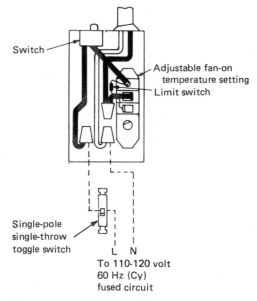

Switch

Adjustable fan-on
temperature setting
Limit switch

Single-pole
single-throw
toggle switch

L   N

To 110-120 volt
60 Hz (Cy)
fused circuit

Fig. 5-32.   Thermostat switch box wiring.

The fan thermostat is located in a control box or switch box which is connected to the fan motor by a flexible conduit (fig. 5-32). For best results the thermostat should be located away from the direct air stream produced by the fan. Attach the thermostat switch box to a rafter above the fan. Use screws instead of nails to mount the box to avoid possible damage to the thermostat.

The thermostat automatically starts and stops the fan motor at preset temperatures. The starting temperature usually ranges from 110° to 160° F., depending on the fan model and temperature conditions.

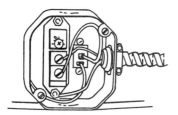

**Fig. 5-33.   Thermostat switch box
with back removed.**

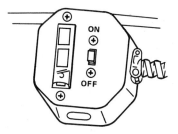

Fig. 5-34. On-off switch on thermostat switch box.

The preset shut-off temperature usually ranges from 70° to 95° F. The high and low temperature settings controlling fan operation can be reset on adjustable thermostats to operate at any temperature within the temperature range of the unit. The temperature settings are changed by removing the back of the switch box and moving the indicator to the preferred setting (fig. 5-33).

The on-off switch in the switch box should be left in the "on" position for normal operation (fig. 5-34). This is a safety switch that prevents the fan from operating when the unit is being serviced. The junction box also contains a limit fuse that will shut off the fan when an excessively high temperature is reached. This fuse cannot be repaired. If damaged, the entire control unit must be replaced.

Always shut off the electric power at the fuse box or circuit breakers before performing any maintenance on the fan. Oil the front and rear fan bearings at the beginning and end of each cooling season. Use four drops of SAE 20 weight nondetergent motor oil or the fan manufacturer's recommended lubricant in each bearing.

ROOF-MOUNTED ATTIC EXHAUST FANS

Externally mounted attic exhaust fans are sometimes called *roof fans, roof-mounted power ventilators, power ventilators,* or *attic exhaust fans.* Their design and operation are identical to the gable-mounted fans except for the housing, which is used to protect the motor and fan blade from the weather, and the self-flashing base (fig. 5-35). The roof-mounted attic exhaust fan is usually located on a roof slope at the back of the structure (fig. 5-36). As shown in table 5-2, the fan draws air into the structure through both gable louvers and the inlet vents in the soffit.

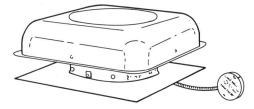

Fig. 5-35. Roof-mounted attic exhaust fan.

Many roof-mounted attic exhaust fans are small enough to fit between rafters spaced the standard 16 inches on center (fig. 5-37). When this is the case, the rafters do not require cutting and roof preparation is the same as for installing turbine ventilators (see fig. 5-23). Other roof-mounted attic exhaust fans may require cutting a rafter and installing headers to frame the opening.

Roof-mounted attic exhaust fans are available in several sizes with fan blade diameters ranging on the average from 10 to 14 inches. They may be operated manually by a switch or automatically by means of a preset thermostat. The operating principle in both the manual and automatic modes is the same as for gable-mounted attic exhaust fans.

The function of a roof-mounted attic exhaust fan is to remove heat and moisture from an attic or attic crawl space closed off from

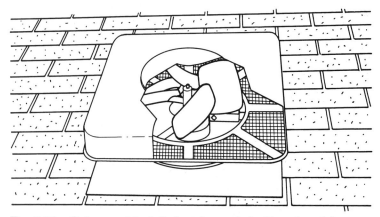

Fig. 5-36. Cutaway of installed roof-mounted attic exhaust fan.

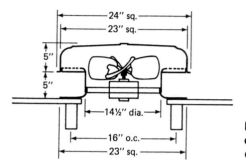

Fig. 5-37. Cross section of roof-mounted attic exhaust fan.

the rest of the structure. It is not used to ventilate the entire structure (see the discussion of whole-house ventilating fans).

Whole-House Ventilating Fans

A whole-house ventilating fan is used with open windows to create a balanced ventilation system. The fan pulls fresh air into the lower level of the structure and discharges stale, warm air from the attic or attic crawl space (fig. 5-38). In this type of ventilation system, both the attic or attic crawl space and the rest of the structure must be open to one another. The air passage between the lower and upper levels may be a stairway, ceiling vent, hatch, or door.

Whole-house ventilating fans are similar to attic exhaust fans in design, construction, and operating principle (figs. 5-39 and 5-40). They are more powerful than the smaller attic exhaust fans, however, having both a larger fan blade diameter and air movement capacity.

The location of a whole-house ventilating fan depends on the design of the structure. In a house having a suitably sized window in an attic endwall or dormer, the best results (and fewest construction problems) can usually be obtained by mounting the fan directly against the window. Whole-house ventilating fans can also be mounted behind louvers in the gable endwall in the same manner as the smaller gable-mounted attic exhaust fan (fig. 5-41).

Whole-house ventilating fans are also frequently installed in the ceiling of the uppermost floor. In one-story houses, the fan can be installed in a ceiling opening at any convenient central location (fig. 5-42). In houses or buildings of two or more stories, the opening is generally located in the ceiling of the top floor hallway. Again, a cen-

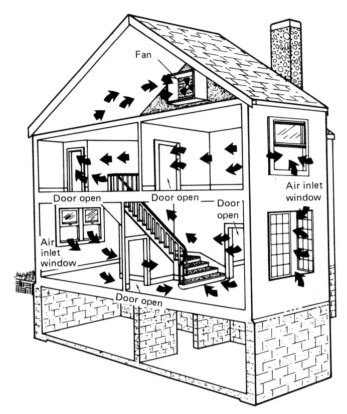

Fig. 5-38. Typical air flow for whole-house ventilation system.

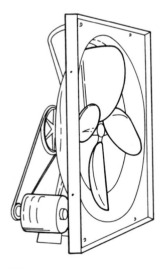

Fig. 5-39. Typical whole-house ventilating fan.

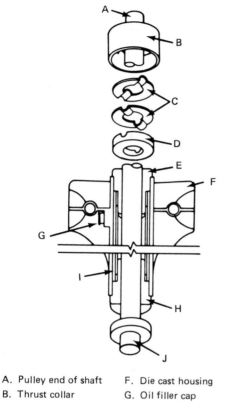

A. Pulley end of shaft   F. Die cast housing
B. Thrust collar   G. Oil filler cap
C. Bead rocker assembly   H. Bronze bearing
D. Thrust disc   I. Large oil reservoir with felt filler
E. Bronze bearing   J. Fan blade end of shaft

Fig. 5-40. Exploded view of typical whole-house ventilating fan.
(Courtesy Sears, Roebuck and Co.)

tral location for the fan works best. The ceiling opening and accompanying grille or shutter must be of sufficient size to avoid excessive resistance to air flow, permitting the air stream to pass through at a moderate velocity.

Both one-speed and two-speed fan systems are available for whole-house ventilation. Single-speed systems will provide effective whole-house air circulation and are available in a variety of different fan sizes. Two-speed systems permit the selection of "high" for rapid air movement during the warm daylight hours and "low" for gentle

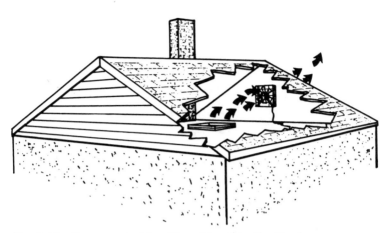

Fig. 5-41. Fan mounted in attic gable endwall with air passage through ceiling vent. *(Courtesy Hayes-Albion Corp.)*

Fig. 5-42. Fan mounted in ceiling with exhaust vents in gable endwalls.

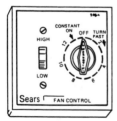

Fig. 5-43. Whole-house ventilation system fan control. *(Courtesy Sears, Roebuck and Co.)*

air during the cooler nights. Two-speed systems also have built-in timers that can be preset to turn off the fan at any time during a 12-hour period. Both the fan speed selector and the timer are contained in a control box or switch box that is connected to the fan motor by wiring and mounted in a conveniently accessible location (fig. 5-43). The size or capacity of a whole-house ventilating fan can be calculated by following these simple steps:

1. Calculate the square footage of the structure by multiplying its width by its length.
2. Consult the map in table 5-3 and determine whether the structure is located in zone 1 or zone 2.
3. Find the square footage of the structure in the figures under the zone in which it is located.
4. Read across to find the recommended fan size. If the structure is not exactly the size shown in table 5-3, select the next larger fan size, even if the smaller size is closer to the area figure of the structure.

Table 5-3. **Fan Blade Sizes, CFM Capacities, and Cutout Opening Dimensions for Different Whole-house Ventilating Fans** (*Courtesy Sears, Roebuck and Co.*)

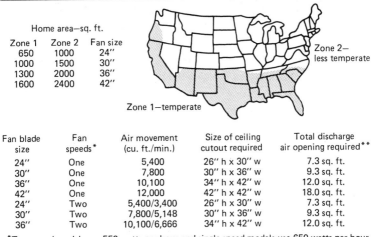

Home area—sq. ft.

Zone 1	Zone 2	Fan size
650	1000	24″
1000	1500	30″
1300	2000	36″
1600	2400	42″

Zone 2—less temperate

Zone 1—temperate

Fan blade size	Fan speeds*	Air movement (cu. ft./min.)	Size of ceiling cutout required	Total discharge air opening required**
24″	One	5,400	26″ h x 30″ w	7.3 sq. ft.
30″	One	7,800	30″ h x 36″ w	9.3 sq. ft.
36″	One	10,100	34″ h x 42″ w	12.0 sq. ft.
42″	One	12,000	42″ h x 42″ w	18.0 sq. ft.
24″	Two	5,400/3,400	26″ h x 30″ w	7.3 sq. ft.
30″	Two	7,800/5,148	30″ h x 36″ w	9.3 sq. ft.
36″	Two	10,100/6,666	34″ h x 42″ w	12.0 sq. ft.

*Two speed models use 550 watts per hour and single speed models use 650 watts per hour.
**Size of openings required vary depending on whether louvers to be used are wood or metal.

Troubleshooting Attic Fans

The three principal problems that occur most frequently when operating attic fans are:

1. Fan motor will not run
2. Not enough air circulating
3. Noisy fan motor or blade

If the fan motor will not operate, check the fuses or circuit breakers first. If the fuse is burned out, replace it. An open circuit breaker should be reset. Both of these problems may result from an energy surge or other factor not related to a malfunctioning fan motor. If the problem continues, check the fan motor for loose wiring at the motor terminals or a broken fan switch. The terminal connections can be tightened, but a broken switch will have to be replaced. The last and most expensive possible cause for the failure of a fan motor to operate is that it is defective. A defective fan motor will have to be replaced.

Insufficient air flow from the fan is usually caused by blocked air openings or passages. Remove any obstructions, adjust the louvers to a more open position, and clean the louver and vent openings. If the problem persists, check the fan belt (if so equipped). A defective or worn fan belt should be replaced; a loose fan belt can be tightened.

A noisy fan motor or blade may be the result of loose mountings, dry motor bearings, or dirty or unbalanced fan blades. If the mounting is loose, tighten the bolts and screws to see if this will eliminate the problem. It may be necessary to use longer bolts or screws, or new holes. Dry fan motor bearings can be lubricated with motor oil. Generally, this is all they will need to eliminate the noise. Dirty fan blades should be cleaned. Unbalanced fan blades should be straightened or replaced.

CHAPTER 6

Skylights and Roof Openings

- Tools, Equipment, and Materials
- Skylights
- Roof Access Openings
- Chimney Openings
- Vent Pipe and Stack Openings
- Roof Ventilators
- Interior Roof Drains
- Expansion Joints
- Other Types of Openings

Dormers, skylights, stairs, chimneys, vent pipes, stacks, and similar types of construction require that openings be cut through the roof deck. Cutting an opening in the roof deck requires basic carpentry skills and some experience because the opening must be carefully planned. All joints must be sealed against rain penetration and any cut rafters or joists must be supported with additional framing to prevent the roof from sagging. Because this type of work affects the external appearance of the roof and offers the possibility of weakening the framework when not done properly, the local building code should be consulted and a building permit obtained if one is required.

Roof openings can be classified as either *major* or *minor*. A major roof opening requires the cutting of one or more rafters, and the installing of headers to which the cut rafters are nailed. Bracing is sometimes added between the rafters near the roof opening to provide additional strength. A minor roof opening can be constructed without cutting any of the rafters. In the case of minor roof openings, the installing of headers or special bracing is often unnecessary.

A roof opening must be carefully flashed to provide protection against leaks. Various methods of applying flashing are described in chapters 8 and 12.

The construction of dormer roof openings is described in chapter 7. This chapter contains descriptions of the framing and construction details for skylights, stairs, chimneys, vent pipes, and other minor roof openings.

Tools, Equipment, and Materials

Basic carpentry tools are required for cutting and framing roof openings. The tools will include a claw hammer, circular or saber saw, straight edge, combination square, measuring tape, and carpenter's apron. A sturdy ladder is required for access to the top surface of the roof.

Roof openings are framed with 2 × 4's and 2 × 6's. Framing lumber can be purchased at a local lumber yard and the amount purchased will depend on the size of the project. Other materials required for the construction of roof openings include common wire nails for the framing, roofing nails, flashing, roofing felt, and roofing cement.

Skylights

Installing a skylight on the roof is an excellent way to introduce more light into halls, baths, and interior rooms or spaces. Studies have shown the skylights provide three to five times more available light than windows.

The use of skylights has increased dramatically since the introduction of window plastic as a substitute for glass. Acrylic is the most commonly used window plastic for skylights, followed by cellulose acetate butrate (CAB) and Lexan polycarbonate sheet. The use of plastic instead of glass has practically eliminated problems of breakage, leakage, and costly maintenance. Window plastic requires no maintenance. If it must be cleaned, dirt can be easily removed with mild soap and water. Fresh paint, grease, and roofing compounds can be removed with rubbing alcohol or Butyl Cellosolve. Before using these or any other substance to clean window plastic, it is always a good idea to test the product first on an unexposed surface prior to use. *Never use a petroleum-based or abrasive cleaner.*

Clear and colorless window plastic admits the maximum amount

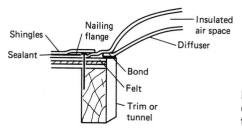

Shingles, Nailing flange, Sealant, Insulated air space, Diffuser, Bond, Felt, Trim or tunnel

Fig. 6-1. Double-dome construction with domes fused together at edges.

of light and heat. When this is not desirable, a translucent white plastic can be used to provide soft, diffused lighting. Window plastic is also available in bronze or gray to further reduce the amount of heat and light passing through the skylight.

Skylight manufacturers have been able to reduce heat loss through window plastic by using double or triple-dome construction. If the domes are fused together at the edges, as in figure 6-1, the air space functions as a thermal insulator. These domes are permanently sealed against dirt, air, or insect infiltration. R-values (thermal resistance) and U-values (overall coefficient of heat transmission) for the different types of plastic are listed in table 6-1.

Condensation may form between the layers of a double or triple dome briefly during extreme weather conditions, but it usually dissipates without any visible sign as weather conditions change. Condensation is most noticeable if all dome layers are clear plastic. Many skylights are designed with condensation gutters and weep holes (or weeping rings) to eliminate moisture (fig. 6-2).

Table 6-1. R-values and U-values for Different Types of Plastic Materials Used in Double-dome Skylight Construction

Plastic Material	U-Factor		R-Factor	
	Winter	Summer	Winter	Summer
Acrylic	.71	.50	1.40	2.00
Cellulose Acetate Butrate (CAB)	.56	.53	1.80	1.88
Lexan Polycarbonate	.288	.281	3.47	3.56

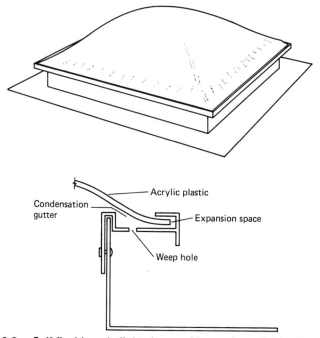

Fig. 6-2. Self-flashing skylight dome with condensation gutter and weep holes.

Types of Skylights

Skylights can be custom-designed or purchased factory-made from the manufacturer. Prefabricated skylights are available in a variety of shapes and sizes for installation on both pitched and flat roofs.

As shown in figure 6-3, skylights are available in many different shapes and styles. The dome, square, and rectangular shaped skylights are the types most commonly used on the pitched roofs of houses. Most skylights are fixed units, but ventilating skylights are also available. Ventilating skylights, which can be operated by hand, pole, or a small electric motor, serve as a source for both additional light and ventilation.

Skylights may be installed over a box frame (or curb) constructed around the roof opening or directly mounted on the plane of a pitched roof. Some curb mounted skylights are manufactured with insulated or noninsulated integral curbs which attach directly to the roof deck (fig. 6-4). Others require that a curb be constructed on site from wood,

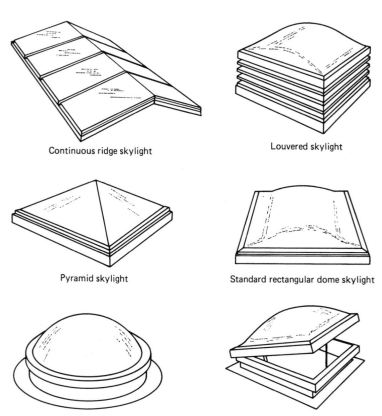

Continuous ridge skylight

Louvered skylight

Pyramid skylight

Standard rectangular dome skylight

Circular skylight

Venting skylight

Fig. 6-3. Examples of different types of skylights.

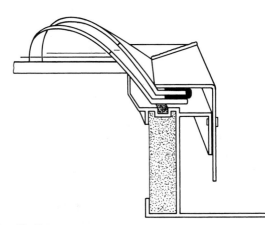

Fig. 6-4. Skylight with insulated integral curb.

151

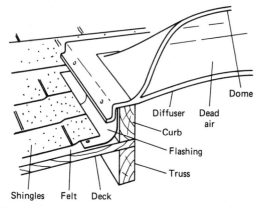

Fig. 6-5. Curb mounted skylight.

concrete, or metal and then secured to the roof deck and flashed (fig. 6-5). The skylight is then installed over the curb and secured to it with corrosion-resistant fasteners. If the skylight is not designed to be used with a curb, it is fastened directly to the roof deck sheathing through factory drilled holes. The self-flashing skylight shown in figure 6-6 is of this type. Installation without a curb gives the skylight a low profile and is less conspicuous on the roof.

Installing a Skylight

Many skylights are designed to fit standard 16- or 24-inch rafter spacings or 24-inch truss spacings (fig. 6-7). Larger units require cutting

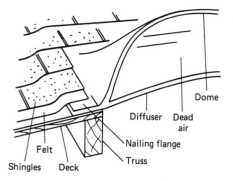

Fig. 6-6. Self-flashing skylight secured directly to roof deck.

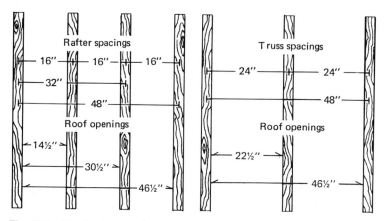

Fig. 6-7. Roof opening dimensions for different rafter and truss spacings.

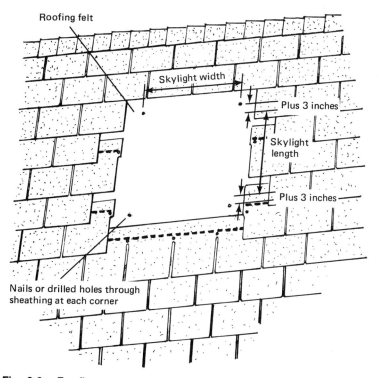

Fig. 6-8. Roofing removed from projected skylight opening.

rafters or trusses and framing the opening. Whether or not rafters or trusses require cutting will depend on the inside dimension (roof opening) listed by the manufacturer for the skylight. For example, a 14¼-inch wide skylight can be used on a roof with rafters spaced 16 inches apart without having to cut any of the rafters. The skylight is simply placed over a pair of rafters, and the required opening is cut out of the roof deck area between them. Similarly, a 22¼-inch wide skylight can be installed on a roof with roof rafters spaced 24 inches apart without cutting any of the rafters. When cutting the rafters is necessary in the installation of the skylight, headers must be added and a curb constructed to serve as a base for the unit.

Figures 6-8 to 6-11 illustrate the procedures used to install a small skylight on a roof with rafters spaced 16 inches on center. Because the skylight is less than 16 inches wide across its inside opening, the rafters do not have to be cut.

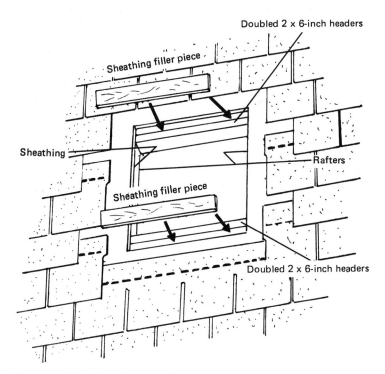

Fig. 6-9. Installing headers.

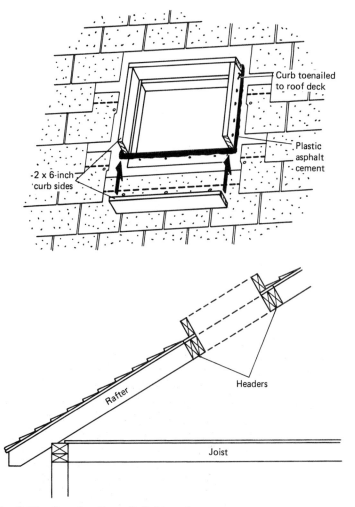

Fig. 6-10. Constructing skylight curb.

Begin by marking the size of the planned opening on the bottom of the roof deck sheathing inside the attic or attic crawl space. The size of the roof opening marked on the sheathing will correspond to the inside opening dimensions of the skylight plus 6 inches added to its length for framing the opening with headers. The 6 inches are distributed so that there are 3 inches at the top of the skylight and

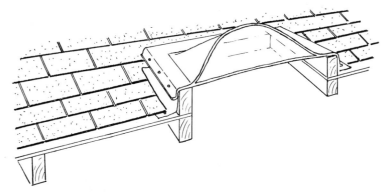

Fig. 6-11. Cutaway view of skylight installed over curb.

3 inches at the bottom (fig. 6-8). Skylight sizes commonly begin at 14¼ × 14¼ inches, which is small enough to fit between rafters without cutting. The first dimension is the width of the skylight inside opening and the second dimension is its length.

Use an electric power drill to drill a hole or drive a tenpenny (10d) nail through the roof at each corner of the square marked on the bottom of the sheathing. Remove the shingles covering the area marked by the four drilled holes or nails plus an additional 5 or 6 inches around them. Do not remove the roofing felt.

Mark the dimensions of the skylight opening on the roofing felt and cut away the roof deck sheathing with a power circular saw. Install doubled 2 × 6–inch headers at both the top and bottom of the roof opening to support the skylight (fig. 6-9). Cut rectangular pieces of plywood thick enough to lie flush with the sheathing and large enough to cover the headers. Nail the plywood pieces in place and construct a curb around the roof opening (fig. 6-10). Curbs may be constructed of 2 × 4's or 2 × 6's depending on the skylight manufacturer's instructions. In either case, the inside surfaces of the curb must be flush with the inside surfaces of the roof opening framework (i.e., headers and rafters). Coat the surface where the curb will lie with plastic asphalt cement, position the curb, and toenail its sides to the framing and roof deck with nails spaced 4 to 6 inches apart. Fit the metal frame of the skylight over the curb and secure it in place with the required number of fasteners (fig. 6-11).

A 48-inch wide or larger skylight requires the construction of a major roof opening. As shown in figure 6-12, it requires the cutting

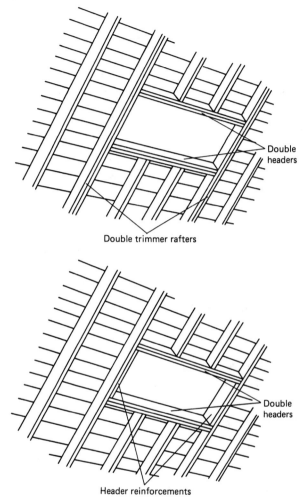

Double headers

Double trimmer rafters

Double headers

Header reinforcements

Fig. 6-12. Two methods of framing major roof openings for skylights.

of two rafters on a roof with rafters spaced 16 inches apart on center. Only one rafter needs to be cut when the spacing is 24 inches apart on center. In any event, the roof must be braced around the planned opening *before* any cutting is done, or the roof deck may sag, causing damage to the sheathing and roofing materials (fig 6-13).

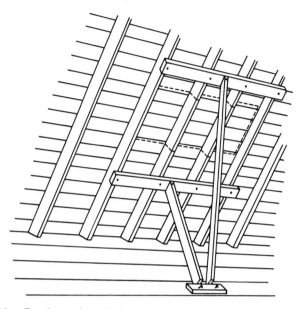

Fig. 6-13. Bracing rafters before cutting them.

As with all types of roof openings, an opening constructed for a skylight must be carefully flashed to provide protection against leaks. Typical flashing methods are described in chapter 8.

Roof Access Openings

A roof access opening is sometimes installed on a flat roof, particularly on the roofs of commercial or industrial buildings. When this is the case, a major roof opening must be constructed in the deck. This requires cutting one or more ceiling joists or rafters and installing headers to reinforce the framing around the opening. Construction details for major roof openings are illustrated in figure 6-14.

The first step in constructing an access opening for a flat roof is to calculate its exact location. The ceiling joists or rafters should then be marked for cutting and the remaining portions braced with 2 × 4's extending down to the floor below. The bracing will prevent the roof from sagging after the opening has been cut through the framing. If stairs are used and they are to be enclosed by partitions, the

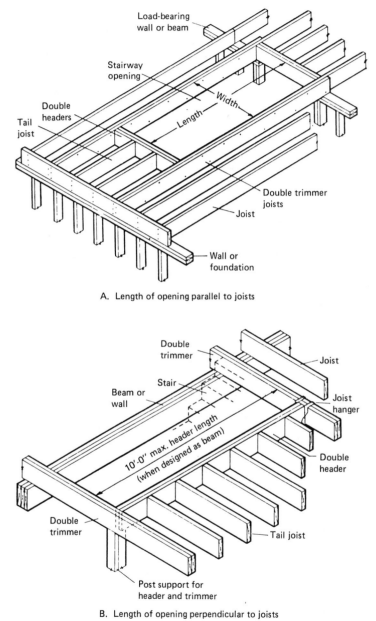

A. Length of opening parallel to joists

B. Length of opening perpendicular to joists

Fig. 6-14. Two methods of framing stair openings.

studs in the partitions will provide excellent support for the ceiling joists or rafters. If no enclosing partitions are planned, permanent braces should be installed from the framing joists or rafters to the floor below.

Roof scuttles and hatches for either ladders or stairs may be placed over the roof opening and secured with nails or bolts through holes provided in the flanges. Roofing material is inserted under the integral cap flashing and caulked with oakum and brush asphalt. Examples of roof scuttles and hatches are shown in figures 6-15 and 6-16. A roof access opening may also be covered by a wood trap door or a small shed-type structure.

Chimney Openings

A chimney that projects through the roof deck usually requires a major roof opening. At least one roof rafter must be cut and headers installed to frame the opening (fig. 6-17).

At least two inches of air space should be provided between the

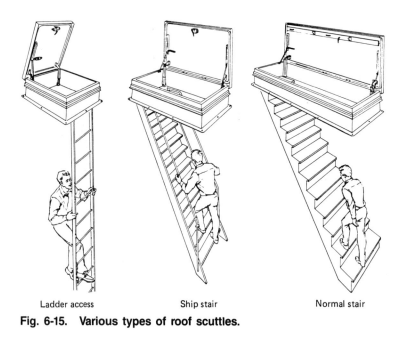

Ladder access Ship stair Normal stair

Fig. 6-15. Various types of roof scuttles.

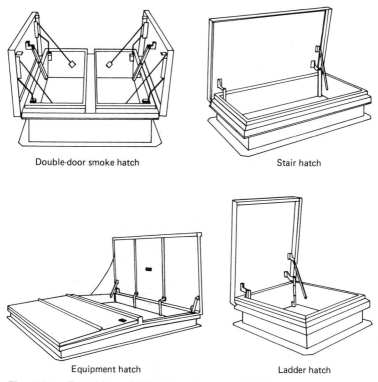

Double-door smoke hatch   Stair hatch

Equipment hatch   Ladder hatch

Fig. 6-16. Examples of different types of roof hatches.

masonry walls of the chimney and the roof framing and deck materials. This is necessary to protect the wood from the heat radiating from the surface of the chimney wall. Always check the local building code for the minimum allowance because it may vary among localities.

Bridging of the same size as the rafters is usually installed in the area near the chimney to provide additional strength for the framing. The bridging runs in the same direction as the headers, that is, at a right angle to the rafters.

Vent Pipe and Stack Openings

Vent and stack openings in the roof deck do not require the cutting of any of the framing members, because the diameter of the opening is invariably less than the 16-inch spacing of the rafters.

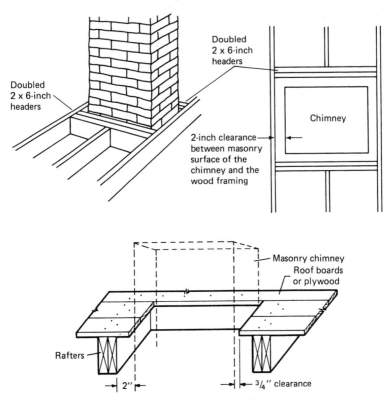

Doubled
2 x 6-inch
headers

Doubled
2 x 6-inch
headers

2-inch clearance
between masonry
surface of the
chimney and the
wood framing

Chimney

Masonry chimney
Roof boards
or plywood

Rafters

2″

3/4″ clearance

Fig. 6-17. Chimney opening framing details.

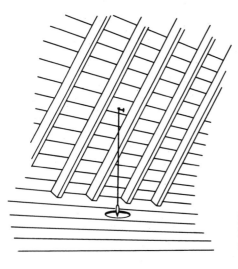

Fig. 6-18. Using plumb bob to mark center of opening on sheathing.

162

If possible, the opening in the roof deck should be placed so that it is directly over the point at which the vent pipe or stack protrudes through the attic floor. The easiest way to mark the opening is to hang a plumb bob from the under side of the roof deck. The point of the plumb bob should hang in the exact center of the vent pipe or stack opening (fig. 6-18).

If the vent pipe or stack protrudes through the attic floor at a point directly under the roof rafter, a bend or elbow should be installed in the piping so that the opening in the roof deck will be located between rafters. This is not as satisfactory an arrangement as a straight pipe, because the bend in the piping offers resistance to the flow of the rising gases, but it does avoid the more complicated and time consuming procedure of cutting a rafter and installing headers and trimmers. The pipe bend or elbow should be installed first and then a plumb bob can be hung from the underside of the roof deck to determine the location of the opening.

Any pipes carrying hot gases and other by-products of the combustion process should have an air space between the surface of the pipe and the wood of the roof deck (fig. 6-19). The size of the air space can be determined by consulting the local building code. The interior of the structure is protected from weather conditions by covering the

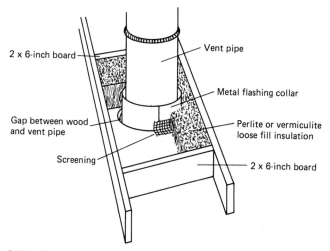

Fig. 6-19. Vent stack details.

air space with flashing and other roof-covering materials. Flashing methods are described elsewhere in this book.

Roof Ventilators

Ventilators are sometimes installed on the roof deck to provide sufficient ventilation in the attic or attic crawl space. These roof ventilators are used on both flat roofs and those with a slope.

As is the case with a vent pipe or stack opening, the opening for a roof ventilator is made between the rafters. Because this is a minor roof opening, there is no need to cut any of the roof rafters in order to install the ventilator.

Roof ventilators are available in a variety of types and designs, ranging from those that are essentially nothing more than a roof opening protected by a metal weather cap to those incorporating a motor-driven fan. The manufacturer of the roof ventilator will provide installation instructions, and these should be read and followed carefully. Additional information covering the installation of roof ventilators is contained in chapter 5.

Interior Roof Drains

Flat roofs on commercial buildings are often equipped with interior roof drains to prevent the ponding of water when it rains, or when ice or snow melt.

Roof drains must be located at the lowest point of the roof to obtain proper drainage. Each roof drain should be provided with a wide flange, expansion joint sleeves, and strainer construction that will allow adjustments for meeting variations in roof level or slope (fig. 6-20).

Expansion Joints

Structural expansion joints are installed on flat built-up roofs of large expanse to protect the roof membrane from stress forces (fig. 6-21). The location of a structural expansion joint should be determined by

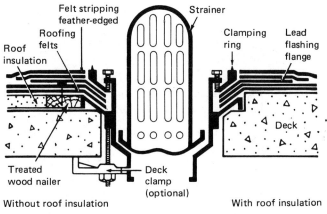

Fig. 6-20. Interior roof drain construction details.

an architect, engineer, or building contractor knowledgeable about the types of stresses found on large roof decks.

Structural expansion joints should be installed at the following locations:

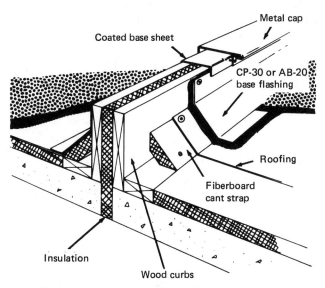

Fig. 6-21. Construction details of typical expansion joint.

1. Where the deck structures change directions as in an *L*-shaped building
2. Where two dissimilar deck materials join
3. Where a new roof has been added to an existing one
4. Where there is a difference of elevation of two adjoining decks
5. Where roof frame members change directions
6. Where building length exceeds 200 feet

The omission of properly placed expansion joints quite often results in roof membrane fracturing caused by stress forces.

Other Types of Openings

Column stubs, sign anchors, railing posts, tank supports, and similar types of projections extending through a flat roof surface should be installed in a metal flashing pan with a 4-inch high collar and a 6-inch wide deck flange (fig. 6-22).

These types of roof deck openings should be made large enough to provide a narrow space between the edge of the deck and the object projecting through it. This space provides an allowance for the expansion and contraction of the roof deck materials.

The flanges of the flashing pan should be embedded in plastic asphalt cement over the roofing felts. Two collars of roofing felt should then be cut to fit snuggly over the flashing pan to seal the flange. The flange should be overlapped 6 inches and 9 inches respectively on all

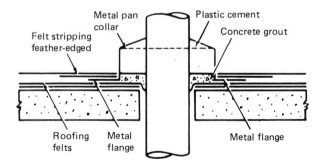

Fig. 6-22. Metal flashing pan with collar and deck flange.

sides with each of the two roofing felt collars embedded in an application of plastic asphalt cement. The roof-covering materials are installed over the collars.

Fill the bottom of the flashing pan with 1 inch of concrete grout, and the remainder with plastic asphalt cement. Slope the plastic asphalt cement layer downward to the outside of the pan lips to provide sufficient drainage.

CHAPTER 7

Dormer Construction

- Types of Dormers
- Tools, Equipment, and Materials
- Attic Preparation
- Dormer Layout

- Cutting and Framing Roof Openings
- Windor Dormer Construction
- Shed Dormer Construction
- Finishing Attics

A dormer is an enclosed room or space with vertical walls that project through the surface of a sloping roof. The word *dormer* is derived from the Latin verb *dormire* (to sleep) and was originally used to designate a room or space used for sleeping. Today, it identifies virtually any type of enclosed room or space with walls rising vertically from a sloping roof.

Types of Dormers

A dormer has one or more windows, depending on its construction. The most common type of dormer used in residential construction is the small window dormer shown in figure 7-1. Two or more of these dormers are generally used on the front slope of a pitched roof. The number used will depend on the length of the roof. When this type of dormer is constructed with a gable roof, it is called a *gable dormer*. Other types of roofs used on the window dormer are the hip roof and shed roof (figs. 7-2 and 7-3). Most window dormers allow both light and ventilating air into the attic. Some dormers, however, are constructed for purely decorative purposes. These nonfunctional dormers

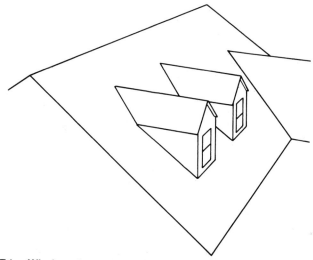

Fig. 7-1. Window dormer with gable roof.

are attached to the roof without cutting an opening through the roof deck.

Some houses have a large dormer that extends almost the entire length of one roof slope. These large dormers may be constructed on

Fig. 7-2. Window dormer with hip roof.

Fig. 7-3. Window dormer with shed roof. *(Courtesy Masonite Corp.)*

either the front slope of the roof, as in figure 7-4, or on the rear slope. Often, two or more window dormers will be constructed on the opposite slope of the same structure. These large roof-length dormers are sometimes called *roof dormers* or, because they are covered with shed roofs, *shed dormers*. In addition to providing light and ventilation, a shed or roof dormer greatly increases the attic area, particularly its head room.

Both the window dormer and the shed or roof dormer require major roof openings. A number of roof rafters must be cut and removed and the opening framed with headers and trimmers in order to accommodate the dormer framing. Construction details for both a gable dormer and a shed dormer are included in this chapter.

Tools, Equipment, and Materials

The tools and equipment required for the construction of a roof dormer are the same as those required for other types of wood frame con-

Fig. 7-4. Shed dormer. *(Courtesy Masonite Corp.)*

struction. The needed tools include a claw hammer, circular or saber saw, straight edge, combination square, measuring tape, plumb bob and string, and carpenter's apron. A sturdy ladder or scaffolding is required for access to the top surface of the roof.

All framing lumber should be well seasoned. Lumber 2 inches thick or less should have a moisture content of not more than 19 percent. Second grades of the various softwoods are normally used for the rafters. Existing rafters cut to form the roof opening for the dormer can also be used on the dormer roof.

Use 2 × 4's for wall studs, top plates, soles, roof rafters, and ceiling joists. A 1 × 6 is recommended for the ridge board of a gable dormer roof. The trimmers and headers are cut from 2 × 6's. The attic and dormer subfloors are usually constructed from ¾-inch thick 4 × 8–foot plywood sheathing panels. The dormer roof and wall sheathing are identical to those used on the rest of the structure. The same holds true for the roofing and siding. Obtain fascia and cap molding to match the trim on the rest of the structure.

Framing lumber is ordered by the nominal size and priced in

Table 7-1. Nominal and Dressed Sizes of Framing Lumber

Nominal Size	Dressed Size
1 × 2	¾ × 1½
1 × 3	¾ × 2½
1 × 4	¾ × 3½
1 × 6	¾ × 5½
1 × 8	¾ × 7¼
1 × 10	¾ × 9¼
1 × 12	¾ × 11¼
2 × 2	1½ × 1½
2 × 4	1½ × 3¼
2 × 6	1½ × 5½
2 × 8	1½ × 7¼
2 × 10	1½ × 9¼
2 × 12	1½ × 11¼

board feet. The nominal size is not the actual lumber measurement or dimension, but instead a designation of the type of frame member. For example, the 2 × 4 commonly used for wall studs actually measures 1½ × 3¼ inches. These size differences should be taken into consideration when ordering lumber for dormer construction. Nominal and actual sizes for framing lumber are listed in table 7-1.

Windows can be purchased made to order or prefabricated complete with exterior and interior trim. The size of the window will determine the size of the rough opening in the dormer.

Purchase 5 pounds of 16d, 8d, and 6d common nails for the job. A 20d nail is recommended for nailing the doubled 2 × 4's of the sole plate to the roof deck sheathing.

Attic Preparation

Most of the work involved in the construction of a dormer is performed from inside the attic. An unfinished attic will usually have exposed floor joists that must be covered before work on the dormer can begin. Covering these joists provides a convenient work surface and protects the ceilings of the rooms below the attic from tools or materials that are invariably dropped during construction. Another important factor to be considered before constructing a dormer or finishing an attic is the type of access involved. Some attics have existing stairways, whereas others are entered through ceiling hatches. If the attic is to be finished as a bedroom, den, or other type of living quarters, the hatch entry should be replaced by a stairway. The stairway opening should be framed and the stairs built before the dormer is constructed.

Attic Subfloor

At this point in the construction of a dormer only a subfloor is nailed to the joists. The finish boards, or floor covering materials such as carpets, tiles, or linoleum, are applied after the dormer has been framed and enclosed and the attic finished (see the discussion of finishing attics in this chapter).

Subflooring commonly consists of wood boards or sheet materials such as plywood, oriented strand board, or waferboard. Oriented strand board—often abbreviated OSB—resembles waferboard but differs in the manner in which its wood strands are arranged. The orientation and length of the strands makes a sheet of OSB strong enough to span greater distances under heavier loads than waferboard.

Subfloor sheets are rated by span index. The numbers of a span index indicate the distance the sheet can span. For example, a sheet of $7/16$-inch OSB has a span index of $24/16$, meaning it can span 24 inches when used as roof sheathing and 16 inches when used as subflooring. To span a greater distance between floor joists, a thicker sheet of OSB must be used. Always check the American Plywood Association (APA) stamp on the product for its span capability.

The size of the plywood panels used to construct the attic subfloor may be a problem. Plywood is usually available in 4 × 8–foot panels and there may be no opening in the attic floor large enough to admit panels this large. If this should be the case, nail four or five 1 × 6–inch boards across the joists in the projected work area. The plywood panels can then be brought into the attic through the dormer roof opening once it is cut.

A more expensive solution to bringing materials into the attic is to use 1 × 6–inch sheathing boards or planks for the subfloor. This method requires more time because far more cutting and nailing is required than when working with plywood panels. Sheathing boards are usually applied so that they run at right angles to the floor joists (fig. 7-5).

Plywood panels are applied with their length running parallel

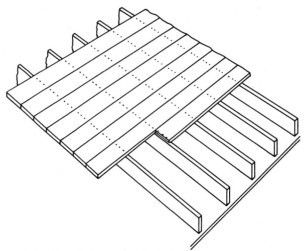

Fig. 7-5. Subfloor of sheathing boards.

to the joists. Butt the panels so that the joints are centered over the joists. Nail them to the joists with common eightpenny (8d) nails placed every 6 to 8 inches (fig. 7-6).

Before laying the subfloor, bridging should be installed between

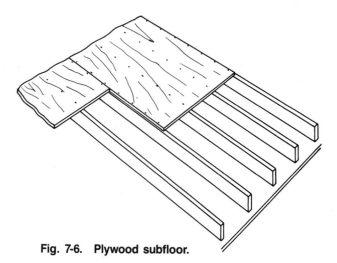

Fig. 7-6. Plywood subfloor.

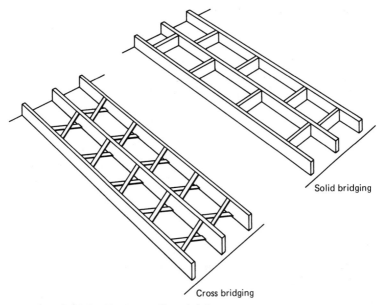

Solid bridging

Cross bridging

Fig. 7-7. Bridging between floor joists.

the joists for reinforcement (fig. 7-7). As the bridging is being installed, rough in all electrical wiring, plumbing, and duckwork required for the finished attic.

Attic Access

Access openings in attic floors are framed out during the construction of the floor system. If the attic is intended only for storage, the access opening will usually be covered by a hatch which may or may not be hinged to the frame. Construction details of two access openings used with hatches are illustrated in figures 7-8 and 7-9. Note the use of doubled headers to provide reinforcement.

The long dimension of stairway openings may be either parallel or at right angles to the joists (fig. 7-10). Framing a stairway opening with its length parallel to the joists is considered the easier of the two construction methods. When the long dimension of the stair opening is at a right angle to the joists, a long doubled header is required for reinforcement. A header under these conditions and without a supporting wall beneath is usually limited to a 10-foot length. A load-

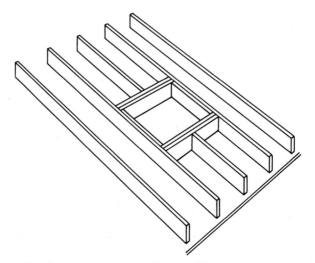

Fig. 7-8. Hatch-type access opening requiring the cutting of one or more floor joists.

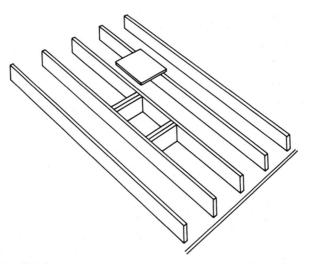

Fig. 7-9. Hatch-type access opening not requiring the cutting of floor joists.

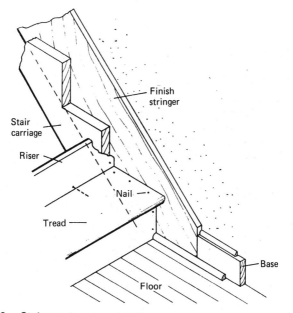

Fig. 7-10. Stairway framing details.

bearing wall under all or part of the opening simplifies framing because the floor joists will then bear on the top plate of the wall instead of being supported at the header by joist hangers or other means. The construction details of a stairway enclosed by a load-bearing wall is illustrated in figure 7-11.

A typical stairway consists of carriages, stringers, treads, and risers. The *stair carriages* carry the treads and support the load on the stairs. Stair carriages are usually made from 2 × 12–inch planks which are notched to receive the treads. Carriages are placed on each side of the stairs. An intermediate carriage is required at the center of the stairs when the treads are 1¹/₁₆ inches thick and wider than 2 feet 6 inches. An intermediate carriage is also required when the treads are 1⁵/₈ inches thick and more than 3 feet wide.

The *treads* are the horizontal boards of a stairway on which the feet are placed. The spaces between the treads are closed with vertical boards called *risers*. The *stringers* function as trim boards and, in some types of stair construction, as support for the treads.

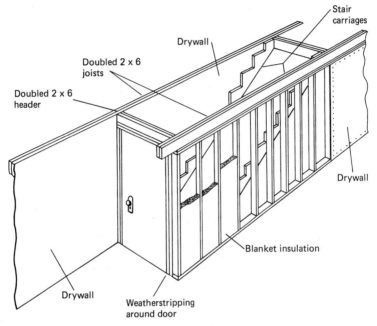

Fig. 7-11. Enclosed stairway construction details.

There is a definite relationship between the height of a riser and the width of a tread in a properly planned stairway. Experience has shown that a riser 7 to 7³/₄ inches high used with a tread about 10 inches wide results in both comfort and safety. To determine whether a proper relationship exists between tread width and riser height, apply either of the two following "formulas" to the dimensions in the stairway layout:

1. Tread width × riser height = approximately 75 inches
2. Tread width = (2 × riser height) = approximately 25 inches

These figures may be adjusted by two or three inches and still fall within the recommended tread width and riser height range required for safety and comfort.

In general, 14 to 15 risers are used for a stairway between a lower floor and an attic. The 8-foot ceiling height of the lower floor plus the floor joists, subfloor, and projected finish floor of the attic should result in a floor-to-floor height of about 105 inches. Each riser will be 7½ inches high if 14 risers are used between the two floors. Fifteen risers used for a 105-inch height would result in 7-inch high risers.

The construction details of a straight-run stairway connecting a lower floor to an attic are illustrated in figure 7-12. It combines a rough notched stair carriage with a finish stringer along each side. The finish stringer is fastened to the wall before the two carriages are fastened in place. Treads and risers are cut to fit snugly between the stringers and are fastened to the rough carriage with finishing nails. The construction may be varied by nailing the rough carriage directly to the wall, as in figure 7-13, and notching the finish stringer to fit over the carriage.

If the attic is used primarily for storage or if no space is available for a stairway to a finished attic, hinged or folding stairs are often used and may be purchased ready to install. They are operated through an opening in the ceiling of a hall and swing up into the attic space out of the way when not in use. Where such stairs are to be installed, the attic floor joists should be reinforced for limited floor loading. One

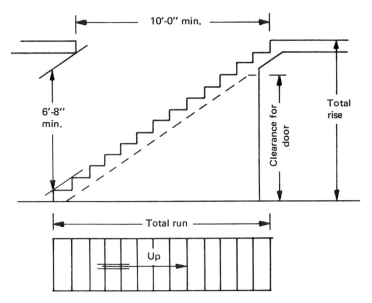

Fig. 7-12. Typical straight-run stairway between attic and lower floor.

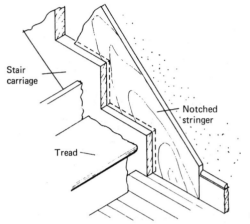

Stair
carriage

Notched
stringer

Tread

**Fig. 7-13. Carriage
nailed directly to wall
with fitted stringer.**

common size of folding stairs requires only a 26 × 54-inch rough
opening. These openings should be framed out as for normal stair
openings.

Dormer Layout

Always check the local building code before beginning any work to
make certain the planned dormer and its specifications meet code requirements. Some building developments or subdivisions also have
restrictions on the types of additions that can be made to a house,
and these will also have to be checked before going ahead with
construction.

A local building permit will have to be obtained before construction begins. A fee is charged for a building permit, which is issued
by the city or county building department after the plans or blue prints
have been approved. The building permit must be posted in a conspicuous place at the construction site. There will probably be at least
one visit from the building inspector while the job is in progress.

Dormer Location

Choosing the type of dormer and determining its location on the roof
are the first two steps in dormer layout. The dormer must be in pro-

portion with the rest of the structure and must be properly centered on the roof.

Working Plan

Draw a working plan of the dormer on graph paper with detailed front and side views. Enter the dimensions of all framing members on the graph paper. The scale drawings of the dormer will be used to estimate the actual (dressed) sizes of the lumber required for the job. The working plan will also mark the exact location of the roof opening (or openings) and will include the angles and locations of the rafter cuts.

Dormer Roof Slope or Pitch

Minimum slope for a shed dormer roof is 1 inch of rise for each foot of run, or 1-in-12. The slope or pitch of a shed dormer roof will always be less than that of the roof on which it is located (fig. 7-14). The slope or pitch of a window dormer roof, on the other hand, will usually be the same as for the house roof (fig. 7-15).

Read the sections in chapters 1 and 2 covering slope and pitch before continuing because these aspects of roof layout must be clearly understood.

Window Dormer Layout

Draw a cross section of the roof to scale with the rafters at the correct angle and proceed as follows:

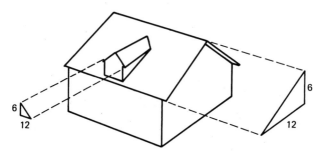

Fig. 7-14. Window dormer roof slope.

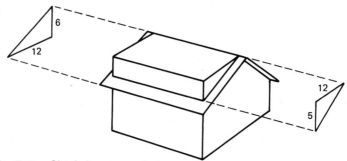

Fig. 7-15. Shed dormer roof slope.

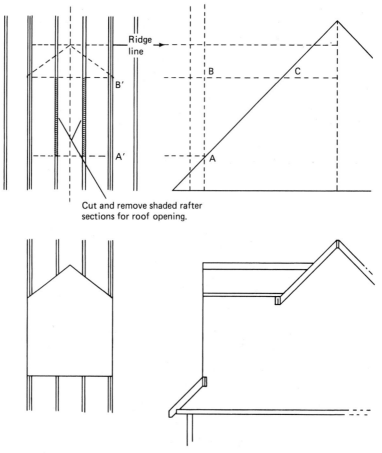

Ridge line →

Cut and remove shaded rafter sections for roof opening.

Fig. 7-16. Window dormer layout.

1. Locate the top surface of the bottom header (point *A* in fig. 7-16) at least 2 feet above the attic floor.
2. Run a vertical line from point *A* to point *B*, a scale distance of 5 feet.
3. Draw a horizontal line at a right angle to point *B* and extend it until it meets the roof at point *C*. The horizontal line *B–C* represents the position of the bottom surface of the dormer sidewall top plate. Point *C* represents the top surface of the top header.
4. Draw a front view of the dormer to scale showing the spacing of the roof rafters and the locations of the doubled rafters framing the roof opening.
5. Transfer point *A* and point *B* from the side view of the dormer to its front view (point *A'* and point *B'* respectively) and draw lines at right angles to the rafters at these points.
6. Run a line perpendicular to the horizontal lines at points *A'* and *B'* in the exact center of the roof opening to locate the position of the dormer roof ridge.
7. Connect point *B'* to the perpendicular line at an angle equal to roof slope or pitch.
8. Transfer the height of the dormer roof ridge in the front view drawing of the dormer to its side view drawing, draw a horizontal line to represent the top of the roof ridge, and mark the point where it intersects the roof.

Shed Dormer Layout

Draw a cross section of the roof to scale with the rafters at the correct angle and proceed as follows:

1. Run a vertical line from the bottom of the ridge board (point *A* in fig. 7-17) to the attic floor (point *B*).
2. Draw a line marking the location of the outside surface of the shed dormer front wall (point *C* if it is flush with the lower exterior wall or point *D* if it is set back from it) perpendicular to the attic floor.
3. Run a horizontal line from point *A* at a right angle to line *A–B* until it intersects with vertical lines from points *C* and *D*, and label the intersecting points *E1 E2* respectively.
4. Measure the length of line *A–E1* (flush dormer wall) or line *A–E2* (set back dormer wall) and enter the amount on the scale drawing in feet.

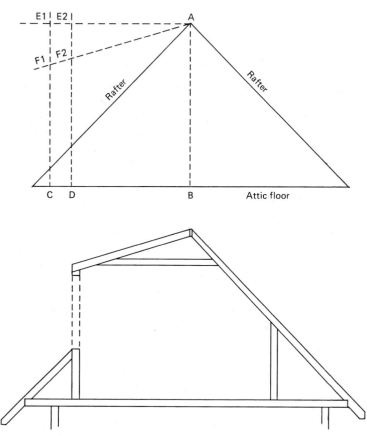

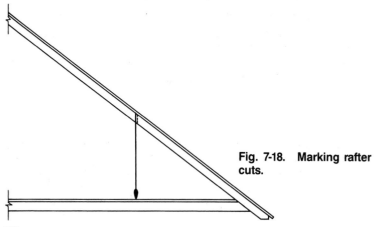

Fig. 7-17. Shed dormer layout.

Fig. 7-18. Marking rafter cuts.

5. Measure down from point $E1$ or point $E2$ a distance in inches *at least* equal to the distance in feet of line A–$E1$ or A–$E2$ to find the minimum slope (1-in-12) of a shed dormer roof. Label these points $F1$ and $F2$ respectively. If, for example, line A–$E1$ was 12 feet, the distance between points $E1$ and $F1$ or between $E2$ and $F2$ would have to be a minimum of 12 inches.

6. Draw a line between point $F1$ or $F2$ and point A to represent the bottom edge of the shed dormer roof rafters.

Cutting and Framing Roof Openings

Never cut the roof opening for the dormer until the floor joists inside the attic have been covered with a subfloor or suitable temporary flooring placed directly below the projected opening in the roof. This precaution will protect the ceiling of the room below from falling debris.

Before cutting through the roof be certain that some sort of protection has been provided for the attic in case it rains, or during those periods when no work is being done. A canvas, plastic sheet, or tarpaulin large enough to extend at least a foot beyond the roof opening on all sides will provide adequate protection. The top of the cover should be nailed to the roof deck on the other side of the roof ridge. The nails are driven through short lengths of 1 × 4–inch boards to prevent the cover from being torn away from the nails in strong winds. When rain threatens, the cover should be drawn down over the opening and nailed at the bottom and sides in the same manner as along the top.

Dormer layout calculations will provide the basis for estimating the dimensions of the roof opening (see the discussion of dormer layout in this chapter).

The procedure for cutting a dormer roof opening may be summarized as follows:

1. Snap a chalk line along the bottom of the rafters to indicate the lower edge of the roof opening (be sure to provide an allowance for the double header and front wall frame).

2. Mark each rafter with a bevel square or chalk line and plumb bob for the exact angle of the cut (fig. 7-18).

3. Install 2 × 4 braces under the roof ridgeboard and under all sections of rafters that will remain after cutting the roof opening.

The rafter and ridge board braces must remain in position until the roof opening is framed and permanently braced. Bracing is required to prevent the roof from sagging. Additional information about the cutting and framing of roof openings is provided in chapter 6.

Window Dormer Construction

The window for a window dormer should be purchased complete with exterior and interior trim. The window manufacturer will provide instructions for installing the window and the trim. Read these instructions carefully before beginning work. If there are any questions, return to the building supply dealer where the window was purchased for clarification of the installation instructions.

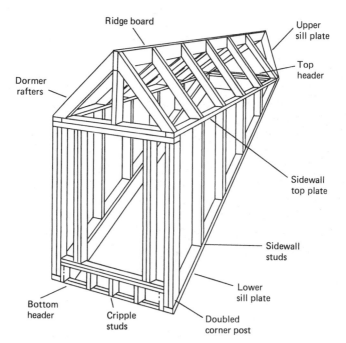

Fig. 7-19. Construction details of typical gable-type window dormer.

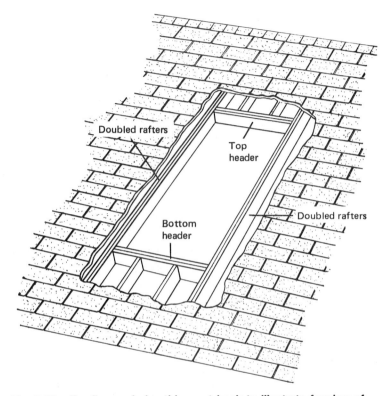

Fig. 7-20. Roofing and sheathing cut back to illustrate framing of roof opening for dormer.

The dormer window should match the style and dimensions of other windows in the structure. An exact match is not necessary, but avoid using window styles and dimensions so different that they make an unpleasant contrast.

Another important consideration is the *number* of window dormers planned for the roof. The main roof of an average size house or building usually has enough space for two or perhaps three window dormers without giving the surface of the roof a crowded appearance. This is usually the case if no minor roof intersects the main roof on the same slope where the window dormers are to be installed. Usually only one or two window dormers should be used on a main roof to which a minor roof is joined.

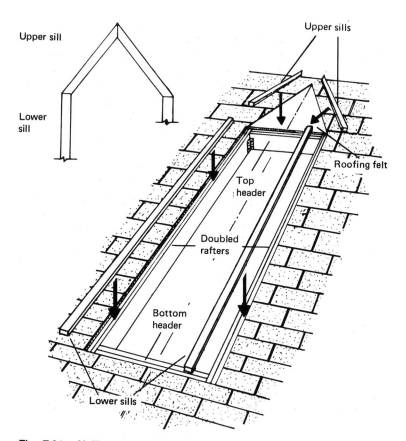

Fig. 7-21. Nailing dormer sill plates to rafters and roof deck.

Window dormers can be installed on a minor roof, but only if the roof is long enough to accommodate them. Once again, a crowded appearance should be avoided. Dormer windows should be spaced the same distance apart as windows would be normally spaced on lower floors. If a lower floor can only accommodate one average size window, then under most circumstances only one window dormer should be used on the roof. These recommendations are offered only as a guide, because there is such a variety in architectural design that one always encounters an exception to the rule.

After the size of the dormer window has been selected, lay out the dimensions of the dormer enclosure. These dimensions will vary widely, depending on the size of the window, the size of the dormer enclosure, and other architectural considerations such as roof design.

In older construction, a rough window opening is ordinarily made 5 inches wider than the width and height of the window glass to be used. Thus, a 24 × 26–inch, double-hung window would require a rough opening of 34 × 36 inches. The extra space is used to accommodate the window frame into which the window sashes (the movable parts) are fitted and hung.

Most windows available today do not require as much space for

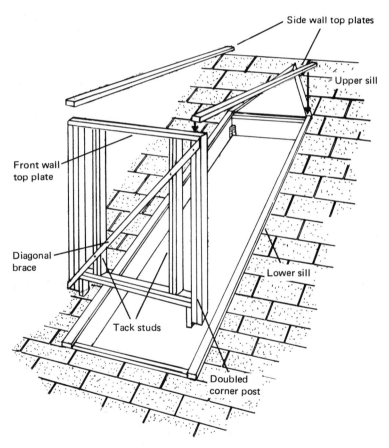

Fig. 7-22. Braced corner posts and window framing supported by sidewall top plates.

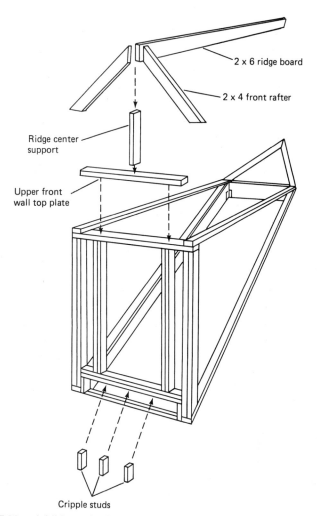

Fig. 7-23. Addition of cripple studs, ridge board, and ridge board support framing.

the window frame. This is particularly true of those with aluminum or other light metal frames. In any event, the size of the rough window opening will be determined by the window dimensions.

Window dimensions also have a direct influence on the width and height of the dormer enclosure. Under normal circumstances, the

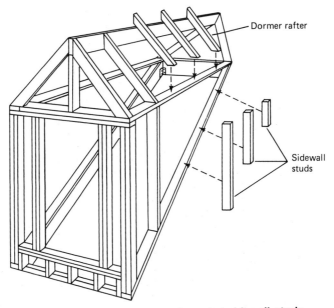

Fig. 7-24. Installation of dormer rafters and sidewall studs.

width of the dormer will not exceed 10 to 12 inches on either side of the window. Thus, a dormer enclosure with a 24-inch wide window should have a maximum width of 44 to 48 inches. Once again, dimensions will vary, depending on the size of the window and other design factors; but the window area should dominate, with the surface of the dormer enclosure functioning simply as a frame for the window.

Figure 7-19 illustrates the construction details of a typical gable-

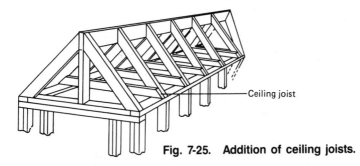

Fig. 7-25. Addition of ceiling joists.

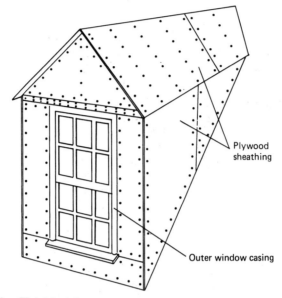

Plywood
sheathing

Outer window casing

Fig. 7-26. Finishing dormer.

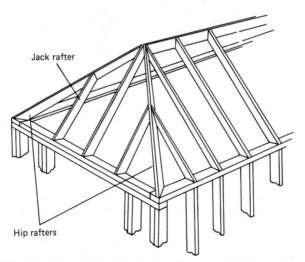

Jack rafter

Hip rafters

Fig. 7-27. Framing details of window dormer with hip roof.

type window dormer used on a roof with rafters spaced 16 inches on center. These construction details are applicable to any size window dormer enclosure; only the dimensions will change.

As shown in figure 7-20, a header usually functions as the base for the window frame as well as a nailing base for the rafter that had to be cut to make room for the dormer opening. Sometimes a plate (or sill header) is nailed across the top of the header, but the bottom of the dormer window usually remains no higher than 5 or 6 inches above the roof surface on roofs with an average slope. However, roofs with steep slopes generally require that lower cripples and a sill header be installed to raise the window opening so that it is centered properly. In larger dormers, it may also be necessary to use cripples above the window opening.

The step-by-step procedure used to construct a gable-type window dormer are illustrated in figures 7-21 to 7-26. Framing details of window dormers covered with hip and shed roofs are shown in figures 7-27 and 7-28.

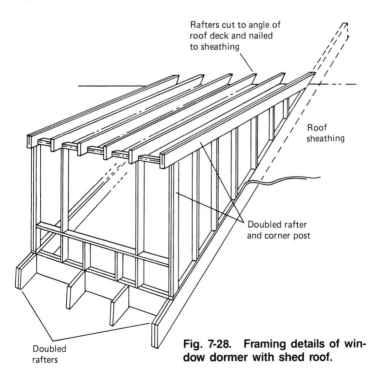

Rafters cut to angle of roof deck and nailed to sheathing

Roof sheathing

Doubled rafter and corner post

Doubled rafters

Fig. 7-28. Framing details of window dormer with shed roof.

The same siding and roofing materials used on the rest of the structure should also be used to cover the window dormer. Installation instructions are included in the various chapters covering specific types of siding and roofing materials. For a description of how to finish the interior, see the discussion of finishing attics in this chapter.

Shed Dormer Construction

A shed dormer is a major project that involves the cutting of most of the rafters along one roof slope. Because its construction is more complicated than the smaller window dormer and represents a major addition to the structure, it should be built by a qualified contractor or a homeowner with above average carpentry skills and experience.

The windows of a shed dormer should be similar in design and size to those in the lower floors. Window spacing is also important because too many or too few windows along the front of the dormer can detract from its appearance. As a general rule, there should never be more windows in the dormer than there are in the floor below.

The construction details of a typical shed dormer are illustrated in figures 7-29 to 7-33. A shed dormer has a flat roof with rafters running from the ridge board of the roof to the top plate of the dormer front wall. The original roof rafters are often recut to serve as the shed dormer rafters. The corner posts are doubled 2 × 4's, as in window dormer construction, and are usually located over the third rafter from

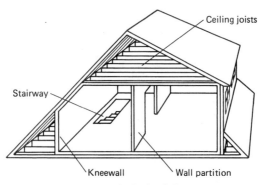

Fig. 7-29. Cross section of typical shed dormer.

1. Mark roof opening dimensions on bottom surface of roof, drive nails through roof deck, and snap chalk lines between nails. Extend to the roof ridge both chalk lines running parallel with roof rafters.

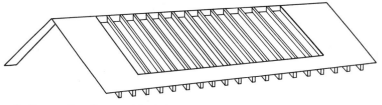

2. Remove all roofing materials within the lines, temporarily brace the roof ridge board and the lower rafter sections, and cut the rafters.

Fig. 7-30.   Marking and cutting roof opening.

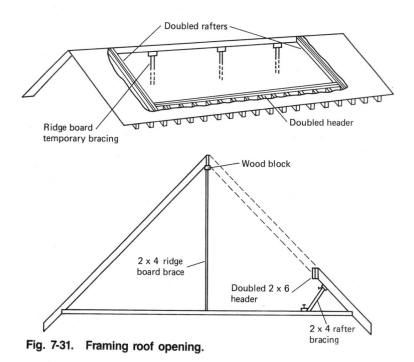

Doubled rafters

Ridge board / temporary bracing

Doubled header

Wood block

2 x 4 ridge board brace

Doubled 2 x 6 header

2 x 4 rafter bracing

Fig. 7-31.   Framing roof opening.

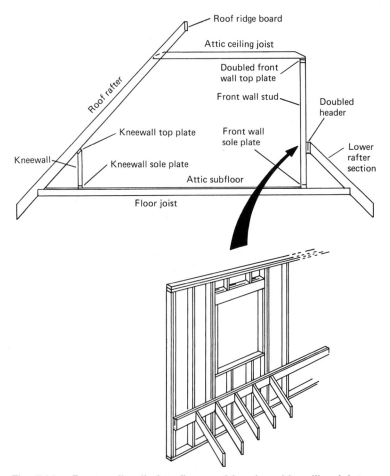

Fig. 7-32. Front wall nailed to floor and header with ceiling joists connecting back roof rafters to front wall top plate.

each end of the roof. The third rafter is always doubled to reinforce the roof opening.

The methods used to finish the interior and exterior of a shed dormer are the same as those described for a window dormer. For a description of how to finish the interior, see the discussion of finishing attics in this chapter.

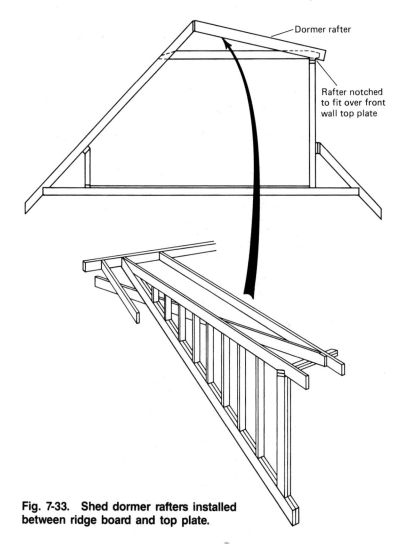

Dormer rafter

Rafter notched
to fit over front
wall top plate

**Fig. 7-33. Shed dormer rafters installed
between ridge board and top plate.**

Finishing Attics

The easiest way to add living space to a house is by finishing the attic. A finished attic can serve as an extra bedroom, den, sewing room, or office. If the attic is large enough, it can be divided into several rooms. Regardless of its intended function, however, finishing the attic will usually increase the resale value of the house.

Before making any remodeling plans, check the amount of headroom in the unfinished attic. There should be a minimum of 8 feet from the bottom of the ridge board to the tops of the floor joists or attic subfloor.

Another important factor to consider is the type of attic access. A finished attic should be reached by a stairway. If there is no stairway, an opening should be framed and stairs constructed before remodeling the attic (see the discussion of attic preparation in this chapter).

Light can be admitted to a finished attic by installing windows

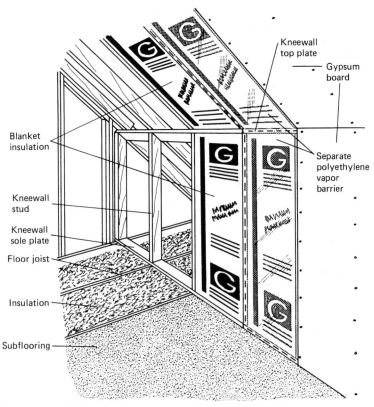

Kneewall top plate

Gypsum board

Blanket insulation

Separate polyethylene vapor barrier

Kneewall stud

Kneewall sole plate

Floor joist

Insulation

Subflooring

Fig. 7-34. Finished attic construction details.

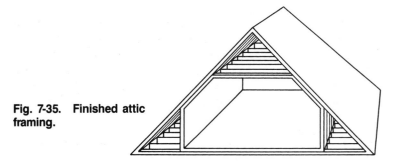

Fig. 7-35. Finished attic framing.

in the gable endwalls. Some window manufacturers produce windows narrow enough to be installed between studs spaced 16 inches apart on center, making it unnecessary to cut the studs. Another method is to add one or more window dormers or a shed dormer. In addition to admitting light, a shed dormer will increase the area of the finished attic by extending the headroom along the roof eave.

Wiring and plumbing must be installed before the framing is covered. The work and materials must be in compliance with local building code requirements. In most cases, a qualified contractor should be hired to do this type of work. The same holds true if heating and cooling ducts are to be installed in the attic and connected to the rest of the duct system.

Construction details common to most finished attics are illustrated in figures 7-34 to 7-36. Ventilating requirements for attics are covered in chapter 5; insulating procedures are described in chapter 4.

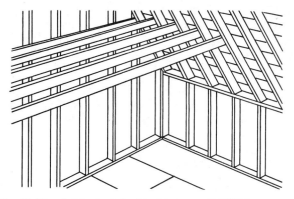

Fig. 7-36. Ceiling joists installed between roof rafters.

Floor Insulation

Insulating the attic floor is optional. Two advantages obtained by insulating the floor are the reduction of heat flowing into the attic from the rooms below and the muffling of sounds from those living spaces. If insulation is added between the floor joists, it must be done before the attic subfloor boards or panels are nailed down. Use 6-inch thick unfaced blanket insulation if there is already some insulation and a vapor barrier between the floor joists.

Subfloor Boards or Panels

The attic subfloor should be laid after the insulation has been installed between the floor joists. When applying plywood, oriented strand board (OSB), waferboard, or other types of sheet material, separate the edges of the sheets by 1/16 inch to allow for expansion. Drive a nail every 8 inches through the panels and into the floor joists.

Roof and Gable Endwall Insulation

Some finished attics have both a flat and sloping ceiling with the latter extending down to the top of a kneewall. The recommended method of insulating this type of construction is illustrated in figure 7-37. Another method is to install blankets between the rafters from the ridge board to the eaves. When this is the case, it is not necessary to insulate the ceiling or kneewalls.

Do not allow the insulation to block the eave vents when installing blankets between rafters. Leave at least a 1-inch clearance between the bottom of the roof deck sheathing and the top of the insulation blanket if ceiling panels or sheets are nailed directly to the bottoms of the roof rafters.

Push blankets between the studs in the gable endwalls and staple the flanges to the framing. Make sure the vapor side is facing toward the attic. The same holds true for the insulation blankets installed between the rafters.

Staple a separate vapor barrier of 4-mil or 6-mil polyethylene material directly to the rafters and endwall studs over the insulation. This separate vapor barrier will prevent moist air inside the attic from condensing on the wood framing and insulation during cold weather.

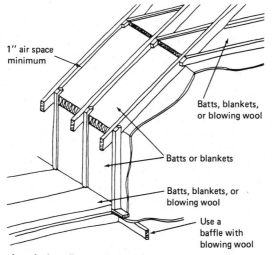

1" air space
minimum

Batts, blankets,
or blowing wool

Batts or blankets

Batts, blankets, or
blowing wool

Use a
baffle with
blowing wool

Fig. 7-37. Insulating flat and sloping ceiling extending down to kneewall. *(Courtesy Johns-Manville Corp.)*

Although the kraft paper covering on blanket insulation will help block the passage of this moisture, it will not be completely effective on its own.

Kneewall Construction

The kneewall is a low wall running parallel to the ridge board (fig. 7-38). Installing a kneewall will extend the floor area of a finished attic and close off the narrow space where the roof rafters meet the floor joists. Drawers and closets can be built into a kneewall for storage. When the area behind the kneewall is used for storage, the spaces between the outer attic rafters and outer attic floor joists must be insulated and the joists covered with subflooring.

A kneewall is framed with 2 × 4's. The studs in the kneewall are usually about 4 feet high or higher, depending on the roof pitch or slope. As shown in figure 7-34, the studs are nailed to a sole plate and top plate which are nailed to the subfloor and rafters respectively. Note that the tops of the kneewall studs are cut at the same angle as the rafters so that the top plate will lie flat against them when nailed in place. As a general rule, the shortest studs are used with steep roofs.

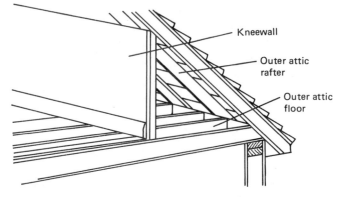

Kneewall

Outer attic rafter

Outer attic floor

Fig. 7-38. Kneewall with outer attic rafter and floor details.

Use a chalk line and plumb bob to locate the position of the kneewall. Construct the kneewall on the attic floor and then raise it into position. Build the kneewall so that each stud is located directly under each rafter. This will place the studs slightly to one side of the floor joists. Wedge the kneewall tightly against the rafters and subfloor, and toenail it in place.

Install a door in the kneewall for access to the outer attic area. Fasten 4-mil or 6-mil polyethylene sheet material as a separate vapor barrier to the inside surfaces of the kneewall studs.

The space behind the kneewall must be properly ventilated or moisture will accumulate and eventually damage the insulation and wood framing. Ventilating procedures are described in chapter 4. Advice may also be obtained from local building supply outlets and ventilation contractors.

Interior Wall and Ceiling Framing

The underside of the original roof may serve as the ceiling of a finished attic or a new ceiling can be installed and supported by joists nailed to the rafters. If a ceiling is built, access must be provided to the area above it.

The interior walls are constructed and erected in the same manner as exterior walls, that is, with a single bottom (sole) plate and double top plates. The upper top plate is used to tie intersecting and crossing walls to each other.

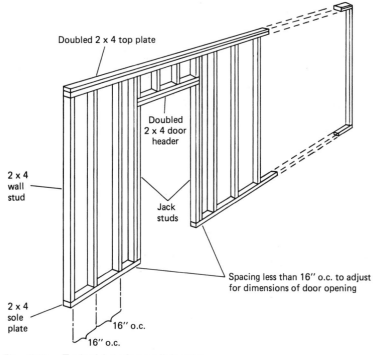

Fig. 7-39. Typical interior wall framing.

The framing of a typical interior wall is illustrated in figure 7-39. The size of the door determines the rough opening dimensions, and these are available at any local building supply outlet. Nail the single top and bottom plates to the ends of the studs, raise the wall framework into the desired position, and nail the bottom plate through the subfloor to the floor joists. Double the top plate, nail the upper one to the lower one, and then toenail through the upper top plate into the ceiling joists. Use sixteenpenny (16d) nails for each step.

Interior Wall and Ceiling Finish

Cover the wall studs with a suitable drywall finish material such as plywood, hardboard, fiberboard, gypsym board, or a similar type of sheet material, or with wood board paneling available in various widths, thicknesses, and patterns. Cover the ceiling joists with gypsum board or fiberboard.

Prefinished plywood for interior walls is available with a variety of different wood veneers or facings. It may be applied vertically or horizontally in 4 × 8–foot and longer sheets. All panel edges require solid backing. Each panel should be at least ¼-inch thick for studs 16 inches on center and ⅜-inch thick for 24-inch spacing. Use 1¼-inch to 1½-inch casing or finishing nails and space the nails 8 inches apart in a line about ⅜ inch from the edge of the panel. Allow a ¹⁄₃₂-inch gap between panels for expansion.

Hardboard and fiberboard panels are applied the same way as plywood panels. Hardboard must be at least ¼ inch thick when used over open framing with studs spaced 16 inches on center. Rigid backing is required for ⅛-inch thick hardboard.

Fiberboard in tongue-and-groove plank or sheet form must be ½-inch thick when studs are spaced 16 inches on center and ¾-inch thick when 24-inch spacing is used. Apply fiberboard with casing or finishing nails slightly longer than those used to apply plywood or hardboard. Fiberboard is also used in the ceiling as acoustic tile and may be nailed to strips fastened to the ceiling joists or installed in 12 × 12–inch or larger tile forms on wood or metal hangers that are suspended from the ceiling joists.

Wood board paneling may also be used to cover the wall. Wood boards may be applied vertically, horizontally, or diagonally. The boards should not be wider than 8 inches except when a long tongue or matched edges are used. Thicknesses should be at least ⅜ inch for 16-inch spacing of frame members and ⅝ inch for 24-inch spacing. Apply the boards with fivepenny (5d) or sixpenny (6d) casing or finishing nails. Use two nails for boards 6 inches wide or narrower and three nails for 8-inch and wider boards. Wood boards are usually stained with a semitransparent stain. Cheaper wood grades may be painted or covered with an opaque stain.

Gypsum board is typically available in 4 × 8–foot and a 4 × 10–foot panels and in ⅜- to ⅝-inch thicknesses. Check the local building code for the required thickness. As a general rule, ½-inch thick gypsum board is sufficient for most types of framing.

Use 1⅝-inch ring-shank drywall nails for ½-inch panels and 1⅞-inch nails for ⅝-inch panels. About 5½ pounds of nails are required for every 1,000 square feet of wall or ceiling surface. Drive the head of each nail below the surface of the gypsum board, but do not break through the paper. Fill the indentation with joint compound. Use 4 × 8–foot gypsum board panels in the attic and apply them horizontally. The panels are heavy and require two workers to install. One worker should hold the panel in place while the other drives in one nail every 8 inches. Handle gypsum board with care because it is fragile and will bend and snap under its own weight.

Gypsum board can easily be trimmed to any required size with a utility knife and straight edge. Cut through the paper on one side and then apply firm, even pressure on the panel from the cut side to break the gypsum core. With the panel either standing up or lying flat on the subfloor surface, cut through the paper on the other side. If the panel has cracked unevenly, smooth it with a medium-tooth wood file.

Gypsum board may be covered with wallpaper or painted. An acrylic-latex type paint is the easiest to use. Before painting gypsum board, apply a primer or an emulsion type sealer to the surface to equalize the difference in absorption between the board paper and the compound used to cover joints and nails. Use of a primer or sealer will assure uniformity of color, suction, and texture over the entire painted wall or ceiling surface. If high-gloss paint is used, the entire gypsum board surface should receive a skim coat of joint compound before a sealer is applied to avoid light shadowing. A sealer application under wallpaper or other wall covering is also recommended to prevent damage to the gypsum board surface should the covering be removed during redecorating. Joint treatment must be thoroughly dry before applying a primer or sealer and the wall finish.

No wall finish is added if the studs are covered with decorative (vinyl-covered) gypsum board panels.

Floor Coverings

The terms *floor covering* and *finish flooring* are used by builders, architects, and interior decorators to refer to the material used as the final wearing surface that is applied to a floor.

There is a wide selection of wood materials that may be used for finish flooring. Hardwoods and softwoods are available as strip flooring in a variety of widths and thicknesses and as random-width planks and block flooring. Carpeting, linoleum, asphalt, rubber, cork, vinyl, and other materials in tile, sheet, or roll form are also available for use as floor coverings.

One of the easiest ways to finish an attic floor is to lay wall-to-wall carpeting over a carpet pad and the subfloor. The advantage of using carpeting over other floor coverings is its ability to absorb sound and its impact resistance.

The floor may also be finished with tile or linoleum, but neither material has the sound-deadening ability of carpeting. Furthermore, both tile and linoleum must be applied with an adhesive over a smooth, level surface on which all cracks have been filled and sanded. Another advantage of using carpeting is that it will conceal surface imperfections.

CHAPTER 8

Roof Flashing Details

- Flashing Applications
- Flashing Materials
- Eave and Rake Flashing
- Valley Flashing
- Chimney Flashing
- Soil Stack and Vent Pipe Flashing

- Vertical Wall Flashing
- Other Types of Roof Flashing
- Roof Juncture Flashing
- Flashing Repairs
- Reroofing

An important preliminary step to the application of the roofing materials is the installation of flashing. The roof must be protected with flashing at every point in its construction where leaks might occur, especially where there may be a particularly heavy concentration of water runoff, such as in roof valleys or along the roof eaves.

Leaks frequently occur at the joint where a minor roof intersects with a main roof or where the roof deck meets a vertical wall. On wood roof decks, the sheathing is subject to a certain amount of expansion and contraction under different weather conditions, which may cause the sheathing to pull away from another surface and provide an entry point for water. The point at which a chimney projects through a roof is also susceptible to leakage. The flashing installed around a chimney must maintain a water-tight seal while at the same time allowing the separate movement of the roof deck and chimney, which are built on separate foundations to avoid any stress and distortion caused by uneven settling between the two. Similar opportunities for leakage occur around projections through the roof deck, such as those made by a vent pipe, soil stack, flue, or roof ventilator.

The various types of flashing materials, applications, and application methods for pitched-roof construction are described in this chapter. The flashing of flat or low-pitched roofs is described in chapter 12.

Flashing Applications

As shown in figure 8-1, the flashing for a typical pitched roof is applied along the roof valleys, rakes, and eaves; around the bases of chimneys, soil stacks, vent pipes, flues, and roof ventilators; over chimney crickets; along the sides of dormers; around skylight curbs and other square or rectangular openings; at the juncture between a roof and a vertical wall; and along roof ridges and the joint formed by the joining of two sloping roof planes.

Flashing is applied in several forms, depending on its location on the roof. Long, wide sheets of flashing material are used in the roof valleys. Narrower widths of material are applied along the roof eaves, rakes, and ridges. Base, step, and counter flashing are used elsewhere on the roof.

Base flashing is applied along the front and sides of a chimney, and along the joint formed between a roof slope and a vertical wall. *Step flashing* may be used instead of base flashing along the sides of chimneys, dormers, and junctures formed by a vertical wall and a roof

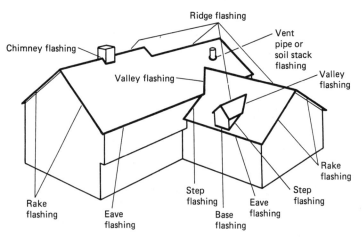

Fig. 8-1. Typical roof flashing applications.

slope. It is sometimes called *shingle flashing* because it is applied in the form of individual, overlapping pieces. *Counter flashing* or *cap flashing* is used to cover base or step flashing where considerable movement can be expected to occur between adjoining surfaces by overlapping (or "capping") it from above.

A flashing flange is usually installed around the base of a soil stack, vent pipe, flue, or other types of small, circular protrusions through the roof surface. A form of base flashing may be used to flash skylight curbs or other types of square or rectangular roof openings. Cricket flashing is a single piece of flashing cut to cover the top of a chimney cricket.

Flashing Materials

Galvanized steel, copper, aluminum, mineral surface roll roofing, and roofing felt are all used as flashing materials on both residential and nonresidential structures. Metal alloys, such as Terne metal or Terne-coated stainless steel, are also used for flashing, primarily on the roofs of public buildings or commercial structures.

In residential construction, metal flashing is generally recommended for roofs covered with wood shingles or shakes, slate, or tile, but it can also be used with asphalt shingles and other types of roofing. Mineral surface roll roofing is commonly used on asphalt shingle roofs. Roll roofing is sometimes used as a flashing reinforcement along roof eaves in combination with a metal drip edge.

Metal flashing is available in rolls or preformed units (fig. 8-2). The preformed units include drip edges and rake edges for mineral fiber (asbestos-cement), asphalt, slate, and wood shingle and shake roofs (fig. 8-3), and metal gravel stops for built-up roofs. They are generally available in 10-foot lengths, which can be cut to size with ordinary tinsnips. Preformed metal valleys with a 1-inch crimp standing seam are produced in 14-inch and 20-inch widths with other sizes available on special order (fig. 8-4). Valley flashing or roll flashing is also available in 12- to 18-inch rolls with most rolls containing 50 linear feet of flashing metal.

Galvanized steel is the most widely used flashing metal. It is cheaper than copper flashing, will not react to contact with masonry or cement, and is easily cut with ordinary tinsnips. The galvanized steel used for flashing is a 26-gauge metal. It must be painted with

Fig. 8-2. Aluminum and galvanized roll valley flashing. *(Courtesy Howmet Aluminum Corp.)*

a protective finish before it is applied or it will rust and deterioriate. Even with a protective finish, a scratch or dent may cause the metal to deteriorate.

Aluminum is an ideal flashing metal for structures located near bodies of water or in areas experiencing high humidity or heavy or frequent rains, because moisture will not rust aluminum or cause it to deteriorate. Aluminum flashing is available unfinished or finished with a thermally set paint applied in full coat to both sides of the metal. The paint colors are either white or brown, which allows the flashing to be used with a wide range of roofing material colors. An unfinished or natural metal finish will weather to a soft gray appearance and will not rust or discolor. Aluminum flashing should have a .019-inch minimum thickness. If it is used in contact with masonry, cement, or dissimilar metals, a coat of bituminous paint should be applied between the surfaces to guard against electrolytic action.

Copper flashing must be a 16-ounce metal with a .020-inch

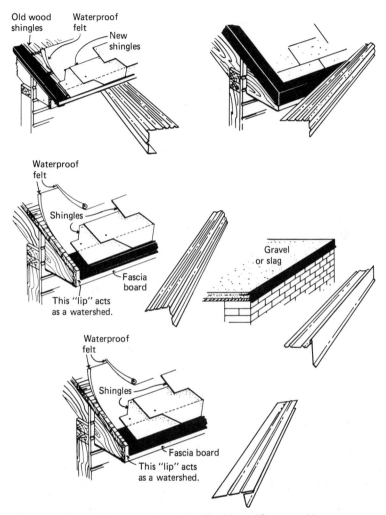

Fig. 8-3. Preformed eave and rake flashing. *(Courtesy Howmet Aluminum Corp.)*

minimum thickness. Copper flashing is expensive and seldom used except in quality construction. On some roofs, copper may be used to flash the chimney with galvanized steel, aluminum, or mineral surface roll roofing used in the valleys, along the eaves, and elsewhere.

Terne metal, which is more commonly known as valley tin or

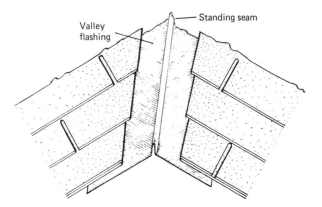

Fig. 8-4. Metal valley flashing with crimped standing seam.

roofer's tin, has been used as a roof flashing metal since the eighteenth century. It is rarely used on houses today, but is still found on the roofs of many public buildings and commercial structures. It consists of copper-bearing steel, which is coated with a lead-tin alloy. One problem with using Terne metal as a flashing is that it must be painted with a protective finish during application or it will deteriorate. The protective finish must be applied by hand on both sides of the metal. A coat of very slow-drying red iron oxide-linseed oil paint is brushed onto the underside of the metal. The exposed surface receives a primer coat of slow-drying red iron oxide-linseed oil paint, which is followed by a good quality linseed oil exterior paint of any desired color.

Terne-coated stainless steel consists of 3–4 nickel chrome stainless steel covered on both sides with Terne alloy (80 percent lead and 20 percent tin). It requires no protective finish and never needs maintenance if properly installed. Its unfinished surface will eventually weather to a uniform dark gray. Terne-coated stainless steel is used as flashing primarily on public buildings or commercial structures, rarely on residential roofs.

Another flashing metal used primarily on nonresidential structures is a zinc-copper-titanium alloy called Econo-zinc (a trademark of Ball Metal and Chemical Division). It is available in 12- or 14-inch wide rolls of .022- or .030-inch thick sheets. It is lightweight, inexpensive, and maintenance-free when applied properly, and it weathers naturally. It is not recommended for use in direct contact with acidic woods or with metals other than aluminum or galvanized steel.

Mineral surface roll roofing has the same composition as asphalt shingles. It will not last as long as metal flashing, but it is far less expensive and easier to install. Mineral surface roll roofing is commonly available in 36-inch wide rolls containing 36 linear feet of material. It is surfaced with mineral granules to produce a wide range of different colors. Some roofing manufacturers also produce an 18-inch wide roll for valley applications.

In order to prevent chemical reactions between dissimilar metals, copper nails should be used with copper flashing, aluminum nails with aluminum flashing, and hot galvanized nails with galvanized steel flashing. Hot galvanized steel roofing nails are also used to apply mineral surface roll roofing and roofing felt.

Eave and Rake Flashing

A metal drip edge should be nailed along each roof eave and rake. The drip edge is normally fastened along the eaves *before* the underlayment is applied and along the rakes *after* it is applied (figs. 8-5 and 8-6). Variations of eave and rake flashing applications are shown in figure 8-3.

Metal drip edges are available in 10-foot lengths of corrosion resistant, nondiscoloring metal. Preformed metal drip edges of galvanized steel, aluminum, or copper are the most commonly used types of eave

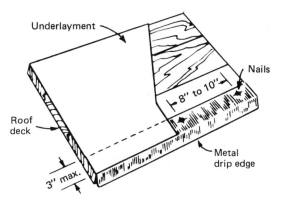

Fig. 8-5. Metal drip edge along eave. *(Courtesy Asphalt Roofing Manufacturers Association)*

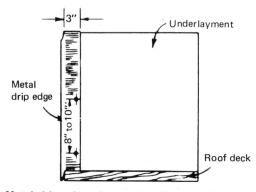

Fig. 8-6. Metal drip edge along rake. *(Courtesy Asphalt Roofing Manufacturers Association)*

or rake flashing used in roofing. Other materials may be used if approved by the roofing manufacturer. As shown in figures 8-5 and 8-6, the metal drip edge is secured to the roof deck with nails spaced 8 to 10 inches apart along the inner edge; it should not extend back from the edge of the roof deck more than 3 inches.

Metal drip edges along the eaves and rakes allow the water to run off the roof without coming into contact with the underlayment or roof deck sheathing. They also prevent or minimize the possibility of rain water being blown back under the roofing. The drip edge along the roof eave is often covered with a layer of mineral surface or smooth roll roofing in areas of the country where the January daily average temperature is 25° F or less, or wherever there is a possibility of ice forming along the eaves and causing a backup of water. The 90-pound mineral surface roll roofing or 50-pound (or heavier) smooth roll roofing is laid along the edge of the roof eave so that it overhangs the underlayment and metal drip edge by ¼ to ⅜ inch (fig. 8-7). The material should extend up the roof deck to a point at least 12 inches beyond the interior wall line of the structure. The mineral surface roll roofing should be applied with its surface face down, and both types of roll roofing should be secured with only enough nails along the top and bottom edges to hold the material firmly and smoothly in place. If the overhang at the eaves requires material wider than 36 inches to meet the minimum 12-inch extension beyond the interior wall line, two courses of material should be laid with the top course overlapping the underlying one at a point midway between the eave

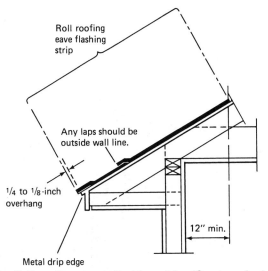

Fig. 8-7. Roll roofing eaves flashing strip. *(Courtesy Asphalt Roofing Manufacturers Association)*

line and the exterior wall line. The lap must be sealed with asphalt cement.

On decks with a slope of less than 4-in-12, but not less than 2-in-12, or in areas of the country where severe ice formation is encountered (fig. 8-8), the second course of material should extend up the roof deck to a point not less than 24 inches inside the inner wall line of the structure (fig. 8-9). The underlaying material should be covered with a layer of asphalt cement before the top layer is applied.

Valley Flashing

A roof valley is formed where two sloping roof planes meet to form an internal angle. During heavy rains, large volumes of water will flow along the joint forming the valley. As a result, it is absolutely necessary to insure that this roof joint is capable of handling the excess water without leaking a portion of it into the structure. This is accomplished by lining the valley with a suitable flashing material.

Metal flashing, mineral surface roll roofing, or smooth surface roll roofing can be used as a flashing material in roof valleys. Because

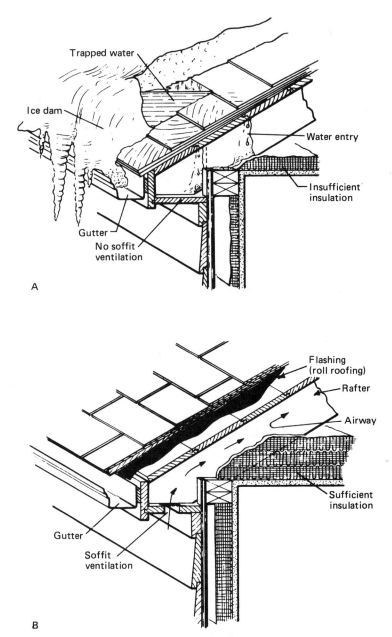

Fig. 8-8. Prevention of ice dam formation along eaves.

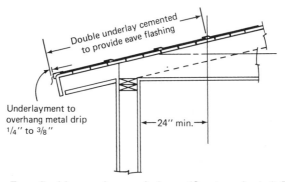

Fig. 8-9. Eave flashing on low roof slope. *(Courtesy Asphalt Roofing Manufacturers Association)*

mineral surface roll roofing is less expensive than metal flashing, the valleys of most asphalt shingle roofs are flashed with this material. Metal flashing is still used in more expensive, higher quality construction, or on roofs covered with wood shingles or shakes, slate, or tile. Smooth surface roll roofing is used as a valley flashing on asphalt shingle roofs constructed with woven or closed valleys.

Valley flashing must be applied after the roofing felt underlayment is nailed to the roof deck. As shown in figure 8-10, a 36-inch

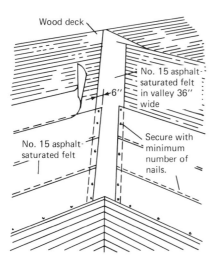

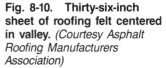

Fig. 8-10. Thirty-six-inch sheet of roofing felt centered in valley. *(Courtesy Asphalt Roofing Manufacturers Association)*

wide sheet of No. 15 asphalt-saturated roofing felt is centered in the valley and attached with enough nails to secure it in place before the underlayment courses are laid across the roof deck.

Metal Flashing

Metal valley flashing can be cut to fit the shape of the roof valley or purchased preformed from a supplier. Preformed metal flashing is available with a 1-inch crimped standing seam down its center. Flashing with a standing seam is required for applications where adjoining roofs have different slopes. The standing seam prevents the heavier volume of water flowing down the steeper slope from overrunning the valley and being forced under the shingles or roofing material on the adjoining slope. The valley should be wider at the bottom to cope with the larger volume of water drainage at lower levels on the roof slope (fig. 8-11).

Metal valley flashing is produced with either flat or crimped edges. Like the standing seam down the center of the valley, the

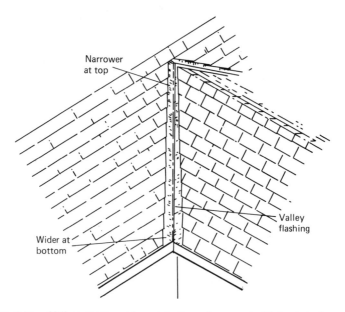

Fig. 8-11. Valley made wider at bottom for more efficient drainage.

crimped edges are also designed to prevent a heavy flow of water from overflowing the valley flashing and running up under the roofing material.

Metal flashing for valleys should not be less than 12 inches wide for roof slopes of 7-in-12 or more, 18 inches wide for slopes of 4-in-12 to less than 7-in-12, or 24 inches wide for slopes less than 4-in-12. The flashing should be secured with nails spaced approximately 12 inches apart in a row 1 inch from each outer edge of flat-edge flashing or driven through cleats attached every 12 inches to crimped-edge flashing. Nails should never be driven near the center of the valley because water may leak through the flashing at the nail holes. Use only nails of a metal compatible with the flashing to prevent corrosion, discoloration, or other reactions. Galvanized steel nails should be used with galvanized steel flashing, aluminum nails with aluminum flashing, and copper nails with copper flashing.

Galvanized steel flashing must be painted before it is applied to protect it from corrosion and deterioration. Clean the metal with a suitable solvent, apply a zinc-based metal primer to both sides of the flashing, and spray the top (exposed) surface with two coats of paint in a color that will match or contrast with the roofing material.

Composition Flashing

The application of a composition flashing material, such as mineral surface or smooth surface roll roofing, depends on whether the roof has open, woven, or closed valleys. If the roof has either woven or closed valleys, a type of construction limited to strip-type asphalt shingle roofs, 50-pound or heavier smooth surface roll roofing is used as the flashing material. A 36-inch wide sheet of the material is nailed over the underlayment so that it extends at least 12 inches beyond the center of the valley on either side (fig. 8-12). The roll roofing sheet should be nailed only along its outer edges (*never* in the center of the valley) with enough nails to hold it firmly in place. The strip shingles are then applied so that they cover the valley flashing either by being woven across the center of the valley or by cutting them to form a seam down the valley center.

An open valley is flashed with two layers of mineral surface roll roofing over the underlayment (fig. 8-13). The first layer consists of an 18-inch wide sheet which is centered in the valley with its surfaced side face down and its lower edge cut to fit flush with the eave lines

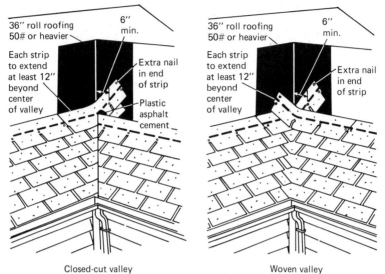

36" roll roofing 50# or heavier

6" min.

Each strip to extend at least 12" beyond center of valley

Extra nail in end of strip

Plastic asphalt cement

36" roll roofing 50# or heavier

6" min.

Each strip to extend at least 12" beyond center of valley

Extra nail in end of strip

Closed-cut valley

Woven valley

Fig. 8-12.   Composition flashing for closed-cut or woven valley.
(Courtesy Asphalt Roofing Manufacturers Association)

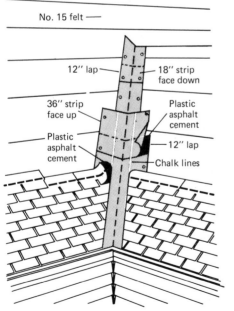

No. 15 felt

12" lap

18" strip face down

36" strip face up

Plastic asphalt cement

Plastic asphalt cement

12" lap

Chalk lines

Fig. 8-13.   Composition flashing for open valley.
(Courtesy Asphalt Roofing Manufacturers Association)

of the two intersecting roofs. The material should be secured to the valley with a row of nails placed 1 inch from each outer edge of the sheet. Use only enough nails to hold the sheet firmly and smoothly in place. Push the material firmly into the center of the valley before nailing the second edge. Nail a second 36-inch wide sheet of mineral surface roll roofing over the first sheet with its surface side up. Make certain that the sheet is centered in the valley and nail it in the same manner as the underlaying sheet.

Whenever it is necessary to use more than one sheet of roll roofing in the valley, make sure the upper sheet overlaps the underlying one by at least 12 inches and seal the lap with a coating of plastic asphalt cement between the two sheets.

Chimney Flashing

A chimney is built on a separate foundation from the one that supports the structure. Because there is always movement between the chimney and the structure, resulting to a large extent from the different rates at which the two separate foundations settle, a chimney requires special flashing treatment. The chimney flashing must be able to absorb movement without breaking its seal and providing an entry point for water. This is accomplished in most cases by allowing the flashing attached to the chimney to overlap or cap the flashing attached to the roof.

Before flashing can be applied, all shingle courses must be completed up to the front face of the chimney. If a cricket or saddle is to be added to the back of the chimney, it must be built before either the roofing or flashing is applied.

The chimney masonry must be cleaned with a stiff wire brush and coated with an asphalt primer before either metal or composition flashing is applied to the surface. Clean an area measuring 6 to 8 inches up from the base of the chimney and then apply the primer. The asphalt primer will produce a much better bonding surface for the plastic asphalt cement.

Cricket Construction

On some pitched roofs, a cricket or saddle is built behind the chimney after the underlayment of roofing felt has been applied to the roof

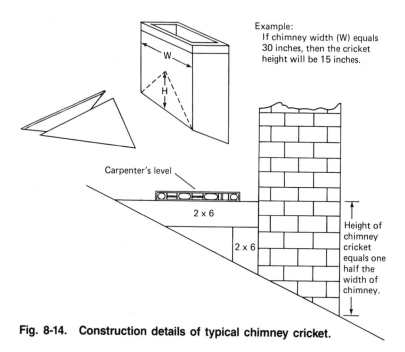

Example:
If chimney width (W) equals
30 inches, then the cricket
height will be 15 inches.

Fig. 8-14. Construction details of typical chimney cricket.

deck. The purpose of the cricket is to channel the flow of water around the chimney and to prevent the buildup of snow and ice behind it. Crickets are used with chimneys 24 inches wide or wider.

The construction of a typical chimney cricket is illustrated in figure 8-14. The frame is used to support the cricket roof, which is made of ½-inch thick plyscord, plywood, or similar material. The cricket roof is cut to fit the angle of the sloping main roof deck.

As shown in figure 8-14, the outer edges of the cricket are designed so that they meet but do not extend beyond the two corners of the chimney. When they do extend beyond, it is usually because a cant strip is installed down both sides of the chimney (fig. 8-15). The cant strip protects the joint formed between the roof and chimney by channeling the water runoff. If a cant strip is used, the cricket will extend no further than the width of the cant strip.

Metal Flashing

Flashing is always applied to the front of the chimney first. A single piece of base flashing is laid across the front of the chimney so that

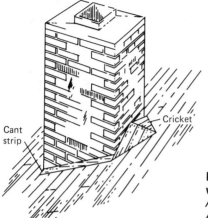

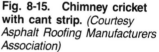

Cant strip

Cricket

Fig. 8-15. Chimney cricket with cant strip. *(Courtesy Asphalt Roofing Manufacturers Association)*

it extends over the last shingle course. A typical base flashing pattern for the front of the chimney is shown in figure 8-16. It is bent along the dotted line in the illustration to conform to the angle formed by the chimney and roof. Lay the lower section over the shingles in a bed of plastic asphalt cement and secure the upper vertical section

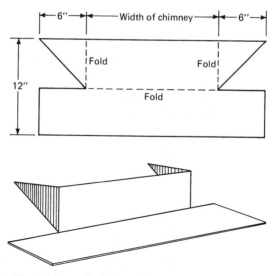

Fig. 8-16. Typical base flashing pattern.

to the chimney surface with plastic asphalt cement and masonry nails. Drive the nails into the mortar joints. Bend the triangular ends of the upper section around the corners of the chimney and secure them with plastic asphalt cement.

Either base flashing or step flashing may be used along the sides of the chimney. Side base flashing is a single, continuous section of metal. The base section pattern shown in figure 8-17 must be bent to conform to the vertical surface of the chimney, the roof slope, and the cant strip attached to the base of the chimney (see dotted lines in fig. 8-17). If a cant strip is not used, the base flashing will only have to be bent once. When a single, continuous piece of base flashing is used, it is always applied to the side of the chimney before the shingle

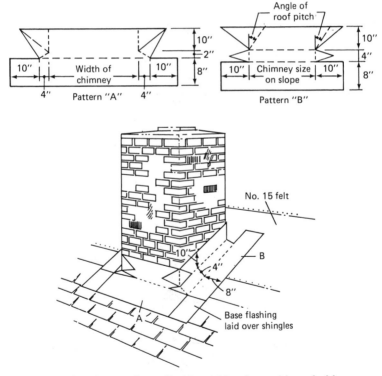

Fig. 8-17. Continuous base flashing strip along sides of chimney.
(Courtesy Asphalt Roofing Manufacturers Association)

courses are laid. Cover the roof deck underlayment next to the chimney with plastic asphalt cement and firmly press the bottom section of the base flashing into the cement. Apply a coating of asphalt plastic cement to the chimney surface where the flashing will cover the masonry, firmly press the upper section of the base flashing against the chimney, and secure it with masonry nails driven into the mortar joint. A tighter fit can be obtained by bending the flashing to a 95° angle before installing it. When the shingle courses are completed up to the base of the chimney, secure the end of the last shingle in each course to the flashing by embedding it in plastic asphalt cement. *Do not nail through the shingle and the flashing.*

If step flashing is used along the sides of the chimney, the shingle courses and the step flashing are alternately applied as the work progresses up the roof slope (fig. 8-18). Step flashing consists of individual, rectangular pieces of flashing metal cut double the shingle exposure dimension and then bent in half to conform to the angle formed by the chimney and roof. For example, a pitched roof with a 5-inch shingle exposure would require cutting 7 × 10–inch flashing metal pieces and then bending them to provide 5 inches on each side. Each piece would be 7 inches wide and would extend 5 inches up the chimney wall and 5 inches onto the roof deck when installed.

The first step flashing piece should be laid so that one edge wraps around the corner of the chimney. Secure the flashing to the chimney and roof deck with a suitable amount of plastic asphalt cement, and then nail the bottom section of the flashing to the roof deck with two

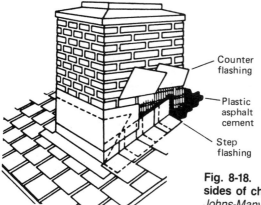

Counter flashing

Plastic asphalt cement

Step flashing

Fig. 8-18. Step flashing along sides of chimney. *(Courtesy Johns-Manville Corp.)*

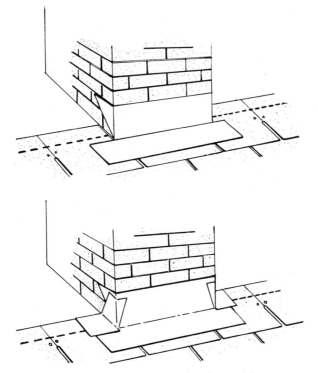

Fig. 8-19. Step flashing pieces installed at front chimney corners.

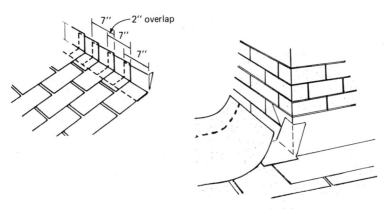

Fig. 8-20. Installing step flashing along sides of chimney.

nails placed so that they will be covered by the overlapping portion of the next piece of step flashing (fig. 8-19). Cover the section of the flashing nailed to the roof deck with plastic asphalt cement and press the end of the last shingle in the course firmly against the flashing (fig. 8-20). *Do not nail through the shingle and flashing.*

Apply the second piece of step flashing so that it overlaps the first one by 2 inches and covers the nails that attach it to the roof deck. Use the same method to secure it to the chimney and roof as described for applying the first piece of flashing. Continue to apply alternate courses of first the flashing and then the last shingle in the shingle course until the work is completed along the side of the chimney. Make sure each piece of flashing overlaps the underlying one by the required 2 inches for a 5-inch shingle exposure. Do not nail the step flashing to the chimney or it will be too rigid to allow for any movement between the roof and the chimney.

If a cricket has *not* been built against the back of the chimney, the base of the chimney will have to be covered with a section of flashing. This is done after the sides of the chimney have been flashed. A suggested pattern for the flashing at the back of the chimney is illustrated in figure 8-21. The lower section is secured to the underlayment with plastic asphalt cement and then nailed to the roof deck. The upper section (cap or counter flashing) is secured to the masonry surface of the chimney with plastic asphalt cement and masonry nails driven through the mortar joints. After the base flashing is installed, the surface of the lower section is covered with plastic asphalt cement, and the shingles are pressed firmly into the cement as each course is laid across the roof above the chimney.

The base flashing at the front of the chimney is covered by a sheet of counter flashing or cap flashing. Counter flashing may also be used to cover the base flashing or step flashing along the sides of the chimney.

Counter flashing is secured to the chimney by bending a 1 1/2-inch lap or reglet along its top edge and inserting it into a mortar joint from which the mortar has been removed. The joint is refilled with either plastic asphalt cement or mortar mixed to the desired consistency (fig. 8-22).

The counter flashing along the front of the chimney is one unbroken piece. Along the sides of the chimney the counter flashing may be either a continuous piece or several overlapping pieces cut to conform to brick dimensions (fig. 8-23).

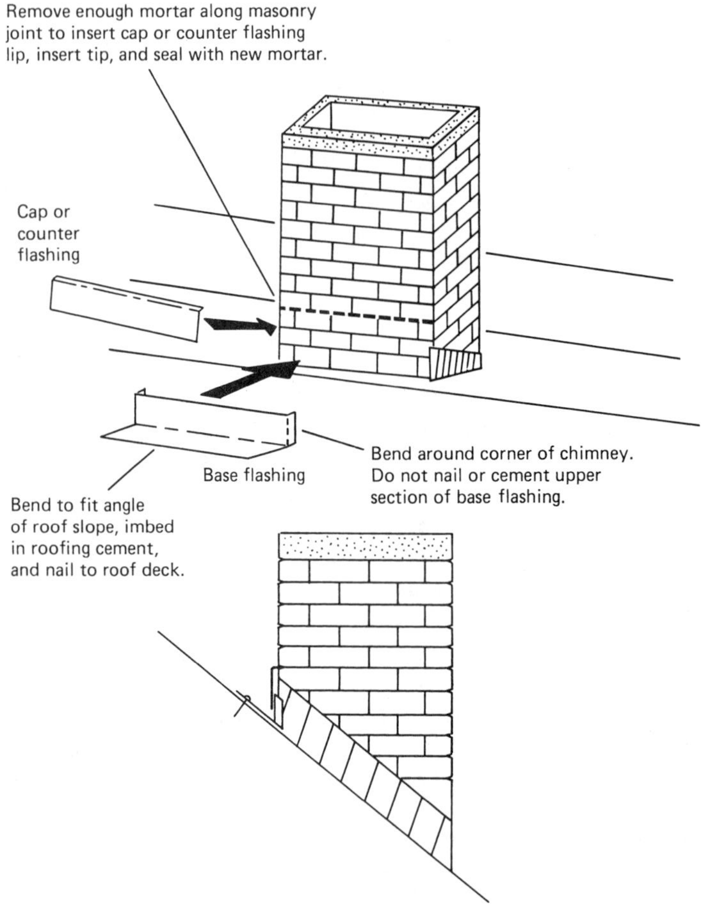

Remove enough mortar along masonry
joint to insert cap or counter flashing
lip, insert tip, and seal with new mortar.

Cap or
counter
flashing

Bend around corner of chimney.
Do not nail or cement upper
Base flashing   section of base flashing.

Bend to fit angle
of roof slope, imbed
in roofing cement,
and nail to roof deck.

Fig. 8-21. Base flashing along back of chimney.

If a single, continuous piece is used along the sides, the bottom end should be bent 2 inches to overlap the counter flashing on the front of the chimney. Embed the 2-inch lap in plastic asphalt cement to bond it to the front counter flashing. Individual counter flashing pieces along the sides of the chimney should overlap at least 3 inches and plastic asphalt cement should be applied between the overlapping sections to seal the joint.

If the chimney has a cricket or saddle, it too must be covered

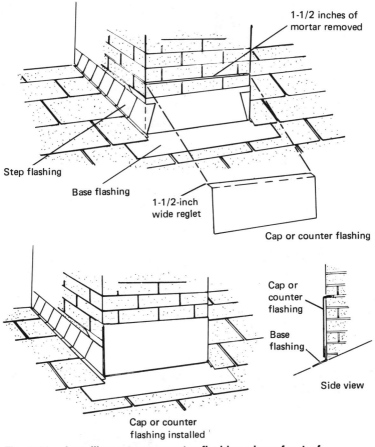

1-1/2 inches of
mortar removed

Step flashing

Base flashing

1-1/2-inch
wide reglet

Cap or counter flashing

Cap or
counter
flashing

Base
flashing

Side view

Cap or counter
flashing installed

Fig. 8-22. Installing cap or counter flashing along front of chimney.

with flashing. Cricket flashing consists of one or more pieces of flashing metal cut to the required dimension and nailed to the cricket roof. If more than one piece of flashing is required to cover the cricket, they should be soldered together after they have been nailed in place.

Applying flashing to a cricket can be made easier by first cutting and fitting a pattern similar to the one shown in figure 8-24. A large rectangular sheet of roll roofing or stiff art paper can be used for the pattern material. Fit the pattern material to the cricket and trim it to the required dimensions. Transfer the pattern to a sheet of

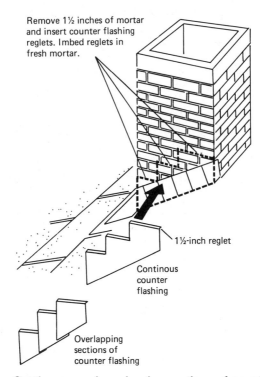

Remove 1 ½ inches of mortar and insert counter flashing reglets. Imbed reglets in fresh mortar.

1½-inch reglet

Continous counter flashing

Overlapping sections of counter flashing

Fig. 8-23. Continuous and overlapping sections of counter flashing along side of chimney.

metal flashing, and then cut and bend the metal flashing to fit the cricket.

The cricket flashing should be large enough to cover the cricket roof and extend a short distance onto the roof deck and up the back surface of the chimney. Cover the cricket with asphalt primer and plastic asphalt cement, and then install the flashing. The edge flap lying against the surface of the chimney is secured in place by embedding it in plastic asphalt cement. The edges lying flat against the roof surface should be embedded in plastic asphalt cement and nailed to the roof deck. Run a ribbon of plastic asphalt cement along the edges of the cricket flashing to provide additional protection against leakage. Counter flashing is sometimes used to cover the back of the chimney after the cricket flashing has been applied.

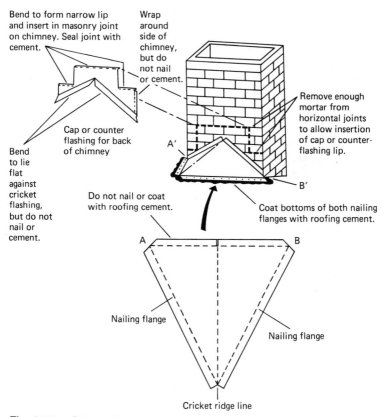

Bend to form narrow lip and insert in masonry joint on chimney. Seal joint with cement.

Wrap around side of chimney, but do not nail or cement.

Remove enough mortar from horizontal joints to allow insertion of cap or counter-flashing lip.

Cap or counter flashing for back of chimney

A'

B'

Bend to lie flat against cricket flashing, but do not nail or cement.

Do not nail or coat with roofing cement.

Coat bottoms of both nailing flanges with roofing cement.

A

B

Nailing flange

Nailing flange

Cricket ridge line

Fig. 8-24. Cricket flashing.

Composition Flashing

Roll roofing can be used for chimney flashing, but it is neither as attractive nor as long lasting as metal flashing. It is, however, inexpensive and easy to install. Its use is limited to asphalt shingle roofs, especially when reroofing them. Before using roll roofing to flash a chimney, check the local building code to determine whether there are any restrictions on its use or if its use as a chimney flashing material is prohibited.

When reroofing, check the condition of the old metal flashing. Reusable counter flashing can be bent up out of the way if the base and counter flashing must be replaced. Apply an 8-inch wide strip of roll roofing at the front and sides of the chimney after the existing roof has been prepared for reroofing. Lay the strip so that one edge butts against the base of the chimney, nail it to the roof with a row of roofing nails along each edge, and run a heavy coating of plastic asphalt cement along the juncture formed by the chimney and the roof (fig. 8-25). When applying the new shingles to the roof, trim the shingle nearest the chimney to leave a gap of approximately ¼ inch between it and the masonry. Coat the top of the roll roofing flashing strip with plastic asphalt cement and firmly press the last shingle of the course down against the flashing strip. A small amount of plastic asphalt cement will squeeze through the gap between the shingle and the chimney.

Use a wire brush to clean the masonry surface of the chimney approximately 6 to 8 inches up from the level of the roof, and then cover the cleaned area with a suitable asphalt primer. Trowel a thick coat of plastic asphalt cement approximately 4 to 6 inches up the chimney surface and 2 inches over the new shingles. Apply a strip of mineral surface roll roofing wide enough to cover the plastic asphalt

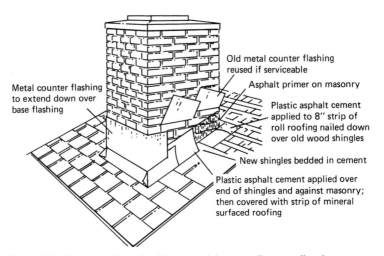

Fig. 8-25. Composition flashing used in reroofing applications.
(Courtesy Asphalt Roofing Manufacturers Association)

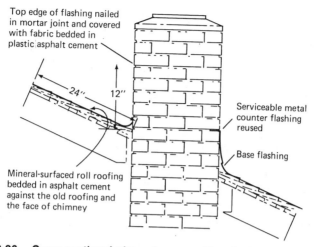

Top edge of flashing nailed in mortar joint and covered with fabric bedded in plastic asphalt cement

24″ 12″

Serviceable metal counter flashing reused

Base flashing

Mineral-surfaced roll roofing bedded in asphalt cement against the old roofing and the face of chimney

Fig. 8-26. Cross-sectional view of composition flashing at chimney. *(Courtesy Asphalt Roofing Manufacturers Association)*

cement on both the chimney and the shingles and long enough to extend 6 inches around both the front and back corners of the chimney. Firmly press the material into the cement.

If there is a cricket at the back of the chimney, cover it with mineral surface roll roofing by the same procedure as used for applying metal flashing. If there is no cricket, apply a single sheet of mineral surface roll roofing so that it extends 12 inches up the chimney and 24 inches up the roof over the existing roofing (fig. 8-26). The material should be embedded in a heavy coat of plastic asphalt cement applied to both the roof and the masonry of the chimney. After firmly pressing the material into the cement, nail its upper edge to the chimney with mortar nails driven into the mortar joints.

If the old counter flashing is reusable, bend it down over the base flashing and embed it in a coat of plastic asphalt cement. If it is not reusable, cover the upper half of the base flashing at the front, back, and sides of the chimney with a strip of wide mesh asphalt-saturated fabric. The strip should be wide enough to extend 2 inches above and below the top edge of the base flashing. It should be embedded in plastic asphalt cement, pressed firmly in place, and then covered with another layer of cement.

Soil Stack and Vent Pipe Flashing

Most roofs have one or more circular pipes that project through the roof deck. These pipes are generally soil stacks, roof ventilators, or vents (exhaust stacks) for fuel-burning equipment, and they require special flashing methods to prevent leakage from occurring where they interrupt the surface of the roof.

Figures 8-27 to 8-30 illustrate a common method of applying flashing around a soil stack. The procedure is outlined as follows:

1. Complete the shingle course through which the pipe will project. As shown in figure 8-27, one of the strip shingles will have to be cut and fitted to the diameter of the pipe.
2. Cut a rectangular flange from mineral surface roll roofing, or 50-pound or heavier smooth surface roll roofing, to fit over the pipe. The roll roofing flashing flange should be large enough to extend 4 inches below the pipe, 8 inches above the pipe, and 6 inches on either side of the pipe.
3. Locate the exact opening on the flange for the circular pipe following the steps illustrated in figure 8-28.
4. Cut out the opening and slip the flashing flange over the pipe, pressing down the flange so that it lies flat against the surface of the roof.

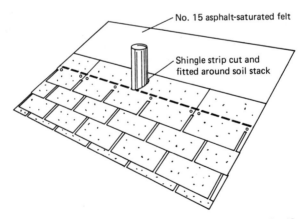

No. 15 asphalt-saturated felt

Shingle strip cut and fitted around soil stack

Fig. 8-27. Shingle strip cut out and fitted around stack. *(Courtesy Asphalt Roofing Manufacturers Association)*

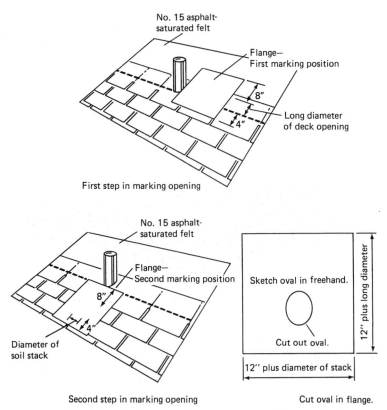

Fig. 8-28. Marking and cutting flange opening. *(Courtesy Asphalt Roofing Manufacturers Association)*

5. Apply plastic asphalt cement around the pipe and over the surface of the flashing flange until a collar extends approximately 2 inches up the pipe (fig. 8-29). Pack the cement down firmly so that all air pockets are eliminated.

6. Complete the next shingle course across the roof, but do *not* nail through the flashing flange. One of the strip shingles may have to be cut to fit around the upper portion of the circular pipe. Any shingle or portion of a shingle overlying the flashing flange must be embedded in plastic asphalt cement (fig. 8-30).

If it is necessary to cut the rectangular flashing flange from two sections of roll roofing, the upper section should overlap the lower one

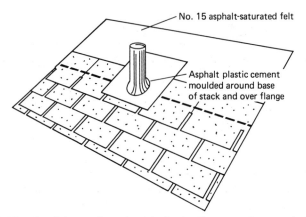

Fig. 8-29. Applying collar of asphalt plastic cement around pipe and over flange. *(Courtesy Asphalt Roofing Manufacturers Association)*

by approximately 2 inches. Cement the two sections together with plastic asphalt cement.

Flashing around a circular pipe located near the roof ridge is applied in the same manner as described in steps 1–6 above with the following exceptions:

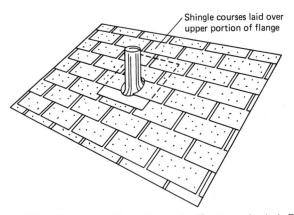

Fig. 8-30. Shingling above the soil stack. *(Courtesy Asphalt Roofing Manufacturers Association)*

1. Bend the flashing flange over the ridge and allow it to lap the roof shingles at all points.
2. Cover the flashing with hip shingles embedded in plastic asphalt cement.

Some soil stacks, vent pipes, and similar kinds of circular pipes are sold with adjustable metal flange collars designed to fit any roof slope. These are as effective as any other kind of flashing when the manufacturer's installation instructions are carefully followed.

Vertical Wall Flashing

On some structures, the rake of a sloped roof may butt against a vertical wall (fig. 8-31). The joint formed where the rake and vertical wall meet must be protected with flashing or leaks may develop. The recommended flashing method is to use individual metal flashing pieces applied with a 2-inch side lap. The application procedure is the same as the one used to apply step flashing to the side of a chimney.

In new construction, the underlayment should extend up the vertical wall about 4 inches to cover the joint formed by the wall and roof (fig. 8-32). If the underlayment has already been applied to the roof deck and the edge of each strip abuts at the vertical wall, lay an 18-inch wide strip of No. 15 asphalt-saturated roofing felt parallel to the rake. Allow 4 inches of the material to extend up the vertical wall, firmly press the roofing felt into the joint, and nail the 14-inch

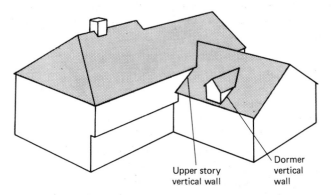

Upper story
vertical wall

Dormer
vertical
wall

Fig. 8-31. Common vertical wall flashing applications.

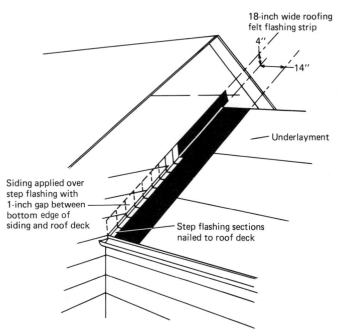

18-inch wide roofing
felt flashing strip

4"

14"

Underlayment

Siding applied over
step flashing with
1-inch gap between
bottom edge of
siding and roof deck

Step flashing sections
nailed to roof deck

Fig. 8-32. Eighteen-inch wide flashing strip applied over underlayment.

wide section to the roof deck with as few evenly spaced roofing nails as possible. Nail the 4-inch wide portion to the vertical wall sheathing in the same way.

Each piece of metal flashing must be provided with a 2-inch side lap. This dimension and the amount of shingle exposure will determine the width of the metal flashing piece. For example, a 5-inch shingle exposure will require a piece of metal flashing 7 inches wide (5-inch exposure plus 2-inch side lap).

Each metal flashing piece should be long enough to extend 4 inches up the vertical wall, and 2 inches onto the roof deck. Taken together, the width and length dimensions require that each metal flashing piece measure 6 × 7 inches.

Begin at the eave and nail the first metal flashing piece to the vertical wall sheathing with a single roofing nail placed in the upper top corner. Use a second roofing nail to nail the 2-inch wide portion of the metal flashing piece to the roof deck. Both nails should be placed where they will be covered by the 2-inch side lap of the next flashing

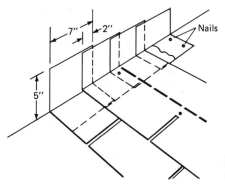

Fig. 8-33. Flashing shingle details.

piece (fig. 8-33). Coat the head of each with plastic asphalt cement to prevent any possibility of a leak developing.

Complete the first course of shingles along the roof eave, and secure the end of the last shingle to the metal flashing piece with plastic asphalt cement. *Do not nail through the shingle and metal flashing.*

Apply the second metal flashing piece to the vertical wall sheathing and the roof deck using the same method described for the first one. Provide at least a 2-inch side lap, but do not allow its edge to extend into the exposure for the first shingle course. On the roof deck, the metal flashing pieces must be completely hidden from view by the shingles. They are hidden from view by the finish siding on the vertical wall.

The remaining metal flashing pieces are applied to the vertical wall sheathing and roof deck in the same manner as described for the first and second pieces. The base flashing along a vertical *masonry* wall is sometimes covered with counter flashing (fig. 8-34). The counter

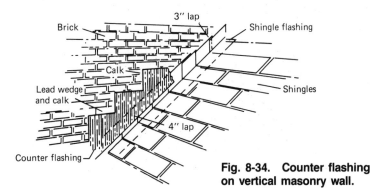

Fig. 8-34. Counter flashing on vertical masonry wall.

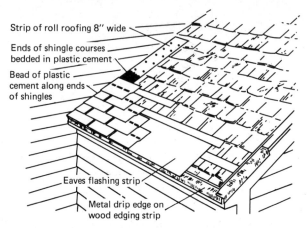

Strip of roll roofing 8" wide

Ends of shingle courses bedded in plastic cement

Bead of plastic cement along ends of shingles

Eaves flashing strip

Metal drip edge on wood edging strip

Fig. 8-35. Vertical wall flashing details when reroofing.

flashing is usually cut in several sections with each section overlapping the underlying one by a minimum of 3 inches. The procedure for applying counter flashing to a vertical masonry wall is the same as the one used to flash the sides of a chimney (see the discussion of chimney flashing). Bring the finish siding down over the flashing after all the flashing pieces have been nailed to the sheathing.

When reroofing, the joint between a vertical wall and the roof rake can be flashed with roll roofing (fig. 8-35). After the existing roof has been prepared for the new layer of roofing materials, nail an 8-inch wide strip of roll roofing along the roof with one edge of the strip butted against the vertical wall surface. Nail along both edges of the roll roofing strip with the nails spaced approximately 4 inches apart in rows 1 inch from each edge. As each course of shingles is laid, cover the roll roofing flashing strip with plastic asphalt cement and firmly press the end of the last shingle of each course into the cement. *Do not nail the shingle to the flashing strip.*

Other Types of Roof Flashing

Flashing is also used around dormers, skylights, scuttles, and other square or rectangular shaped roof openings; along the joint of the external angle or juncture formed by a change of pitch on a gambrel roof; and sometimes along the roof ridge. Flashing details for skylights,

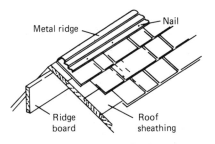

Metal ridge

Nail

Ridge
board

Roof
sheathing

Fig. 8-36. Metal ridge flashing.

scuttles, and similar types of roof openings are provided in chapter 6. Dormer flashing is discussed in chapter 7.

Ridge flashing is not used on most roofs because they are adequately protected by a row of ridge shingles that protect the joint by deflecting the water. However, ridge flashing is recommended for a wood shingle or shake roof with Boston-type ridge construction to prevent water entry. The ridge is flashed with a strip of roll roofing extending 3 inches down from both sides of the ridge and nailed along its outer edges. The nails should be driven in rows 1 inch from each edge and spaced approximately 4 inches apart. The wood shingles or shakes are installed over the ridge flashing.

A metal flashing strip can be used along the ridge of either a wood shingle or shake roof, or an asphalt shingle roof instead of the usual row of ridge shingles (fig. 8-36). The 6-inch wide metal flashing strip is installed over the ridge *after* the roofing has been applied. Use nails of a compatible metal driven in rows 1 inch from each outer edge of the strip and placed approximately 6 inches apart. The flashing should extend 3 inches down from each side of the ridge and should be embedded in a thick coat of plastic asphalt cement.

Roof Juncture Flashing

A gambrel roof will have one or two changes of pitch on each slope, depending on its design. These changes of pitch form roof junctures that must be covered with a flashing material. The flashing details of a typical gambrel roof are shown in figure 8-37. Note that a cant strip required under the shingle course immediately above the roof juncture. The shingle course immediately below the juncture is doubled, and the flashing is inserted between them.

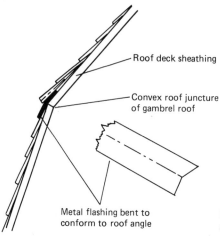

Roof deck sheathing

Convex roof juncture
of gambrel roof

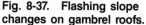

Metal flashing bent to
conform to roof angle

**Fig. 8-37. Flashing slope
changes on gambrel roofs.**

Flashing Repairs

Carefully inspect metal flashing for cracks in the plastic asphalt seals, loose or missing nails, small holes, or small rust spots. Cracked seals can be repaired by scraping away the old material and recoating the surface with plastic asphalt cement or a suitable caulking compound. Cracks around the embedded edges of chimney flashing should be freed of loose mortar, cleaned with a stiff wire brush, and then sealed. Never apply the cement or caulking compound when temperatures are below 40° F. If the seal *beneath* the flashing has deteriorated and is causing the leak, carefully pull the flashing back from the surface, remove the old cement, and apply a thick coat of new plastic asphalt cement with a putty knife.

Small holes in metal flashing can be repaired with a spot of plastic asphalt cement or a patch covered with cement. Replace the flashing if the holes are larger than ½ inch across.

Small rust spots should be cleaned down to the bare metal with a suitable solvent and wire brush and then covered with a metal patch and plastic asphalt cement.

Composition flashing can be repaired by first cleaning the surface and then covering the damaged area with plastic asphalt cement or other suitable sealant. Do not attempt to repair composition flashing if the material shows signs of extensive deterioration.

Reroofing

Old metal flashing can be left in place and reused when reroofing if it is still in a serviceable condition. If it is badly deteriorated, it should be removed and replaced with new flashing. If the old flashing is asphalt-saturated roofing felt and the roof covering material (asphalt strip shingles, individual shingles, etc.) is to be completely removed, the general rule is to remove and replace the flashing as well. If the old roofing felt flashing is retained, it should be carefully inspected for holes or tears. Any damage found must be repaired before the new roof covering materials are applied.

CHAPTER 9

Applying Strip Shingles

- Asphalt Strip Shingles
- Tools and Equipment
- Roof Deck Preparation
- Shingling Patterns and Exposures
- Nailing Shingles

- Applying Strip Shingles
- Special Shingling Procedures
- Roof Repair and Maintenance
- Reroofing with Asphalt Shingles

Asphalt shingles are the most popular type of roofing material in both new construction and reroofing applications. It is estimated that asphalt shingles have been used for roofing on almost three fourths of the houses in the country (fig. 9-1). They have also been used on commercial and, to a lesser extent, industrial structures. The popularity of asphalt shingles may be attributed to the fact that they are relatively inexpensive, widely available, and produced in a variety of colors, shapes, and styles. They are also easy to install, maintain, and repair, and they have a service life of at least 15 years.

Asphalt shingles are available in strip form or as individual shingles. The application of individual and hex strip shingles is discussed in chapter 10. The application of asphalt strip shingles, the most widely used form of roofing material, is described in this chapter.

Asphalt Strip Shingles

Asphalt shingles are sometimes called composition shingles because they are composed of more than one material. Each shingle has a base

Fig. 9-1. Asphalt shingle roof.

covered on both sides with an asphalt coating to provide a waterproofing seal (fig. 9-2). Ceramic or mineral granules of various colors are embedded in the top coating of asphalt to provide an attractive, protective surface. The shingle base may be either an organic mat of wood and paper fibers or a fiberglass mat. Shingles formed around an organic mat are frequently called *asphalt shingles* If a fiberglass mat is used, they are called *fiberglass shingles*. Fiberglass shingles are slightly more expensive than organic base shingles, but they provide greater fire and wind resistance, and have a longer service life. The application instructions provided in this chapter can be used for both asphalt and fiberglass shingles.

Asphalt strip shingles are generally available in standard 12 × 36–inch lengths, but some roofing manufacturers also produce strip

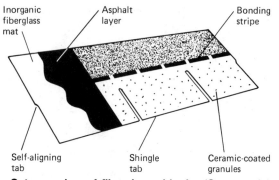

Inorganic fiberglass mat

Asphalt layer

Bonding stripe

Self-aligning tab

Shingle tab

Ceramic-coated granules

Fig. 9-2. Cutaway view of fiberglass shingle. *(Courtesy Johns-Manville Sales Corp.)*

shingles in slightly narrower or wider widths and longer lengths to provide a rustic or textured roof appearance, or deep shadow lines.

The most popular type of asphalt strip shingle is the self-sealing three-tab unit (fig. 9-3). These are sometimes called *three-tab shingles* or *square tab shingles*. Each self-sealing shingle has a strip of special thermoplastic adhesive that is spaced at intervals on each shingle. When the adhesive is activated by the warmth of the sun, it bonds

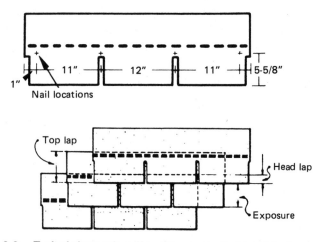

Fig. 9-3. Typical three-tab self-sealing asphalt strip shingles. *(Courtesy Asphalt Roofing Manufacturers Association)*

shingle to shingle to hold them in place against strong winds, rain, and snow.

Asphalt strip shingles are also available with no tabs and cutouts, with staggered or random edge butts, with butts of random width, and other design variations to give a distinctive roof appearance. Some are also manufactured with additional layers of asphalt and other materials to produce a thicker, heavier shingle. Examples of different types of asphalt strip shingles are illustrated in figure 9-4.

Asphalt strip shingles are sold in bundles with approximately 27 shingle strips to the bundle. In most applications, three bundles will cover a roofing square (100 square feet of roof surface). Bundles should be stored flat so that the strips will not curl when they are opened. When possible, store the bundles or shingles indoors until they are to be applied to the roof. If the bundles or shingles cannot be stored in-

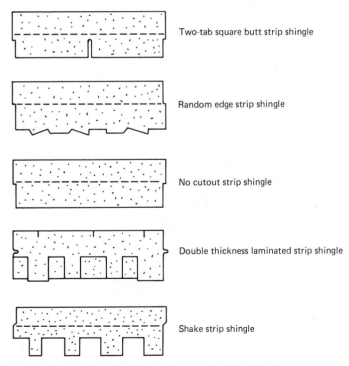

Two-tab square butt strip shingle

Random edge strip shingle

No cutout strip shingle

Double thickness laminated strip shingle

Shake strip shingle

Fig. 9-4. Examples of different types of strip shingles.

doors, stack them on wood planks or 2 × 4's laid close together and clear of the ground. When not working, cover the bundles or shingles with a large sheet of polystyrene or other equally suitable waterproof material for protection from rain or snow. Avoid working when it is extremely cold because the shingles are brittle and will not bend easily. If working in cold weather cannot be avoided, warm the shingles before bending them.

Tools and Equipment

One or more sturdy ladders or scaffolding will be needed when shingling a roof. The tools required for this type of work include a claw hammer, a carpenter's apron to hold the roofing nails, a chalk line, a utility knife, a tape measure or folding rule, and a putty knife or pointed trowel for applying roofing cement. Additional information about the tools and equipment used in asphalt shingling is presented in chapter 3. Be sure to read and follow the work safety rules included in that chapter before beginning work.

Roof Deck Preparation

A wood roof deck must be completely dry and tightly built of good quality, well-seasoned lumber or plywood. It must also have adequate support and the rafters should meet the requirements of their span in size and spacing. If these conditions have been satisfactorily met, the deck can then be covered with the required underlayment and flashing (new construction), the existing roof can be repaired to provide a smooth, flat surface for the new roofing materials (reroofing), or the existing roofing materials can be stripped and new materials applied directly to the roof deck (reroofing). Detailed instructions on all aspects of roof deck preparation are provided in chapter 3.

Underlayment

In new construction or when old roofing materials have been removed, wood decks or roofs with slopes of 4-in-12 to 7-in-12 should be covered with overlapping strips of No. 15 asphalt-saturated roofing felt. The roofing felt should be lapped 2 inches horizontally and 4 inches at the ends. It should also be carried 6 inches from both sides around hips, ridges, and corners. Low-pitched roofs and roofs with slopes greater than 7-in-12 require special treatment. Detailed instructions for the application of the underlayment are included in chapter 3.

Flashing

Chimneys are usually flashed with galvanized steel. Metal drip edges are used along the rakes and eaves. On most asphalt shingle roofs, the valleys are flashed with mineral-surface roll roofing. See chapter 8 for detailed instructions on installing flashing.

Shingling Patterns and Exposures

Asphalt shingles are laid with the upper portion of each shingle covered by an overlapping one in the next course. Because the uncovered portion of each shingle from its bottom edge to the bottom edge of the overlapping shingle is exposed to the weather, it is sometimes called the weather exposure of the shingle or the shingle exposure. On most roofs, asphalt shingles are laid with a 5-inch exposure.

When applying asphalt shingles, the first course is started with one full size shingle at either rake. The shingle sizes used to begin the second and succeeding courses will depend on the shingling pattern desired for the roof. The 4-inch, 5-inch, and 6-inch patterns are used when shingling with standard three-tab asphalt strip shingles.

The 6-inch shingling pattern is the one used on most roofs and its application is described in the next section (see the discussion of applying strip shingles). The 6-inch shingling pattern requires that the second course be started with a shingle from which 6 inches (one half tab) have been removed. Succeeding courses are started with shingles from which 12 inches, 18 inches, 24 inches, and 30 inches have been removed. The seventh course is started with a full size shingle and the pattern is repeated until the ridge is reached (fig. 9-5). The starting shingles of this pattern are trimmed in 6-inch increments and are applied so that their cutouts break joints on halves. The cutouts of each shingle are centered on the tabs of the underlying shingles. Vertical chalk lines must be snapped on the underlayment as work progresses to keep the cutouts aligned up the roof. The shingle edges or cutouts in every other course must be aligned within ¼ inch. Because

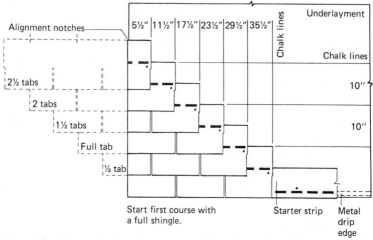

Fig. 9-5. Six-inch shingling pattern. *(Courtesy Johns-Manville Sales Corp.)*

of minor differences in shingle dimensions, this alignment is difficult to maintain without the use of vertical chalk lines to serve as guides.

The 4-inch shingling pattern is recommended for low-pitched roofs with slopes of 2-in-12 to less than 4-in-12. When using the 4-inch

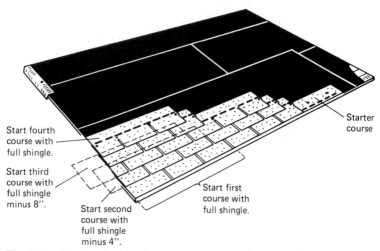

Fig. 9-6. Four-inch shingling pattern repeated in the fifth, sixth, and seventh courses. *(Courtesy Asphalt Roofing Manufacturers Association)*

shingling pattern, the second course is started with a shingle from which 4 inches have been removed. Succeeding courses are started with shingles from which 8 inches, 12 inches, 16 inches, 20 inches, 24 inches, 28 inches, and 32 inches have been removed. The tenth course is started with a full-size shingle and the pattern is repeated until the roof ridge is reached. With this shingling pattern, the cutouts break joints on thirds and create a diagonal alignment. A variation of the 4-inch shingling pattern is to repeat the 4-inch, 12-inch, and 16-inch offsets in the fifth, sixth, and seventh courses (fig. 9-6).

Like the 4-inch shingling pattern, the 5-inch pattern also produces a diagonal shingle alignment. The second course is started with a shingle from which 5 inches have been removed. The first shingles in succeeding courses are trimmed in 5-inch increments (10 inches, 15 inches, 20 inches, etc.) until only a 6-inch section remains (30 inches removed from a 36-inch shingle). The pattern is then repeated to the roof ridge.

A 9-inch shingling pattern is recommended for two-tab and no cutout asphalt strip shingles (fig. 9-7). The second course starts with a shingle from which 9 inches have been removed. The first shingles in succeeding courses are trimmed in 9-inch increments. A full-size shingle is used to start the fifth course and the pattern is repeated until the roof ridge is reached.

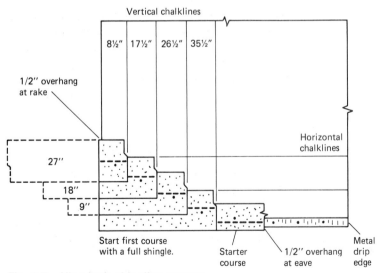

Fig. 9-7. Nine-inch shingling pattern.

Some strip shingles, particularly those designed to produce a rustic, thatch, or other distinctive appearance, must be applied in accordance with the roofing manufacturer's suggested shingling pattern. Always read and carefully follow the roofing manufacturer's application instructions for these shingles.

Shingles may also be applied in random patterns by varying shingle offsets (see the discussion of random shingle spacing in this chapter).

Nailing Shingles

Properly nailing down the shingles is fundamental to any good roofing procedure. This requires the selection of the right nails, using enough nails, and placing them in the most suitable pattern. The shingle manufacturer will recommend the best nailing pattern to use and will include these recommendations on each bundle of shingles. Typical nailing patterns for strip shingles are shown in figures 9-8 and 9-9.

The nails used to apply asphalt strip shingles are 11- or 12-gauge hot galvanized roofing nails with heads at least 3/8 inch in diameter. The nails should be long enough to penetrate at least 3/4 inch into

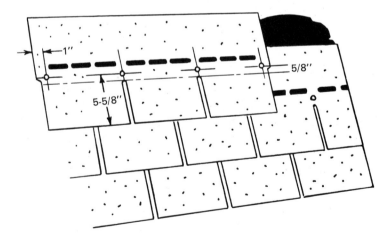

Fig. 9-8. Correct nailing pattern for typical three-tab self-sealing strip shingle with standard 5-inch exposure. *(Courtesy Asphalt Roofing Manufacturers Association)*

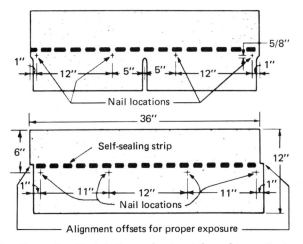

Fig. 9-9. Correct nailing patterns for two-tab and no cutout self-sealing strip shingles. *(Courtesy Asphalt Roofing Manufacturers Association)*

solid wood board sheathing and through plywood sheathing. Recommended nail lengths for different roofing applications are listed in table 9-1.

On most pitched roofs, four roofing nails are used to secure each strip shingle. Use six nails on roofs with steep slopes or roofs located in areas where strong winds are common and place a spot of roofing cement under each shingle tab for additional holding power.

Carefully align each shingle before nailing it and check to make certain that no end joint or cutout is less than two inches from any nail in an underlying course. Check the first course to make sure it has been laid perfectly straight before applying any of the succeeding

Table 9-1. Recommended Nail Lengths for Different Roofing Applications

Application	Nail Length
Over new deck	1¼ inches
Over roll roofing	1¼ inches
Over asphalt shingles	1½ inches
Over wood shingles	1¾ inches

courses. Check the alignment of shingles and shingle courses at regular intervals against the horizontal chalk lines. Make certain that all cutouts are aligned from the roof eaves to the ridge.

Nail each shingle at the end closest to the shingle just laid and then proceed to nail across the shingle to its opposite end. Do not nail both ends of the shingle first and then the middle. Doing so may cause the shingle to buckle in the middle. If a self-sealing strip shingle is being used, do not nail above or through the adhesive strip. The nails should be driven through the shingle at a point $5/8$ inch above the cutouts on tab-type shingles or 6 inches up from the bottom edge on no cutout shingles.

Drive the nails as straight as possible. If a nail is driven crookedly or at a slant, the edge of the head may cut into the shingle. Drive the nails until their heads contact the surface of the shingle. Do not drive the nail heads into the shingle material.

Shingles with the standard 5-inch exposure should be nailed $5\frac{5}{8}$ inches above the bottom edge so that the nail heads are covered by the shingles in the next course (see fig. 9-8). The nails at either end of the shingle should be located approximately 1 inch up from the long side edge. If an exposure other than 5 inches is used, the nail heads must be placed so that they are covered by the shingles in the next course.

Applying Strip Shingles

Gable roofs shorter than 30 feet long are generally shingled by starting at either rake and shingling toward the opposite one. Figure 9-5 illustrates the shingling procedure used when starting at a left-hand rake with three-tab asphalt strip shingles.

In new construction, after the underlayment, drip edge, and any other required flashing have been nailed or otherwise secured to the roof deck, snap vertical and horizontal chalk lines to provide an alignment guide for shingles and shingle courses during application. The vertical chalk lines serve as guides for the shingle cutout. As shown in figure 9-5, the vertical chalk lines should be spaced at $5\frac{1}{2}$-inch, $11\frac{1}{2}$-inch, $17\frac{1}{2}$-inch, $23\frac{1}{2}$-inch, $29\frac{1}{2}$-inch, and $35\frac{1}{2}$-inch intervals parallel to the roof edge when using three-tab, square-butt asphalt strip shingles applied with a 6-inch offset.

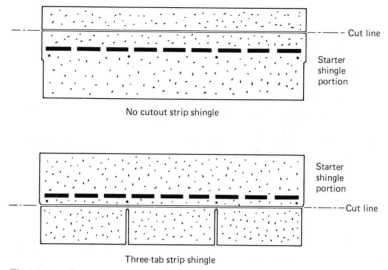

No cutout strip shingle

Three-tab strip shingle

Fig. 9-10. Starter course shingles.

Applying Starter Course

Lay a starter course along the edge of the roof with a ½-inch overhang at both the rakes and the eaves. The starter course normally consists of a row of shingles with the tabs cut off (fig. 9-10). If tabless shingles are used, the upper 3 inches of each shingle should be cut off in a straight line parallel with its bottom edge. Nail the shingle down on the roof deck, allowing for a ½-inch overhang along the rakes and eaves of the roof. The nails should be placed approximately 4 inches above the eave line (fig. 9-11).

Two alternative methods of applying the starter course are to use full shingles with the tabs reversed or a continuous starter strip of 9-inch wide or wider mineral surfaced roll roofing (fig. 9-12).

A stiffer backing for the shingles along the roof eaves can be provided by using a course of wood shingles under the starter course (fig. 9-13). Each shingle is laid edge to edge and secured with nails driven 1 inch from each of its side edges and approximately 6 inches up from its bottom edge. The course of wood shingles is then covered by the starter course and first course of asphalt strip shingles.

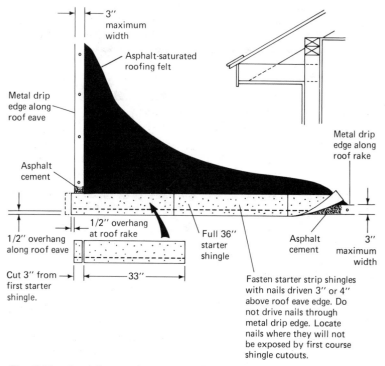

Fig. 9-11. Applying starter course shingles.

Applying First Course

The first course of shingles is started with a full strip shingle laid directly over the starter strip. The butt edges of the first course shingles should be laid flush with the bottom edge of the starter strip at the left-hand rake of the roof. Align the end of the 36-inch shingle with the 35½-inch chalk line on the roof deck (fig. 9-14). The ½-inch difference is accounted for by the overhang at the rake.

Applying Second and Succeeding Courses

Cut 6 inches, or one half of a tab, from a shingle strip and begin the second course with it. Align one end of the shingle strip with the 29½-inch chalk line and allow the other end to overlap the rake by ½ inch. The amount of shingle exposure allowed on the first course

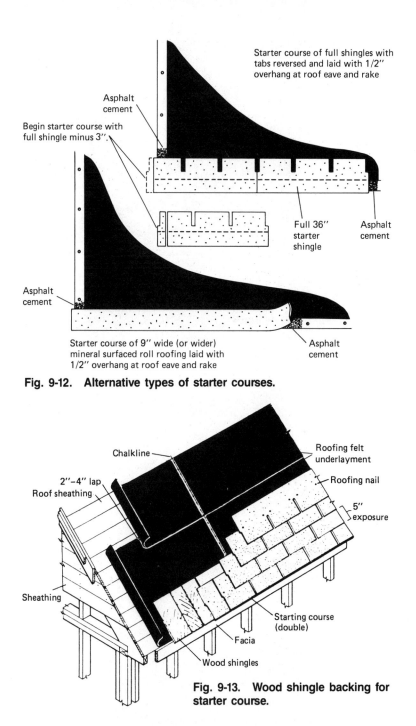

Starter course of full shingles with tabs reversed and laid with 1/2" overhang at roof eave and rake

Asphalt cement

Begin starter course with full shingle minus 3".

Full 36" starter shingle

Asphalt cement

Asphalt cement

Starter course of 9" wide (or wider) mineral surfaced roll roofing laid with 1/2" overhang at roof eave and rake

Asphalt cement

Fig. 9-12. Alternative types of starter courses.

Chalkline

Roofing felt underlayment

Roofing nail

2"–4" lap
Roof sheathing

5" exposure

Sheathing

Starting course (double)

Facia

Wood shingles

Fig. 9-13. Wood shingle backing for starter course.

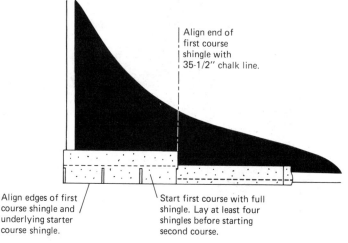

Align end of
first course
shingle with
35-1/2" chalk line.

Align edges of first
course shingle and
underlying starter
course shingle.

Start first course with full
shingle. Lay at least four
shingles before starting
second course.

Fig. 9-14. First shingle course.

will depend on the shingle manufacturer's recommendations. For this type of shingle it will generally be about 5 inches (fig. 9-15).

Start the third course by cutting off 12 inches, or a full tab, from the shingle strip. Nail the shortened shingle strip in position with the

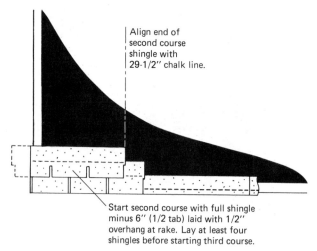

Align end of
second course
shingle with
29-1/2" chalk line.

Start second course with full shingle
minus 6" (1/2 tab) laid with 1/2"
overhang at rake. Lay at least four
shingles before starting third course.

Fig. 9-15. Second shingle course.

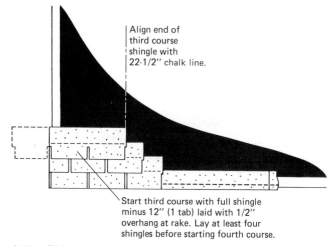

Align end of third course shingle with 22-1/2" chalk line.

Start third course with full shingle minus 12" (1 tab) laid with 1/2" overhang at rake. Lay at least four shingles before starting fourth course.

Fig. 9-16. Third shingle course.

far edge of the shingle aligned with the 23½-inch chalk line (fig. 9-16). Allow the same ½-inch overhang at the rake and the same exposure for the second course of shingles as was provided for the first course.

Cut off 18 inches, or a tab and a half, from a shingle strip and use it to begin the fourth course. One end of the shingle should be aligned with the 17½-inch chalk line, and the other end should extend ½ inch beyond the edge of the rake (fig. 9-17). Allow the same shingle exposure in the third course as you did for the first and second courses.

Cut 24 inches (two tabs) from a shingle strip and begin the fifth course with the remaining 12-inch section of shingle. One edge of the shingle should overhang the rake edge by ½ inch. Align the other edge with the 11½-inch chalk line, and provide the same exposure for the fourth course as in the previous shingle courses (fig. 9-18).

Start the sixth course with a 6-inch wide shingle (half tab). Align one end with the 5½-inch chalk mark, and allow a ½-inch overhang at the rake. Provide the required exposure for the fifth shingle course (fig. 9-19).

Now that you have laid the starter shingles for each of the first six courses, complete each course all the way to the rake at the opposite end of the roof. Be sure to allow a ½-inch overhang at this

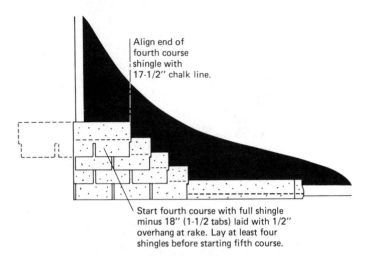

Align end of
fourth course
shingle with
17-1/2" chalk line.

Start fourth course with full shingle
minus 18" (1-1/2 tabs) laid with 1/2"
overhang at rake. Lay at least four
shingles before starting fifth course.

Fig. 9-17. Fourth shingle course.

rake, and cut off any excess shingle beyond the ½-inch overhang. Try
to make this line as straight as possible. Some roofers cut away the
excess after the shingle is applied; others prefer doing it before it is
nailed in position.

The seventh course begins with a full shingle at the rake, and
is aligned with the shingles of the first course. The first shingle of each

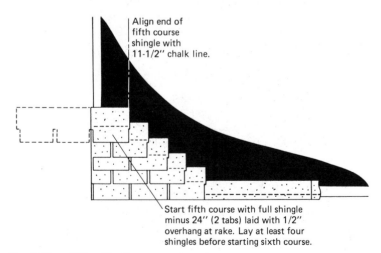

Align end of
fifth course
shingle with
11-1/2" chalk line.

Start fifth course with full shingle
minus 24" (2 tabs) laid with 1/2"
overhang at rake. Lay at least four
shingles before starting sixth course.

Fig. 9-18. Fifth shingle course.

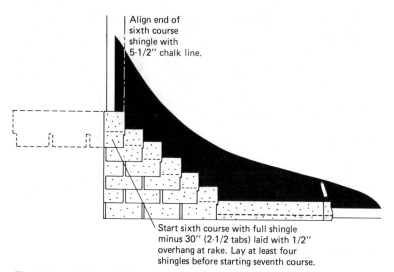

Align end of sixth course shingle with 5-1/2" chalk line.

Start sixth course with full shingle minus 30" (2-1/2 tabs) laid with 1/2" overhang at rake. Lay at least four shingles before starting seventh course.

Fig. 9-19. Sixth shingle course.

succeeding course is 6 inches shorter, thereby repeating the shingling pattern of the first six courses. Repeat the pattern with every seventh course until the roof ridge is reached. Go to the other side of the roof and apply shingles in the same manner. The roof ridge is shingled last (see the discussion of shingling hips and ridges).

The dimensions given in the preceding paragraphs are for shingling with 36-inch self-sealing three-tab, square-butt asphalt strip shingles applied with a 6-inch offset and a ½-inch overhang at the rakes and eaves. Changes in these specifications or the type of offset will require modifications of each of the shingling steps. Whenever shingles without a self-sealing adhesive strip are used, each tab should be embedded in a quarter-size spot of roofing cement.

Roofing manufacturers will provide application instructions with the shingles, but the recommended specifications will vary among manufacturers. For example, although a 5-inch shingle exposure is fairly standard, recommendations for overhang at the eave and rake range from ¼ to ¾ inch.

Valley Shingling

Roof valleys are formed where two roofs join at an angle (fig. 9-20). The joint formed by the two roofs is a potential weak point in the

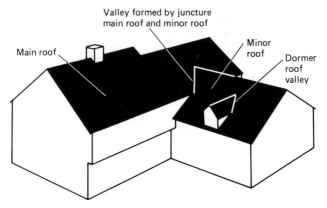

Fig. 9-20. Roof valleys.

structure because it provides a channel for water runoff. In order to prevent leakage, the joint formed by the two roofs must be adequately covered with roofing materials. The three common methods of covering a roof valley are

1. Open valley shingling
2. Woven valley shingling
3. Closed-cut valley shingling

Both strip shingles and individual shingles can be used in open valley shingling. *Only* strip shingles can be used in woven or closed-cut valleys.

OPEN VALLEYS

In open valleys, the flashing along the valley joint is left exposed (fig. 9-21). The shingles ending at the valley are cut to form an edge running in a straight line parallel to the joint. The valley must be covered with flashing before the shingles can be applied. Valley flashing instructions are described in chapter 8. The procedure for constructing an open valley after the flashing has been applied is as follows:

1. Snap two chalk lines along both sides of the valley. These lines should be 6 inches apart at the ridge, or 3 inches up from the joint on either side of the valley, and must diverge at the rate of ⅛ inch per foot in the direction of the eave.

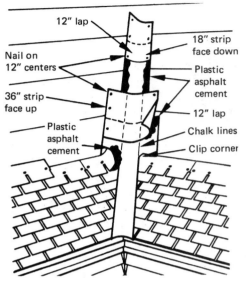

12" lap

18" strip
face down

Nail on
12" centers

Plastic
asphalt
cement

36" strip
face up

12" lap

Plastic
asphalt
cement

Chalk lines

Clip corner

Fig. 9-21. Open roof valley. *(Courtesy Asphalt Roofing Manufacturers Association)*

2. Run the shingle courses from each roof slope to the chalk lines in the valley and cut the last shingle of each course so that it ends parallel to the line.
3. Cut approximately 1 inch off the upper corner of each shingle touching the chalk lines to guide water runoff into the valley. As shown in figure 9-21, a diagonal cut should be made.
4. *Cement* the shingles to the surface at the chalk lines. Do *not* nail within 7 inches of the chalk lines because the nail holes will penetrate the flashing strips and may cause leaks. Press the shingles firmly into the asphalt cement before nailing.

WOVEN VALLEYS

A woven valley is completely covered with shingles. Alternate shingle courses are laid across the valley and woven together as shown in figure 9-22. Because of the additional coverage provided by the shingles, it is not necessary to cover the valley joint with a double thickness of roll roofing. Only a single 36-inch wide strip of mineral surface roll roofing is required, and it is applied according to the instructions given for constructing an open valley. The flashing strip is laid *over* the underlayment (see chapter 8).

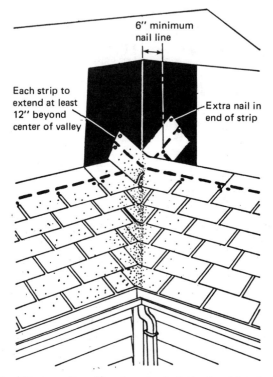

Fig. 9-22. Woven valley. *(Courtesy Asphalt Roofing Manufacturers Association)*

In woven valleys, the roof shingles are first laid to a point approximately three feet from the center of the valley on each roof slope. The remainder of the procedure is as follows:

1. Snap a chalk line down each side of the valley 6 inches from the valley joint.
2. Run the first course of shingles (from the main roof) across the valley and onto the adjoining roof for at least 12 inches. Nail the shingles down, but keep all nails at least 1 inch from the chalk line on each side of the valley
3. Run the first course of shingles from the adjoining roof across the valley to the main roof and nail it in position. Remember to keep the nails at least 1 inch outside the chalk lines.

4. Run the remaining shingle courses alternately back and forth across the valley, weaving the shingles together and nailing them in position as described in steps 2 and 3 above.

CLOSED-CUT VALLEYS

In closed-cut valleys, the valley is completely covered with shingles, but the shingle courses are not woven together (fig. 9-23). A single 36-inch wide flashing strip of mineral surfaced roll roofing is laid down the length of the valley before the shingles are applied. The method used for applying the flashing strip of roll roofing is included in the instructions describing the construction of an open valley.

After applying the roll roofing flashing strip, the remainder of the procedure for constructing a closed-cut valley is as follows:

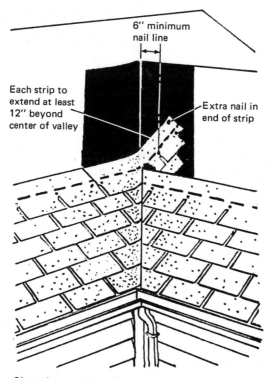

6" minimum
nail line

Each strip to
extend at least
12" beyond
center of valley

Extra nail in
end of strip

Fig. 9-23. Closed-cut valley. (*Courtesy Asphalt Roofing Manufacturers Association*)

1. Snap a chalk line down each side of the valley 6 inches from the valley joint.
2. Run each course of shingles from the main roof across the valley and onto the adjoining roof for at least 12 inches. Nail the shingles to the roof, but keep all nails at least 1 inch outside the chalk line on each side of the valley. Bring all the shingle courses across from the main roof, before completing the courses on the adjoining roof.
3. Snap a chalk line 2 inches from the valley joint on the unshingled (adjoining) roof.
4. Run each course of shingles from the adjoining roof to the 2-inch chalk line and trim away the excess. Cut a diagonal piece off the upper corner of each shingle ending in the valley to provide better drainage.
5. Nail the shingles to the roof, but keep all nails at least 1 inch outside the chalk lines snapped down on each side of the valley. The ends of each shingle course are secured to the valley with a 3-inch wide strip of asphalt roofing cement.

Hips and Ridges

Most roofing manufacturers provide special shingles to cap the roof ridge and hips. Ridge and hip shingles can also be made from the same ones used to cover the roof.

The ridge and hip shingles for a roof covered with square-butt strip shingles should be at least 9 × 12 inches. Any tabs must be cut off the shingles. The common 12 × 36–inch three-tab shingle strip is large enough to supply three ridge or hip shingles.

All shingling on the flat surfaces of the roof must be completed

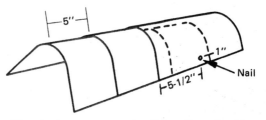

Fig. 9-24. Standard exposure and nail locations for hip and ridge shingles. *(Courtesy Asphalt Roofing Manufacturers Association)*

Fig. 9-25. Applying ridge shingles.

before the ridge or hip shingles are applied. After the ridge or hip shingles have been cut to the required size and shape, bend each shingle lengthwise down the center so that each side of the ridge or hip will have the same amount of overlap. Avoid splitting or cracking the shingle as you bend it. If the weather is cold, warm the shingle until it is flexible enough to bend without cracking.

Each shingle is secured to the ridge or hip with a nail on each side of the roof juncture. Position the nails 1 inch up from the bottom edge and 5½ inches back from the exposed edge (fig. 9-24). These nail positions are for standard shingle exposure.

Ridge shingles are applied by beginning at either end of the roof and working toward the opposite end (fig. 9-25). Begin hip shingles at the bottom of the hip and work toward the ridge. On hip roofs, the hips should be shingled before the ridge. Allow 5 inches of exposure for both ridge and hip shingles when standard exposures are being used.

Avoid using any kind of metal ridge roll with asphalt shingles. After a period of time the metal may corrode and discolor the shingles.

Special Shingling Procedures

A gable roof with a straight, unbroken surface is usually shingled by starting at the left-hand rake, or the most visible rake, and working across to the opposite one. Many gable roofs also have at least two valleys where a minor roof intersects with a major or main roof. The special shingling procedures required for roof valleys have already been described in this chapter (see the discussion of valley shingling). Special shingling procedures are also required for a variety of other types of roof applications including hip roofs, roofs with spans of more than 30 feet, steep roofs, low-pitched roofs, and roofs broken by dormers, skylights, and other roof openings. These special shingling procedures are described in this section.

Shingling from Roof Center

The recommended method of applying strip shingles to hip roofs, roofs with two equally visible rakes, or roofs longer than 30 feet, is to start at the center and work in both directions toward the two roof rakes.

Begin by snapping a vertical chalk line in the exact center of the roof. Cut off 3 inches from the bottoms of as many shingles as necessary and apply a starter strip running in each direction from the center to the two rakes. The starter strip is applied with the mineral surface coating of each shingle facing down against the roof deck, and should overhang both eaves and rakes of the roof by approximately 1/2 inch. The amount of overhang must be maintained with each shingle course at the rakes or the roof will have an uneven appearance.

Begin the first course of shingles by aligning the first full shingle with its butt edge even with the starter course edge and its center tab centered on the vertical chalk line (fig. 9-26). Succeeding courses are laid from the centerline with the required exposure and offset. The starter strip and each shingle course must be trimmed to fit the rake or hip ridge line. The ends of each of the two shingles terminating a course should be secured to the deck with asphalt roofing cement.

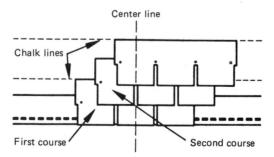

Fig. 9-26. Shingling from roof center. *(Courtesy Asphalt Roofing Manufacturers Association)*

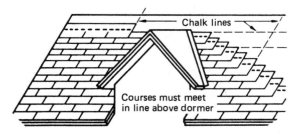

Fig. 9-27. Shingling around dormers. *(Courtesy Johns-Manville Sales Corp.)*

Shingling Around Dormers

Shingling around a dormer requires special care because the shingle courses must be laid exactly parallel to the eave line or their pattern will not continue evenly above the dormer roof. The courses can be kept parallel by snapping horizontal and vertical chalk lines on either side of the dormer as guides for shingle alignment (fig. 9-27).

When the roof shingle courses reach the dormer, flashing must be installed along its base. Dormer base flashing serves the same function as the flashing installed along the base of a chimney. Base flashing installation procedures are described in chapters 7 and 8.

After the base flashing is installed, the shingle courses are continued up the roof to a point slightly above the lower edge of the dormer valley (fig. 9-28). As each course is laid, step flashing should be installed where the last shingle contacts the dormer wall (fig. 9-29).

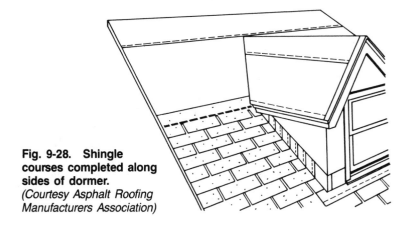

Fig. 9-28. Shingle courses completed along sides of dormer. *(Courtesy Asphalt Roofing Manufacturers Association)*

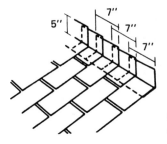

Fig. 9-29. Step flashing details along dormer. *(Courtesy Johns-Manville Sales Corp.)*

If the dormer has a gable roof, two valleys will be formed at the point it joins the main roof. Flashing must be applied to these valleys before any shingle courses are laid. Mineral surface roll roofing is the most common type of flashing material used on asphalt shingle roofs.

The flashing and shingling of an *open* dormer valley are illustrated in figures 9-30 to 9-33. The application procedure may be outlined as follows:

1. Complete the main roof shingle courses to a point just above the lower end of the dormer valley (see fig. 9-28).
2. Cut an 18-inch wide strip of mineral surface roll roofing long enough to extend the length of the dormer roof valley *plus* a 2-inch allowance at the bottom and a 12-inch allowance at the top.
3. Bend and fit the 18-inch wide strip of roll roofing with its mineral surface face down to the angle of the valley. Allow 6 inches of material to cover the dormer roof deck and 12 inches to cover the main roof deck (fig. 9-30).
4. Cut the bottom of the flashing strip so that there is a ¼-inch overhang along the eave of the dormer roof deck and a 2-inch projection below the point where the two roofs join.
5. Nail the roll roofing flashing strip to the valley with the mineral surface face down. Locate the nails approximately 1 inch from the edges of the roll roofing strip. Make certain the roll roofing fits tightly into the valley before nailing.
6. Cut and fit a 36-inch wide strip of roll roofing to the valley with the mineral surface side facing up.
7. Center the wider strip in the valley and cut the bottom edge so that it is flush with the edge of the underlying 18-inch wide strip (fig. 9-31).

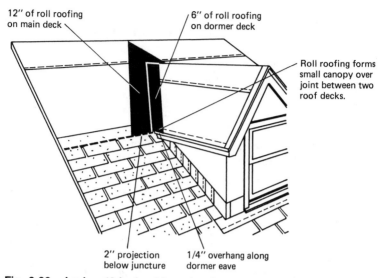

12" of roll roofing on main deck

6" of roll roofing on dormer deck

Roll roofing forms small canopy over joint between two roof decks.

2" projection below juncture

1/4" overhang along dormer eave

Fig. 9-30. Laying 18-inch wide strip of roll roofing in dormer valley.

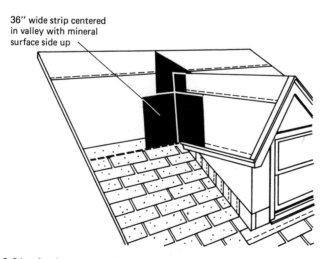

36" wide strip centered in valley with mineral surface side up

Fig. 9-31. Laying second or upper strip of roll roofing.

8. Nail the 36-inch wide strip to the valley with nails positioned 1 inch in from the edges. Keep all nails at least 7 inches from the valley joint. Additional protection against leakage can be provided by running a strip of asphalt roofing cement along the edges of the roll roofing.

9. Snap two chalk lines along both sides of the valley. These lines should be 3 inches on either side of the center of the valley where the dormer roof ridge intersects with the main roof and must diverge at the rate of ⅛ inch per foot in the direction of the dormer roof eave (fig. 9-32).

10. Complete all shingle courses on the main roof to the chalk line in the valley and cut the last shingle to conform to the angle of the line (fig. 9-33).

11. Embed the end shingles of each course in plastic asphalt cement and press them down firmly onto the valley flashing. Do not nail through the flashing.

12. Repeat steps 1–10 on the other side of the dormer.

13. Shingle the dormer roof by starting at the eave at each rake and running the courses to the chalk lines in the valleys. Cut the ends of the shingles to fit the angle of the valley and embed them in plastic asphalt cement. Do not nail through the flashing. The shingling method is the same as that used for shingling a main roof.

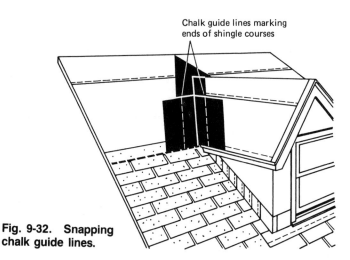

Chalk guide lines marking
ends of shingle courses

Fig. 9-32. Snapping chalk guide lines.

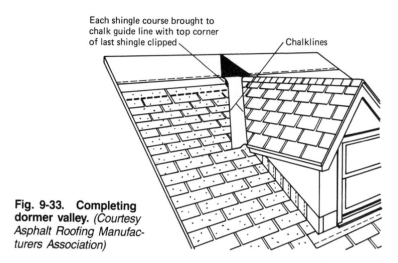

Each shingle course brought to chalk guide line with top corner of last shingle clipped

Chalklines

Fig. 9-33. Completing dormer valley. *(Courtesy Asphalt Roofing Manufacturers Association)*

14. Apply shingles to the dormer ridge by starting at the rake and working toward the main roof. Secure the shingles with one nail on each side of the shingle 5⅛ inches up from its bottom edge (for a 5-inch exposure) and 1 inch up from each side edge. Split the last dormer ridge shingle and nail it to the roof deck with two nails in each half (fig. 9-34).

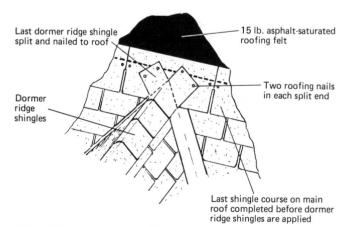

Last dormer ridge shingle split and nailed to roof

15 lb. asphalt-saturated roofing felt

Two roofing nails in each split end

Dormer ridge shingles

Last shingle course on main roof completed before dormer ridge shingles are applied

Fig. 9-34. Shingling dormer ridge.

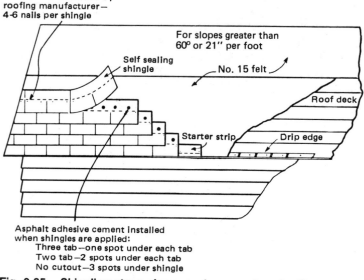

Fig. 9-35. Shingling steep slopes and mansard roofs. *(Courtesy Asphalt Roofing Manufacturers Association)*

Shingling Steep Slopes

The maximum roof slope considered suitable for normal shingle application is 7-in-12 or approximately 60°. On steeper roof slopes, the factory applied self-sealing adhesives on the backs of the shingle tabs show a tendency to pull free. A strong wind will frequently bend these loose shingles back, which not only ruins the appearance of the roof, but also exposes the roof to possible leaks.

Mansard roofs and others with slopes exceeding 7-in-12 or 60° require that the shingles be applied with certain modifications of standard shingle application procedures. These modifications are illustrated in figure 9-35 and are outlined as follows:

1. Nail the shingles to the roof deck according to the shingle manufacturer's instructions and recommended nailing pattern.
2. Nail self-sealing strip shingles so that the nails are not positioned in the adhesive strip.
3. Apply a quick-setting asphalt adhesive cement to the back of the bottom section of each shingle immediately upon installation. One spot of cement approximately 1 inch in diameter should be applied to the back of each tab of a three-tab

shingle; two spots for each tab of a two-tab shingle; and three evenly spaced spots for a tabless (no cutout) shingle. The roofing manufacturer will generally recommend a suitable asphalt adhesive cement.

4. Reduce the possibility of moisture entrapment behind the sheathing by providing adequate through ventilation in the space immediately below the roof deck (see chapter 5).

Shingling Low-Pitched Roofs

Low-pitched roofs with slopes of 2-in-12 to just under 4-in-12 should be covered with square tab shingle strips. Because the rate of water runoff is so much slower on a low slope, there should be a double layer of underlayment along the eaves to protect the sheathing of the roof deck. The method for applying underlayment on low slopes is described in greater detail in chapter 3. An eaves flashing strip of roofing felt should be cemented to the roof edge before any shingles are applied.

If self-sealing shingles are not used, each tab or lower section of the shingle should be cemented down by applying at least three quarter-size spots of roofing cement.

Random Shingle Spacing

Random shingle spacing is an application method in which the shingle courses are started with irregular tab widths. It is used on roofs with long and unbroken horizontal roof surfaces of 30 feet or more where perfect vertical alignment of shingle cutouts is extremely difficult. Many different patterns are possible depending on the sequence of the tab widths. In random shingle spacing, no rake tab should be less than 3 inches wide, the centerlines of shingle cutouts in any one course should be located at least 3 inches laterally from the cutout centerlines of the underlying and overlapping courses, and the rake tab widths should not be repeated closely enough to draw attention to the cutout alignment.

A typical random shingle spacing pattern is shown in figure 9-36. Note that the first course along the eave starts with a full 12-inch tab. The second course starts with a half or 6-inch tab, the third course with a 9-inch tab, and the fourth course with a 3-inch tab. The pattern is repeated in the sixth course, eleventh course, and so on until the roof ridge is reached.

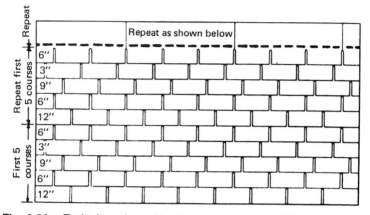

Fig. 9-36. Typical random shingle spacing pattern. *(Courtesy Asphalt Roofing Manufacturers Association)*

Asphalt strip shingles without cutouts should be applied so that each course starts with a random length shingle. The courses should be offset at least 3 inches to eliminate the possibility of shingle joints coinciding with the nails in the underlying course.

Ribbon Courses

A distinctive look can be achieved on asphalt shingle roofs by using ribbon courses at the eaves and every fifth course up the roof. As shown in figure 9-37, each ribbon course results in a triple thickness of shingles that emphasizes the horizontal roof line. The application method is as follows:

1. Cut a 4-inch wide strip from the top of a 12 × 36–inch asphalt strip shingle.
2. Nail the 4 × 36–inch strip to the eave either flush with the flashing, or with the same overhang provided the first course of shingles. Complete this 4-inch wide course the length of the eaves.
3. Nail the 8 × 36–inch strip directly over the 4 × 36–inch strip with the bottom edges of both strips flush (fig. 9-38). Complete this 8-inch wide course the length of the roof.
4. Lay the first course of 12 × 36–inch shingles along the eaves with its bottom edge flush with the edges of the underlying

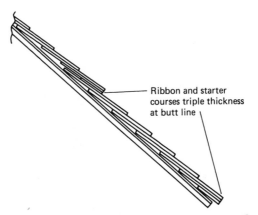

Fig. 9-37. Cross section of roof with ribbon courses. *(Courtesy Asphalt Roofing Manufacturers Association)*

strips. Offset the cutouts of the first course so that they are not aligned with those of the underlying 8 × 36–inch strip.

5. Lay ribbon courses at every fifth shingle course up the roof using the method described in steps 1–4 above.

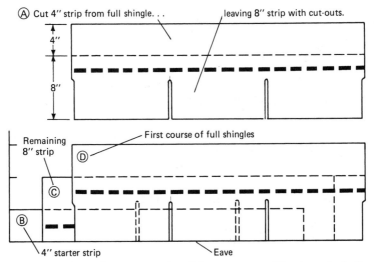

Fig. 9-38. Offsetting cutouts for ribbon courses. *(Courtesy Asphalt Roofing Manufacturers Association)*

Roof Repair and Maintenance

The roof surface is exposed to a variety of potentially harmful weather conditions. Over a period of time the heat of the sun causes asphalt shingles to blister, curl or cup, fade and lose their color, or dry out and become brittle. Temperature changes cause dried out, brittle shingles to split or crack. Strong winds blow away portions of split or cracked shingles. They also lift shingles and loosen the roofing nails, or create bare spots by by blowing away the mineral granules. Rain can be blown under loose, curled, or cupped shingles and leak through the sheathing of the roof deck. Rain can also rust galvanized steel flashing, and the sun can dry out and crack the flashing cement.

The service life of an asphalt shingle roof can be extended by proper repair and maintenance. This can be accomplished by periodically checking the roof for signs of damage or deterioration and making the necessary repairs. The check list should include the following:

1. Roof deck and framing
2. Shingle courses
3. Hip and ridge shingles
4. Flashing
5. Roof openings
6. Gutters and downspouts

Roof Deck and Framing

Go into the attic or attic crawl space and check for leaks. The best time to do this is when it is raining. A minor leak can sometimes be

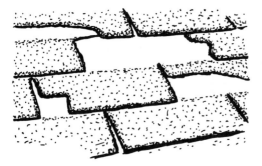

Fig. 9-39. Shingle damage.

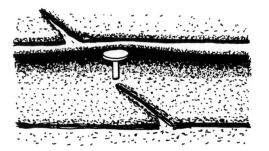

Fig. 9-40. Popped nails.

stopped by patching the underside of the sheathing, but most require the removal of some or all of the roofing materials before repairs can be made.

Check the roof framing for adequate support and the condition of both the rafters and roof deck sheathing. Replace warped, split, or broken rafters, and warped or damaged sheathing. If the framing is inadequate for the weight of the roof, it will have to be reinforced. See chapter 3 for detailed instructions concerning the inspection and repair of roof decks and roof framing.

Shingle Courses

Check the roof surface for damaged or missing shingles, missing portions of shingles, cracked, split, or blistered shingles, cupped or curled shingles, lifted shingles, and loose or missing nails (figs. 9-39 to 9-41).

Fig. 9-41. Bare spots.

If these conditions are extensive, it is an indication that the roof is in an advanced stage of deterioration and must be replaced or reroofed. If there are relatively few occurrences of these conditions, only minor repairs are required. Before making any repairs, however, check the condition of the underlayment. If it is dried out and brittle, both the shingles and underlayment should be replaced regardless of the condition of the former.

REPLACING DAMAGED SHINGLES

Lift the overlapping shingle in the next course to provide access to the nails of the damaged shingle. Carefully remove the nails from the damaged shingle by prying them up with a chisel or pry bar. Lift out the damaged shingle and fill the old nail holes with roofing cement or a sealer. Insert the replacement shingle carefully so that it

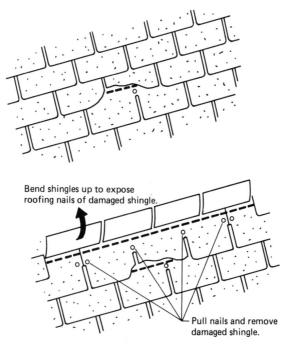

Fig. 9-42. Removing damaged shingle.

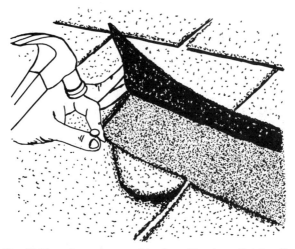

Fig. 9-43.   Nailing down loose shingles. *(Courtesy Celotex Corp.)*

does not damage the underlayment. Lift the overlapping shingle and nail the new one to the roof with the required number of nails and nailing pattern (fig. 9-42). *Do not drive the nails into the old nail holes.* Cover the nail heads with roofing cement and run a strip of cement around the underside of the edges of both the new and overlapping shingles and press them down firmly.

REPAIRING CRACKED OR LOOSE SHINGLES

If a portion of a shingle has cracked off but has not blown away, nail down both sections (fig. 9-43). Cover the heads of the roofing nails with roofing cement and run a strip of cement over the crack between the two shingle sections. If the shingle is only loose, insert dabs of roofing cement under the shingle, press it down firmly, and nail it to the roof.

REPLACING MISSING SHINGLES

Follow the procedure described for replacing damaged shingles. Do not use the old nail holes when nailing the replacement shingle to the roof.

CUPPED OR CURLED SHINGLES

Apply roofing cement under the cupped or curled section, press it down firmly, and nail it to the roof. Cover the nail heads with dabs of roofing cement.

Hip and Ridge Shingles

Hip and ridge shingles are exposed to higher wind velocities than those on other sections of the roof. As a result, they are more prone to lifting or tearing (fig. 9-44). A hip or ridge shingle with a minor tear is not difficult to repair. Embed the shingle in roofing cement and press it down firmly against the hip or ridge. Nail on each side of the tear, cover the nail heads with a dab of roofing cement, and run a strip of roofing cement down the line of the tear. Badly torn or damaged hip and ridge shingles should be replaced. Lifted shingles can be nailed down, but the nails must not be driven into the old nail holes.

Flashing

Check for rusted metal flashing, flashing with holes, torn felt flashing, and cracked or missing flashing cement. Small rust spots can be scraped or wire-brushed from the surface of metal flashing. The surface should then be painted to protect it from corrosion. Where rust has penetrated through the metal, the flashing should be replaced. Small pin holes in the metal can be covered with roofing cement after the area around the hole has been cleaned with a wire brush. Replace roofing felt used as flashing when it is badly torn or brittle. Cracked or brittle flashing

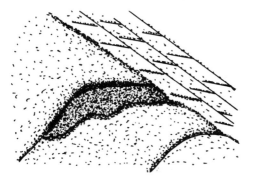

Fig. 9-44. Torn hip and ridge shingles.

cement should be removed and the joint cleaned and then recemented. Cracks around the embedded edges of chimney flashing should be freed of loose mortar, cleaned with a wire brush, and completely sealed with a suitable sealer. See chapter 8 for additional information on the inspection, repair, and replacement of flashing.

Gutter Systems

GUTTERS

Clean leaves and other debris from the gutters. Check metal gutters for rust spots and rusted out areas, and make the necessary repairs (see chapter 17). Leaks usually develop at seams and end caps of gutters. The points to check for leaks on a metal gutter system are illustrated in figure 9-45. Scrape away the old sealer, clean the seam, and apply a suitable sealer along all connecting edges and joints.

GUTTER HANGERS

Check to see if the gutter hangers are properly fastened beneath the shingles. If the gutter hanger straps are nailed to the tops of shingles, check for signs of deterioration and repair as necessary. Deterioration at this point may require replacement of some or all of the starter course shingles.

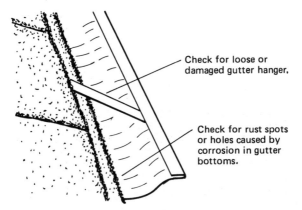

Check for loose or damaged gutter hanger.

Check for rust spots or holes caused by corrosion in gutter bottoms.

Fig. 9-45. Points to check for leaks on metal gutters.

DOWNSPOUTS

Check downspouts and downspout fasteners for looseness and repair as necessary. Replace badly rusted metal downspouts.

SPLASH BLOCKS

Check the splash block at the bottom of each downspout for accumulations of mineral granules. The heavy loss of mineral granules from the shingles is usually a sign that the roof requires reroofing. Deposits of mineral granules will also collect in the gutters.

Roof Openings

Inspect the flashing and flanges around skylights to make certain that they are secure and unbroken. The putty or synthetic sealer around the skylight glass should be unbroken and tight. These are all potential sources of water leaks. Check the condition of stack or vent pipe flashing and repair as necessary. Check the roof for possible leaks where television antennas or antenna guy wires have been anchored to the roof (fig. 9-46). If the television mast has been secured to a vent pipe, strong winds can cause it to loosen the caulking around the vent pipe flashing. Check for these problems and make the necessary repairs. Detailed instructions describing the inspection and repair of flashings and flanges for roof openings are provided in chapters 6 and 8.

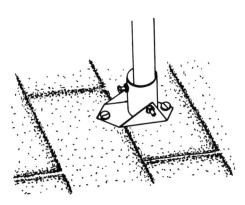

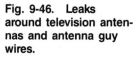

Fig. 9-46. Leaks around television antennas and antenna guy wires.

Reroofing with Asphalt Shingles

Asphalt shingles can be used in reroofing to cover any type of roofing material except tile, slate, mineral fiber (asbestos-cement) shingles, and metal roofing. These roofing materials are too hard and brittle to serve as a suitable nailing base. Asphalt shingles are most commonly applied over an existing asphalt shingle roof or a wood shingle (or shake) roof. Under certain conditions they can also be applied over roll roofing and built-up roofs. *Never reroof over more than one existing roof.* If the existing roof has already been reroofed once, all the roofing materials will have to be removed before the new roofing can be applied.

Reroofing Preparations

ROOFING OVER ASPHALT SHINGLES

Prepare the existing roof so that it provides a smooth, flat nailing surface for the new shingles. Remove loose nails and renail shingles where necessary. Buckled, cupped, or curled shingles as well as surface rolls should be split with a utility knife and nailed flat. Warped sheathing boards should be nailed flat or replaced. See chapters 2 and 3 for additional details on preparing the roof for reroofing.

ROOFING OVER WOOD SHINGLES

Remove all loose and protruding nails, and renail the shingles. *Do not drive the new nails into the old nail holes.* Split warped or curled wood shingles and nail the sections flat. Replace missing shingles. If the existing shingles are badly curled or turned at their butts, nail beveled wood feathering strips along the butts (fig. 9-47). If the existing shingles along the rakes and eaves are badly deteriorated, cut them back far enough to install 1 × 4–inch or 1 × 6–inch wood strips with their edges overhanging the eaves and rakes the same distance as the old shingles. Using these strips along the eaves and rakes improves wind resistance, provides better alignment of the asphalt shingles, and serves as a suitable nailing base for the new roofing.

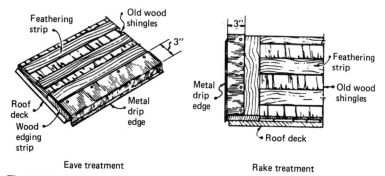

Fig. 9-47. Preparation of existing wood shingle roof for reroofing with asphalt strip shingles. *(Courtesy Asphalt Roofing Manufacturers Association)*

Flashing

Old flashing should be saved whenever possible. If it is too deteriorated to be saved, it can still be used as a pattern for new flashing. Use metal drip edges along the rakes and eaves. The length of the roof along the eaves should also be covered by a strip of roll roofing. Additional flashing details are discussed in chapter 8.

Reroofing

After the roof surface has been properly prepared for reroofing, the procedure used to apply asphalt strip shingles is the same over either an existing asphalt or wood shingle roof.

STARTER COURSE

Use full-length shingles with their tabs removed as the starter course. Each starter course shingle should be cut equal in width to the exposure of the old shingles (usually 5 inches). Trim 6 inches off the length of the first shingle and position it at the left-hand rake with its mineral surface up and its adhesive strip next to the edge of the roof (fig. 9-48). Nail it to the roof so that it overlaps the drip edge ¼ inch to ⅜ inch

Fig. 9-48. Applying starter course. *(Courtesy Owens-Corning Fiberglas Corp.)*

at the rake and eave. Continue the starter course with full-length (36-inch) shingles across the bottom edge of the roof to the opposite rake.

FIRST COURSE

Cut 2 inches from the top edge of a full-width new shingle, overlap the starter course, and nail it to the roof (fig. 9-49) with its cut edge aligned with the butt edge of the old shingle. Complete the first course by continuing across the roof to the opposite rake.

SECOND COURSE

Use full-width shingles with their top edges aligned with the butt edges of the existing shingles in the next course (fig. 9-50). Cut the first shingle

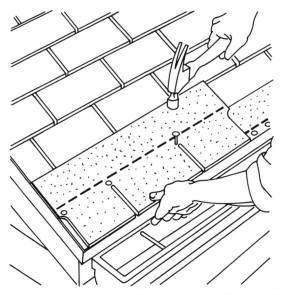

Fig. 9-49. Applying first course. *(Courtesy Owens-Corning Fiberglas Corp.)*

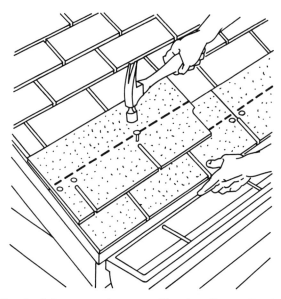

Fig. 9-50. Applying second course. *(Courtesy Owens-Corning Fiberglas Corp.)*

Fig. 9-51. Applying third course. *(Courtesy Owens-Corning Fiberglas Corp.)*

Fig. 9-52. Shingling woven valley. *(Courtesy Owens-Corning Fiberglas Corp.)*

Fig. 9-53. Shingling closed-out valley. *(Courtesy Owens-Corning Fiberglas Corp.)*

Fig. 9-54. Applying hip and ridge shingles. *(Courtesy Owens-Corning Fiberglas Corp.)*

in the second course to meet the required offset of the shingling pattern. The full size of the second course shingles will reduce the exposure of the first course. This reduction of exposure occurs only in the first course, which is partially concealed by the gutter system.

THIRD AND SUCCEEDING COURSES

Full-width shingles with their top edges aligned with the butts of the existing shingles are used in the third course and all succeeding courses (fig. 9-51). Shingle exposure for the new roofing will coincide with that of the existing roofing.

VALLEY DETAILS

Woven or closed-cut valleys are recommended for reroofing with asphalt strip shingles (figs. 9-52 and 9-53). Application details for both types are described in this chapter.

HIPS AND RIDGES

Hips and ridges are completed as in new construction (fig. 9-54). The hip and ridge shingles will have the same exposure as the shingle courses.

CHAPTER 10

Applying Specialty Shingles

- Tools and Equipment
- Roof Deck Preparation
- Hex-Type Shingles
- Individual Hex Shingles
- Giant Individual Shingles
- Interlocking Shingles
- Roof Repair and Maintenance

Specialty shingles are designed for the specific purpose of forming a roof that stays flat and secure regardless of weather conditions. When applied to the roof, specialty shingles produce a distinctive woven or hexagonal pattern in each shingle course (figs. 10-1 and 10-2). Most

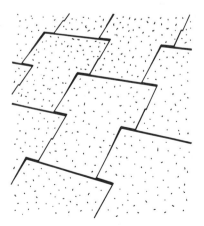

Fig. 10-1.   Woven shingle pattern.

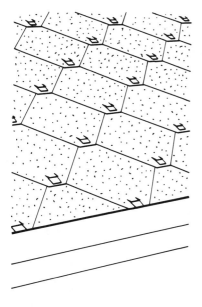

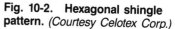

Fig. 10-2. Hexagonal shingle pattern. *(Courtesy Celotex Corp.)*

are individual shingles, but strip types are also available. Some are manufactured with locking devices to interlock adjacent shingles and thereby provide greater resistance to strong winds.

Specialty shingles are similar in composition to standard asphalt strip shingles. They are made from roofing felt saturated with asphalt and coated with a layer of mineral granules. They are available in a variety of colors, shapes, and sizes, and carry the same UL class C fire rating as standard asphalt strip shingles.

In many areas of the country, specialty shingles are being replaced by the self-sealing, three-tab asphalt strip shingles which provide the same protection against strong winds. As a result, their availability is limited to certain geographical areas.

Tools and Equipment

Specialty shingles are applied with the same types of tools and equipment as are used for applying standard asphalt strip shingles. Descriptions of these tools and equipment are found in chapter 2.

Roof Deck Preparation

Specialty shingles can be applied to the same types of roof decks as asphalt strip shingles. Roof deck preparation instructions are covered in chapter 3. Specific instructions for some types of reroofing applications are included in this chapter. Flashing details are described in chapter 8.

Hex-Type Shingles

Hex-type shingles can be used in new construction or when reroofing over an existing roof. In either case, roofing with hex-type shingles produces a distinctive hexagonal pattern in each shingle course.

Hex Strip Shingles

Hex strip shingles are used primarily in new construction for roof slopes of 4-in-12 or greater. They are not recommended for slopes under 4-in-12. These low-pitched roofs require special roof deck preparation and are covered more effectively with self-sealing square-tab asphalt strip shingles.

Before hex strip shingles can be applied, the roof deck must be properly prepared by applying an underlayment, roofing felt (flashing) along the roof eaves, metal drip edges at the eaves and rakes, and other flashings where required. Either two-tab or three-tab hex strip shingles may be used to finish the roof.

Each two-tab or three-tab hex strip shingle is nailed to the roof with 11- or 12-gauge hot-dipped galvanized steel roofing nails with heads at least 3/8 inch in diameter (fig. 10-3). The nails must be long enough to penetrate at least 3/4 inch into 1-inch thick wood board sheathing or through plywood sheathing. They are driven through the shingles in a line 5¼ inches above the exposed butt edge. On two-tab shingles, the nails are located 1 inch from each end of the strip and 3/4 inch from the angled sides of the cutout. Three-tab shingles are secured with nails driven 1 inch from each end of the strip and one nail centered over each cutout.

Each course is applied with the bottom edge of each shingle tab aligned with the top edge of the cutout of the shingle in the underlying course. Free-type tabs should be cemented to the underlying shingle with a spot of roofing cement (fig. 10-4).

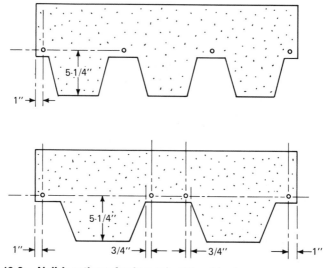

Fig. 10-3. Nail locations for hex-tab strip shingles.

TWO-TAB SHINGLE APPLICATION

The construction details of a roof covered with two-tab hex strip shingles are shown in figure 10-5. Begin by nailing a starter course of strip shingles along the roof eave with their mineral surface side up and the tabs pointing toward the ridge line (fig. 10-6). Cut one half the tab off the first shingle in the starter course and apply it at the right-hand corner of the roof. Nail it to the sheathing so that it

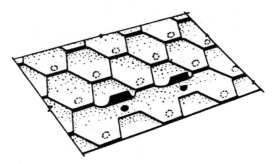

Fig. 10-4. Cement spot locations under hex shingle tabs. *(Courtesy Asphalt Roofing Manufacturers Association)*

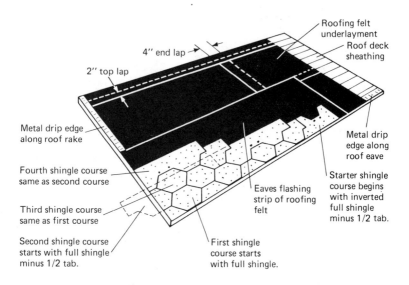

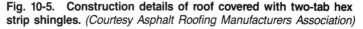

Fig. 10-5. Construction details of roof covered with two-tab hex strip shingles. *(Courtesy Asphalt Roofing Manufacturers Association)*

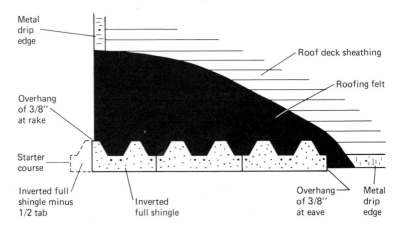

Fig. 10-6. Starter course of two-tab hex strip shingles.

extends about ³/₈ inch beyond the rake and roof eave. Nail the rest of the shingles in the starter course with the same ³/₈ inch overhang at the eaves as the first one. Make certain that each shingle is in perfect alignment before nailing it to prevent buckling. *Drive the nails straight to avoid cutting the shingle with the edge of the nail head, and drive the nail head flush with the surface of the shingle.*

Start the first course of shingles at the left-hand rake of the roof with a full shingle. The vertical edge of the shingle should extend ³/₈ inch beyond the rake and the bottom edge of its tabs should overhang the roof eave by ³/₈ inch. Finish the first course with a ³/₈-inch overhang at the eave for the bottom edge of the tabs of each shingle.

Start the second course and all remaining even-numbered courses with full shingles minus a half tab. Each even-numbered course will end with a full shingle extending ³/₈ inch beyond the edge of the rake. Start the third course and all remaining odd-numbered courses with a full shingle. End each odd-numbered course with a full shingle minus a half tab and its edge extending ³/₈ beyond the rake.

THREE-TAB SHINGLE APPLICATION

The roofing details of a roof covered with three-tab hex strip shingles are shown in figure 10-7. Begin the starter course with a full shingle from which 6 inches have been cut. Start shingling from the left-hand corner of the roof, with its mineral surface up and its tabs pointed toward the ridge line. Nail it to the sheathing so that it extends about ³/₈ inch beyond the rake and roof eave (fig. 10-8). Complete the starter courses by nailing full-size shingles to the sheathing with the same ³/₈-inch overhang as the first shingle. Start the first course of shingles and all remaining odd-numbered courses with a full-size shingle at the left-hand rake. Complete the first shingle course across the roof by cutting a half tab from the last shingle and nailing it to the sheathing so that it overhangs the rake by ³/₈ inch. Start the second and all remaining even-numbered shingle courses with a full-size shingle from which a half tab has been removed. Nail the shingle to the roof deck with the required ³/₈-inch overhang at the left-hand rake. Complete the second course across the roof to the opposite rake. The last shingle in the second and all even-numbered courses should be a full-size shingle extending ³/₈ inch beyond the rake.

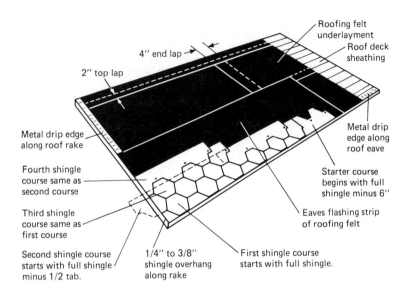

Fig. 10-7. Roof covered with three-tab hex strip shingles. *(Courtesy Asphalt Roofing Manufacturers Association)*

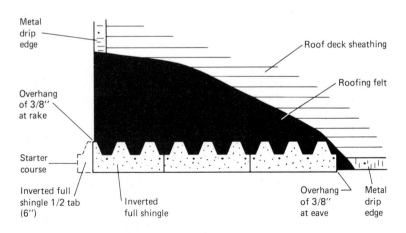

Fig. 10-8. Starter course of three-tab hex strip shingles.

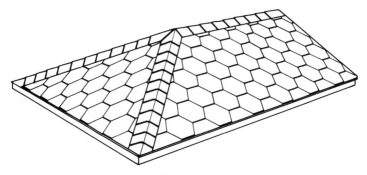

Fig. 10-9. Hip and ridge details for hex-tab strip shingle roof. *(Courtesy Asphalt Roofing Manufacturers Association)*

HIP AND RIDGE SHINGLE APPLICATION

Hips and ridges can be formed by applying special preformed hip and ridge shingles supplied by the roofing manufacturer or by cutting them to size from 12 × 36–inch square-butt asphalt strip shingles (fig. 10-9). If the hip and ridge shingles are formed on site, they should be cut to approximately 9 × 12 inches and bent lengthwise to provide equal exposure on both sides of the hip or ridge. Apply the shingles to the hips or ridge with a 5-inch exposure. Secure them with one nail on

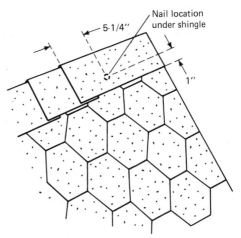

Fig. 10-10. Hip and ridge shingle nailing instructions. *(Courtesy Asphalt Roofing Manufacturers Association)*

each side of the shingle located 1 inch up from the long edge parallel to the hip or ridge and 5½ inches back from the bottom edge of the exposed end of the shingle (fig. 10-10). The next shingle applied to the hip or ridge will overlap the nails of the underlying shingle. Hips are shingled by beginning at the bottom and working toward the roof ridge. Ridges are shingled by beginning at either end of the roof and working toward the opposite end.

Individual Hex Shingles

Individual hex shingles are used primarily for reroofing roofs with slopes of 4-in-12 or more. There are two types of individual hex shingles: the staple-applied shingle and the lock-down shingle. The staple-applied shingle is secured to the roof with 2 nails and one staple fastener. Lock-down shingles have tabs and slots or other devices so that the shingles can be locked together across the roof. Interlocking the shingles in this manner results in better than average wind and weather resistance.

Snap horizontal and vertical chalk lines across the roof to insure proper shingle and course alignment. The top center of each shingle should be aligned with a vertical chalk line. The horizontal chalk lines are used for course alignment.

STAPLE-APPLIED HEX SHINGLES

A staple-applied hex shingle is secured to the roof with one nail driven through each tab and a staple driven through the lower corner (fig. 10-11). The nails are located 1 inch in from the side edge and 1 inch

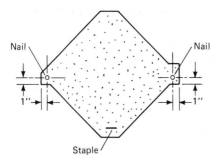

Fig. 10-11. Staple-applied hex shingles. *(Courtesy Asphalt Roofing Manufacturers Association)*

Table 10-1. **Minimum Nail Lengths for Different Roofing Applications** (*Courtesy Asphalt Roofing Manufacturers Association*)

Type of Application	1″ Wood Board	³⁄₈″ Plywood
Over new wood decks	1¼″	Nails should be long
Over old asphalt shingles	1½″	enough to penetrate
Over old wood shingles	1¾″	through plywood deck

up from the bottom edge of each tab. Use either 11- or 12-gauge hot galvanized nails with heads at least ³⁄₈ inch in diameter. Minimum nail lengths for different roofing applications are listed in table 10-1. The nails must be long enough to penetrate at least ³⁄₄ inch into solid deck sheathing or through plywood sheathing. The lower corner of the shingle is stapled to the tabs of adjacent shingles in the course immediately below.

The construction details of a wood shingle roof reroofed with individual staple-applied hex shingles are illustrated in figure 10-12.

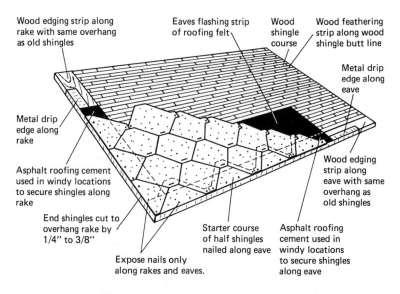

Fig. 10-12. Construction details of wood shingle roof reroofed with individual staple-applied hex shingles. (*Courtesy Asphalt Roofing Manufacturers Association*)

The starter course consists of half shingles laid along the roof eaves. The half shingles for the starter course are cut so that the tabs are retained. The recommended method of applying staple-applied hex shingles is to start at the center of the roof and work in both directions toward each rake. Begin by nailing the first half shingle of the starter course at the exact center of the roof with about a ³/₈ inch overhang at the eave. Drive one nail through each tab and drive two more evenly spaced nails across the bottom of the shingle (fig. 10-13). Lay the rest of the shingles of the starter course in both directions toward the rakes. Make certain that the bottoms of the shingles are aligned and that there is a ³/₈-inch overhang at the eave. Secure each shingle with four nails, one in each tab and two more along the bottom. Cut the end shingles of the starter course to overhang each rake by ³/₈ inch. In areas of the country where strong winds are common, embed the shingles along the eaves and rakes in a 6-inch wide strip of asphalt roofing cement before nailing them to the roof.

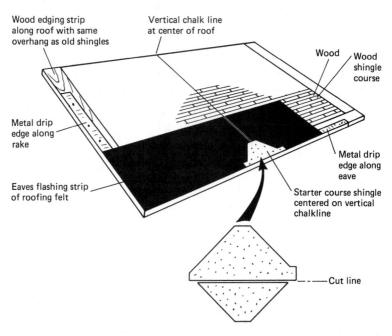

Fig. 10-13. Installing staple-applied shingles with the center starting method.

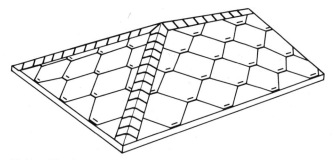

Fig. 10-14. Hip and ridge construction details. *(Courtesy Asphalt Roofing Manufacturers Association)*

Begin the first course by nailing a full-size shingle to the center of the roof with its bottom edge aligned with the edge of the eaves flashing strip of roofing felt or extended beyond the eave for a ³⁄₈-inch overhang and centered over the shoulder tabs of the starter course shingles. Apply full-size shingles from the center of the roof toward each rake. Secure each shingle with its tabs butted and aligned with the tabs of adjacent shingles and its top aligned with a vertical chalk line. Nail the shingles to the roof with one nail in each shoulder tab and two exposed nails in the lower corner, as in figure 10-12. Trim the shingles at the rake for a ³⁄₈-inch overhang. All shingles along the rakes are also nailed with exposed nails.

The shingles in the second and succeeding courses are secured with nails driven through the shoulder tabs, but the lower corner of each shingle is stapled to the adjacent tabs of the underlying shingles instead of being nailed to the roof as is the case with the shingles in the first course.

Hip and ridge construction details for a roof shingled with individual staple-applied hex shingles are shown in figure 10-14. Factory-made hip and ridge shingles are available from the roofing manufacturer or they may be cut and formed on site. A hip or ridge shingle can be made by cutting a 9 × 12–inch rectangular shingle from 90-pound mineral-surfaced roll roofing or from one of the hex shingles used on the roof. Bend the shingle lengthwise for an equal amount of exposure on each side of the hip or ridge. Beginning at either end of the ridge or at the bottom of the hip, position the shingle over the top edge and secure it on each side with a nail located 5½ inches up from the exposed end and 1 inch up from each edge (fig. 10-15). All

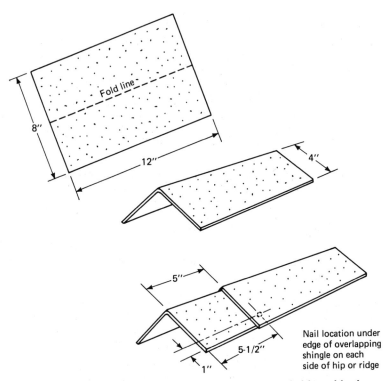

Fig. 10-15. Exposure and nail locations for hip and ridge shingles.

succeeding courses on the hip or ridge should be applied with a 5-inch exposure. Trim the ends of the last shingles at either end of the hip to fit the angle of the eave or juncture with the ridge shingle.

LOCK-TYPE HEX SHINGLES

Lock-type hex shingles are designed with a built-in locking device that ties or locks adjacent shingles to one another. Several different types of lock-type hex shingles are available. The shingle shown in figure 10-16 is secured to the roof with one nail in each shoulder. The lower corner of each shingle is locked by inserting the locking tabs under the sides of shingles in the underlying course (fig. 10-17). The lock-type shingles shown in figure 10-18 are nailed to the roof with three nails instead of two. Two nails are driven through the shingle tab and one through the center of the left shoulder. When the next course of

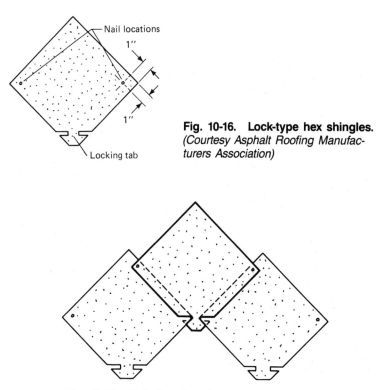

Fig. 10-16. **Lock-type hex shingles.** *(Courtesy Asphalt Roofing Manufacturers Association)*

Fig. 10-17. **Shingle edges inserted in locking tab.**

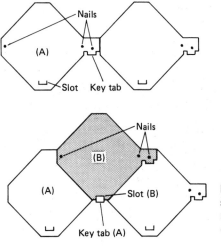

Fig. 10-18. **Lock-type hex shingles secured with three nails.** *(Courtesy Celotex Corp.)*

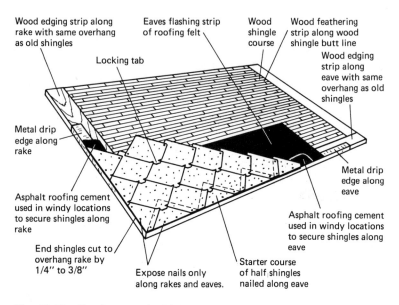

Wood edging strip along rake with same overhang as old shingles

Eaves flashing strip of roofing felt

Wood shingle course

Wood feathering strip along wood shingle butt line

Wood edging strip along eave with same overhang as old shingles

Locking tab

Metal drip edge along rake

Metal drip edge along eave

Asphalt roofing cement used in windy locations to secure shingles along rake

Asphalt roofing cement used in windy locations to secure shingles along eave

End shingles cut to overhang rake by 1/4" to 3/8"

Expose nails only along rakes and eaves.

Starter course of half shingles nailed along eave

Fig. 10-19. Roof covered with lock-type shingles. *(Courtesy Asphalt Roofing Manufacturers Association)*

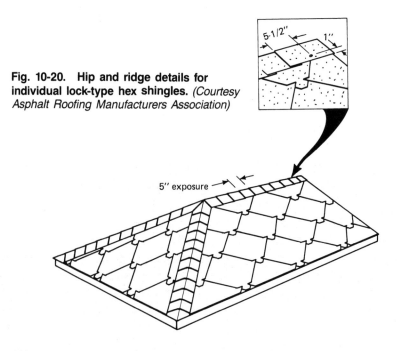

Fig. 10-20. Hip and ridge details for individual lock-type hex shingles. *(Courtesy Asphalt Roofing Manufacturers Association)*

5-1/2"   1"

5" exposure

shingles is applied, the shingle tabs in the lower course are inserted through the slots in the lower corner of the shingles in the upper course. Construction details of a roof covered with lock-type hex shingles are shown in figure 10-19. As with staple-applied shingles, the shingles along both the eaves and rakes should be embedded in a 6-inch wide strip of asphalt roofing cement to provide additional protection against strong winds (fig. 10-20). The methods used to shingle the hips and ridges are the same as those used when shingling with individual staple-applied hex shingles (see fig. 10-12).

Giant Individual Shingles

Large rectangular-shaped 12 × 18–inch individual shingles are used in either new construction or reroofing applications to produce a single coverage roof. Either the Dutch lap or American-application method may be used to apply giant individual shingles.

Dutch Lap Method

The *Dutch lap method*, or *Scotch method* as it is also sometimes called, is a shingling method used to apply giant individual shingles over any kind of old roofing that provides a smooth surface. Although used primarily in reroofing, the Dutch lap method can also be used to apply shingles to a new roof deck if an adequate underlayment has first been applied.

In new construction, roof deck preparation and the use of underlayment and flashing follow the same methods used with standard asphalt strip shingles (see chapters 3 and 8).

The Dutch lap application method produces a single coverage roof, and the shingles are laid from either rake (from left to right or from right to left), but *never* from the center of the roof deck toward both rakes. This application method is not recommended for roof slopes of less than 4-in-12.

Both nails and metal fasteners are used in the Dutch lap method. Each shingle is secured with two nails and one fastener. The nails used are 11- or 12-gauge hot galvanized roofing nails with large heads of at least 3/8-inch diameter. In reroofing, nails with a minimum length of 1 3/4 inches should be used when applying shingles over an existing wood shingle roof. Reroofing over an existing asphalt shingle roof requires nails with a minimum length of 1½ inches. Slightly shorter nails (1¼-inch minimum length) may be used when applying shingles by the Dutch lap method to a new wood deck.

Some shingle manufacturers provide fasteners in the form of special metal clips. Others recommend using noncorrodible wire staples and a stapling device. If fasteners are unobtainable or undesirable, a spot of asphalt roofing cement applied under the shingle will serve as well.

When applying the shingles from the left rake to the right one, the nails should be located in the upper left-hand and lower right-hand corners of the shingle. The fastener (or spot of asphalt roofing cement) is located in the exposed lower left-hand corner of the shingle. The fastener is used to secure the shingle to the overlapped portion of the adjacent shingle in the same course. *Never* use nails to secure the exposed corner of a shingle to the roof deck *except* along the rake and eave. The locations of the nails and fasteners are reversed when the shingles are applied from the right roof rake to the left one.

The Dutch lap method of applying giant individual shingles to

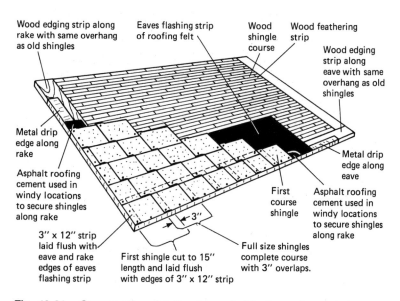

Fig. 10-21. Construction details of wood shingle roof reroofed with giant individual shingles (Dutch lap method). *(Courtesy Asphalt Roofing Manufacturers Association)*

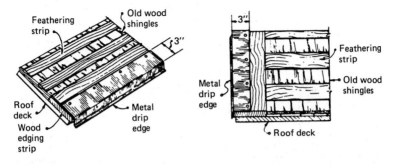

Fig. 10-22. Feathering strips, edging strips, and metal drip edges. *(Courtesy Asphalt Roofing Manufacturers Association)*

an existing wood shingle roof is illustrated in figure 10-21. *A starter course is not required when using the Dutch lap application method.*

Begin by cutting back the old wood shingles approximately 6 inches from the eaves and rakes, and nail a wood edging strip to the wood deck. Nail feathering strips along the bottom edge of each course of wood shingles and metal drip edges to the rakes and eaves (fig. 10-22). Install an eaves flashing strip of roofing felt extending from the eave line up the roof to a point 12 inches inside the interior wall line of the structure (fig. 10-23).

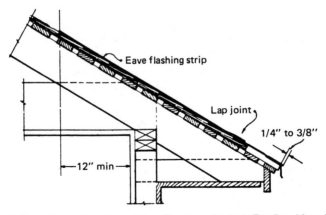

Fig. 10-23. Eaves flashing strip. *(Courtesy Asphalt Roofing Manufacturers Association)*

Cut a shingle 3-inches wide by 12-inches long and nail it to the roof deck, allowing a ³/₈-inch overhang at the eaves and rakes. In areas where strong winds are common, embed the strip in roofing cement before nailing it. Cut a second shingle to a length of 15 inches and nail it over the first 3 × 12–inch shingle so that its edges are flush with it. Complete the course by applying full-size shingles with a 3-inch side lap. Trim the last shingle in the course so that it overlaps the opposite rake ¹/₄ to ³/₈ inches.

Start the second course with a shingle cut to a 12-inch length. Allow a 2-inch top lap over the first course and a ³/₈-inch overhang at the rakes. Complete the second course with full shingles applied with a 3-inch side lap.

The third course begins with a shingle cut to a 9-inch length, and the fourth with a shingle cut to a 6-inch length. The fifth course begins with a full shingle, and the pattern repeats itself. Each course is completed with full-size shingles with a 3-inch side lap and a 2-inch top lap over the shingles in the previous course.

Hip and ridge shingles are made by bending each 12 × 18–inch giant individual shingle lengthwise down the middle. This allows a 6-inch overlap of shingle on either side of the hip or ridge. Beginning at the bottom of the hip or at either end of the ridge, apply the shingles with a 5-inch exposure. The nails should be placed 1 inch up from the edge and 5¹/₂ inches back from the exposed edge (see fig. 10-15).

American Method

When the American method is used to apply giant individual shingles, the shingles are laid with their long 18-inch dimension vertical to the roof eaves and parallel to the rakes. It is primarily this feature that distinguishes the American method from the Dutch lap method.

The American method may be used to apply shingles in either new construction or reroofing applications. When reroofing, no underlayment is required if the existing roof surface is smooth, flat, and in good condition. If the American method is used to apply shingles over an existing wood shingle roof, the same preparations used in the Dutch lap method must be used here. In other words, the wood shingles should be cut back about 6 inches along the roof rakes and eaves and a wood edging strip nailed to the roof deck in their place. Feathering strips should also be nailed to the bottom edge of each course of shingles to provide a smooth and even deck surface, and metal drip edges must be nailed along the roof rakes and eaves.

As with the Dutch lap method, roof deck preparation and the application of the underlayment and flashing in new construction follow the same methods used with standard asphalt strip shingles.

The shingles in the starter strip are laid horizontally along the roof eave with their edges touching. Each shingle is nailed to the roof deck with three evenly spaced roofing nails located 2 inches up from the eave. The bottom edge of the starter strip may be either flush with the edge of the eave flashing or extend beyond it for a suitable overhang of about 1/4 inch.

Start the first course with a full shingle placed so that its long dimension is vertical and with its edges flush with the rake and eave edges of the starter course. The remaining shingles in the first course should be separated by a 3/4 inch gap and nailed to the roof deck with two roofing nails. Each nail is located 6 inches from the bottom edge of the shingle and 1½ inches in from the sides (fig. 10-30). The 3/4 shingle gap and nailing pattern is used in each shingle course.

The way in which the second and succeeding courses are laid will depend on whether the shingle joints break on thirds or on halves. Both patterns are illustrated in figures 10-24 and 10-25. In either case, all shingles are laid with the long dimension vertical.

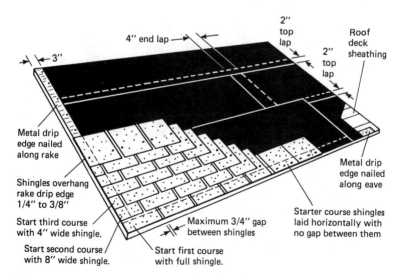

Fig. 10-24. Giant individual shingles applied by the American method with joints breaking on thirds. *(Courtesy Asphalt Roofing Manufacturers Association)*

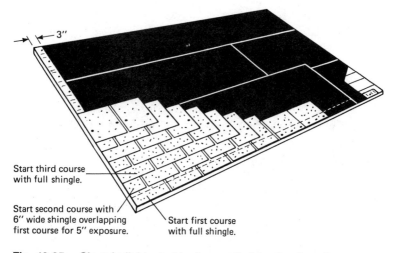

Fig. 10-25. Giant individual shingles applied by the American method with joints breaking on halves. *(Courtesy Asphalt Roofing Manufacturers Association)*

If the shingle courses are to break on thirds, the second course should begin with an 8-inch wide shingle. Provide the same overhang at the rake as was provided for the first course, and allow a 5-inch exposure for the latter (see fig. 10-25). Apply full shingles for the remainder of the second course, and make certain that the 3/4-inch gap between shingles is maintained.

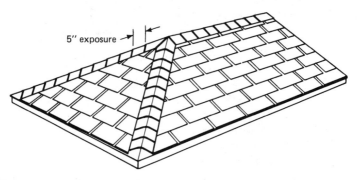

Fig. 10-26. American method hip and ridge details. *(Courtesy Asphalt Roofing Manufacturers Association)*

Fig. 10-27. Hip and ridge shingle nail locations.

Begin the third course with a 4-inch wide shingle, and complete the course with full shingles. The entire pattern is repeated in the fourth course with a full shingle.

If the shingle courses are to break on halves, begin the second course and every *even*-numbered course (fourth, sixth, eighth, and so on) with a 6-inch wide shingle, and the odd-numbered courses with a full shingle. The remaining shingles in every course are full shingles except where they have to be cut at the opposite rake line.

The roof ridge and the hips are covered with shingles bent lengthwise down their center. This allows 6 inches of shingle on either side of the ridge or hip.

Begin shingling at either end of the roof ridge or at the bottom of the hip, allowing a 5-inch exposure for each shingle (fig. 10-26). Secure each shingle to the roof with two nails. Each nail should be located 5½ inches from the exposed end and 1 inch up from the edge (fig. 10-27).

Interlocking Shingles

Interlocking or lock-down shingles provide the same resistance to high wind damage as the hex lock-down shingles and are applied in a similar manner. They can be used in new construction or for reroofing over an existing roof. The principal difference between an interlocking shingle and a hex shingle is that the former does not have a hex shape. The interlocking shingle produces a woven pattern across the roof instead of a hexagonal one (see fig. 10-1).

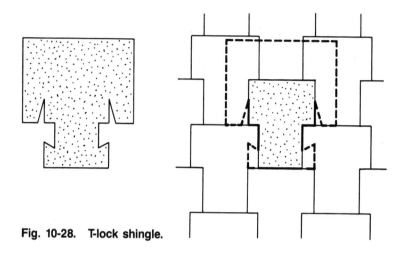

Fig. 10-28. T-lock shingle.

Some interlocking shingles are shaped like a *T* and are called T-lock shingles (fig. 10-28). Many other shapes are also available, however, and they differ primarily on the design and location of the locking device (fig. 10-29). Most are individual shingles, but there is also a strip-type shingle available.

Interlocking shingles are usually applied to roofs with slopes of 4-in-12 or more. They can also be applied to roofs with slopes less than 4-in-12 if the roof deck is properly prepared in accordance with the roofing manufacturer's instructions.

To apply the shingles use 11- or 12-gauge hot galvanized roofing nails with heads at least ⅜ inch in diameter. Minimum nail lengths for different roofing applications are listed in table 10-1. Nail locations are specified by the roofing manufacturer for each type of shingle, and they will vary depending on shingle design.

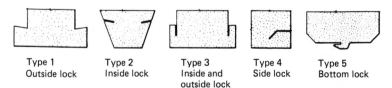

| Type 1 | Type 2 | Type 3 | Type 4 | Type 5 |
| Outside lock | Inside lock | Inside and outside lock | Side lock | Bottom lock |

Fig. 10-29. Typical locking devices for interlocking type shingles.
(Courtesy Asphalt Roofing Manufacturers Association)

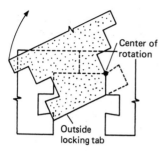

Type 1. Outside lock

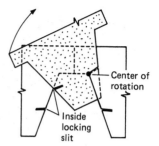

Type 2. Inside lock

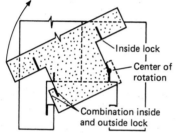

Type 3. Inside and outside lock

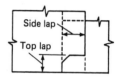

Type 4. Side lock

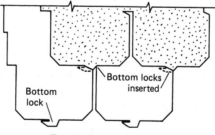

Type 5. Bottom lock

Fig. 10-30. Typical locking methods for interlocking type shingles. *(Courtesy Asphalt Roofing Manufacturers Association)*

Closely follow the roofing manufacturer's application instructions when applying interlocking shingles. The many different types make it difficult to generalize about their application methods. Examples of different locking devices and the methods used to interlock shingles are illustrated in figure 10-30. The locking device must be carefully and correctly engaged to insure a smooth and flat roofing surface. Although interlocking shingles are basically self-aligning, shingling above dormers and other roof openings and protections is made easier by snapping horizontal and vertical chalk lines. If the application instructions require the removal of the locking device or a portion of it for the starter and first course shingles, embed each shingle in a 6-inch wide strip of roofing cement along the roof eaves and rakes.

Hips and ridges can be finished with preformed shingles available from the manufacturer or rectangular shinges cut to a minimum 9 × 12-inch dimension from 90-pound mineral-surfaced roofing felt or from interlocking shingles if large enough. The hip or ridge shingle should be bent lengthwise to provide an equal 4½-inch exposure on both sides of the hip or ridge. Hips are shingled by starting at the bottom and working toward the point at which the hip meets the roof ridge. Ridges are shingled after the hips are completed by starting at either end of the ridge and working toward the opposite one. Each hip or ridge shingle is laid with a 5-inch exposure and secured with two nails located 5½ inches up from the bottom of the shingle and 1 inch up from each side edge (see fig. 10-27).

Roof Repair and Maintenance

A roof covered with specialty shingles requires little maintenance and will last about as long as one covered with standard asphalt strip shingles. The shingles should be periodically checked for a dried out and brittle condition. This is usually an indication that reroofing is in order. If the underlayment is also dried out and brittle, strip the entire roof and install a new underlayment and shingles.

Repairs to roofs covered with specialty shingles are generally similar to those made on roofs covered with standard asphalt strip shingles (see chapter 9). This is particularly true of the hip and ridge shingles which are easily replaced. Because of the interlocking design of the interlocking type specialty shingle, it is unlikely that any will be missing from the roof. This is not the case with hex strip shingles which share the same problems as standard asphalt strip shingles, and are repaired, removed, or replaced in much the same manner.

Most leaks will occur at or around the roof flashing and can be repaired by following the instructions covered in chapter 8.

If the shingles have been embedded in roofing cement along the roof eaves and rakes, check for looseness and use a putty knife to insert fresh roofing cement under the shingles where required. New roofing cement may also have to be inserted under loose hex strip shingle tabs if they are loose.

Nail down any loose nails on the roof deck and cover the exposed nail heads with a spot of roofing cement to prevent leaks.

Nail down any torn shingles and use roofing cement to cover the nail heads and the edges of the torn material.

CHAPTER II

Roll Roofing

- Tools and Materials
- Roof Deck Preparation
- Applying Roll Roofing
- Roll Roofing Repairs
- Reroofing

The easiest and least expensive method of covering a roof is by applying layers of smooth or mineral-surface roll roofing. This roofing material is recommended for use on flat or low-pitched roofs with a minimum slope of 1-in-12 or more. Because it is a quick and uncomplicated roofing method, roll roofing is frequently used in rural areas to cover the roofs of sheds, garages, and other types of farm buildings. It is also sometimes used to cover the roofs of houses. Among its disadvantages are its short service life (about 5 to 12 years depending on weather conditions) and its drab, unattractive appearance.

Tools and Materials

Table 11-1 lists some common types of roll roofing and their specifications. Most roll roofing is produced in 36-inch wide rolls with each roll containing 36 feet of material. Pattern edge roll roofing is available in longer lengths.

Roll roofing is made by impregnating an organic felt material with an asphalt or coal tar saturant and then coating it with a viscous, weather-resistant asphalt. It is manufactured in 36-inch wide rolls in weights per roofing square of 45 to 120 pounds.

Table 11-1. Roll Roofing Weights and Sizes

Type of Roll Roofing	Shipping Weight Per Roofing Square	Length	Width	Top Lap	Exposure
Smooth	65 lb.	36'	36"	2"	34"
Surface	55 lb.	36'	36"	2"	34"
	45 lb.	36'	36"	2"	34"
Mineral	90 lb.	36'	36"	2"	34"
Surface	90 lb.	36'	36"	3"	33"
	90 lb.	36'	36"	4"	34"
Pattern	105 lb.	42'	36"	2"	16"
Edge	105 lb.	48'	32"	2"	14"
19" Selvage	110 lb.	36'	36"	19"	17"
Double Coverage	120 lb.	36'	36"	19"	17"

The principal types of roll roofing are shown in figure 11-1. Smooth-surface roll roofing is the least expensive and the easiest to apply. It is used on sheds and other farm buildings where economy and utility outweigh considerations about appearance. The exposed nail application method is generally used when applying smooth-surface roll roofing.

Mineral surface roll roofing is made by embedding colored granules in the surface of the material while the asphalt coating is still hot. The other surface of the roll roofing sheet is identical in composition to smooth-surface roll roofing.

Pattern-edge roll roofing is produced in 36-inch wide rolls with diagonal shaped tabs along the bottom edge. The pattern edge or tabs resemble shingles except that the tabs are not staggered in each course.

Nineteen-inch selvage double coverage roll roofing is a roof covering material produced specifically for double coverage applications. Each roll is 36 inches wide and is divided horizontally into two sections. The lower 17-inch wide section is intended for exposure and is generally covered with an embedded mineral surface. The upper, or 19-inch wide selvage section, has a smooth surface and is covered by the mineral-surfaced lower half of the next overlapping course.

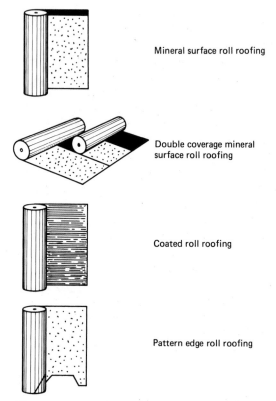

Mineral surface roll roofing

Double coverage mineral surface roll roofing

Coated roll roofing

Pattern edge roll roofing

Fig. 11-1. Types of roll roofing. *(Courtesy Asphalt Roofing Manufacturers Association)*

Roof Deck Preparation

No underlayment is required with roll roofing. Before laying the roll roofing, the roof deck should be swept with a broom. Knot holes should be covered with a piece of tin secured to the deck with mastic. The roof sheathing must be completely dry before the roll roofing can be applied.

Edging boards should be applied to the roof eaves and rakes before applying roll roofing over an existing asphalt shingle roof (see the discussion of reroofing in this chapter).

Flashing Details

Roll roofing flashing details are similar to those used with asphalt shingling. Only composition materials (saturated felt) should be used as flashing with roll roofing. For a description of the types of flashing and the methods used to install them, read chapter 8. Additional information about flashing is contained in this chapter.

Applying Roll Roofing

Roll roofing should not be applied to a roof deck when the temperatures are below 45° F unless special precautions have been taken, because there is the danger of the coating on the roll roofing sheet cracking as the material is unrolled. If the roll roofing must be applied at temperatures before 45° F, then the roll should be warmed before it is used.

One problem with roll roofing is that the material has a tendency to curl up. This problem can be dealt with by cutting the material into 12- to 18-foot lengths and piling them on a flat surface. The weight of the pile will eventually flatten out the roll roofing.

In new construction, 11- or 12-gauge hot galvanized or aluminum roofing nails with large diameter heads should be used to nail the roll roofing to the roof deck. The nail heads should measure 3/8 inch in diameter, and the shanks should be 7/8 to 1 inch long.

End laps and top laps should be cemented with a lap cement or quick setting cement recommended by the manufacturer of the roll roofing material. Always store the cement in a warm place until you are ready to use it. If you must work in temperatures below 45° F, warm the cement container by placing it in a pan of hot water. *Never heat a can of lap cement over a flame. The cement is highly flammable and could explode.*

Roll roofing is most commonly applied with each course running parallel to the roof eaves. However, if smooth surface roll roofing is used, the courses may also be applied so that each course runs parallel to the roof rakes (figs. 11-2 and 11-3). The horizontal application method offers the best protection, because each course overlaps the underlying one on the downward slope of the roof.

In addition to the direction in which the courses are laid, roofs

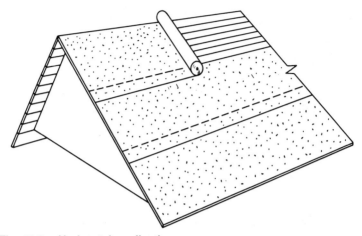

Fig. 11-2. Horizontal application.

of this type can also be distinguished by the amount of top lap, the nailing method, and the type of roll roofing material used.

The *top lap* is the amount of material in a course of roll roofing that overlaps the underlying course. The amount of top lap is often recommended by the roll roofing manufacturer (see table 11-1), but the minimum roof pitch is also a strong determining factor. In any event, the top lap should never be less than 2 inches. A 2-inch top

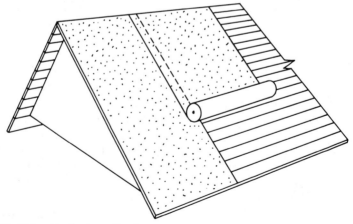

Fig. 11-3. Vertical application.

lap is recommended for roofs with pitches down to 2 inches when the exposed nail method of application is used. On roofs with pitches as low as 1 inch and where the roll roofing is applied with the concealed nail method, a 3-inch top lap should be used.

Both the exposed nail and concealed nail methods are used in applying roll roofing. The advantages of the concealed nail application method are its appearance and the fact that the nail heads are protected from weather conditions.

Smooth-Surface Roll Roofing

The easiest and quickest way to apply smooth surface roll roofing is to use the exposed nail application method. This application method can be used either when applying the courses parallel to the eaves, or parallel to the rakes. The procedure for applying the roll roofing courses parallel to the eaves is illustrated in figure 11-4, and can be outlined as follows:

1. Lay the first course of roll roofing parallel to the eave of the roof so that approximately ¼ to ⅜ inch of material overlaps both the eave and rake.
2. Nail the material along the eave, rake, and end (lapping sections) of the course with nails spaced 2 inches apart in rows 1 inch from all edges. The nails in the rows should be slightly

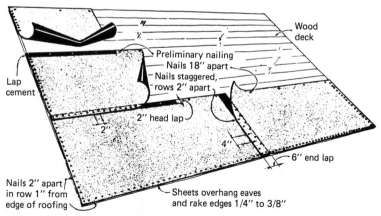

Fig. 11-4. Horizontal exposed nail application method. *(Courtesy Asphalt Roofing Manufacturers Association)*

staggered to prevent splitting the wood sheathing boards of the roof deck.

3. Nail the top edge of the first course with nails spaced 18 inches apart and about ½ inch down from the edge. These nails will be covered by the 2-inch lap of the second course.

4. Apply the second course of roll roofing so that it extends ¼ to ⅜ inch over the rakes at each end of the roof, and laps the first course by at least 2 inches. Space the nails 18 inches apart along the top edge of the sheet.

5. Lift the lower edge of the second course and apply lap cement along the top 2 inches of the underlying (first) course. Press down the bottom edge of the second course and nail it to the roof deck, staggering the nails so that they are 2 inches apart and at least ¾ inch above the bottom edge of the material.

6. Apply the succeeding courses in the same manner as the second course, and trim the top of the last course even with the ridge. The top edges of the courses on either side of the roof ridge should butt, not overlap.

7. Cut a 12-inch wide length of roll roofing to cover the ridge.

8. Snap a chalk line 5½ inches down from the ridge on either side of the roof deck. These two chalk lines *must* run exactly parallel to the ridge.

9. Apply a 2-inch wide strip of lap cement along each chalk line. Do not allow the lap cement to extend below either chalk line.

10. Bend the 12-inch wide strip of roll roofing along its center, lay it along the ridge, and embed it in the cement. Nail the roll roofing strip to the ridge with nails staggered 2 inches apart in rows ¾ inch up from the bottom edges of the strip.

Double coverage smooth surface roll roofing is possible with the exposed nail application method by allowing each course to overlap the underlying one by 19 inches. This permits an exposure of 17 inches when using 36-inch wide rolls.

As shown in figure 11-5, the exposed nail application method can also be used to lay roll roofing courses parallel to the rake (i.e., vertical to the roof eave). The first course or strip of roll roofing should be started at the roof ridge at either rake. This first course is nailed along the rake and eave with nails spaced on 2-inch center and ap-

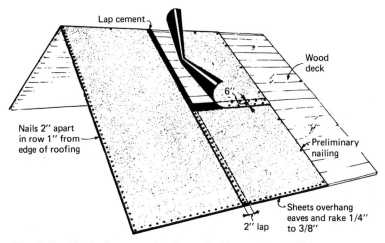

Fig. 11-5. Vertical exposed nail application method. *(Courtesy Asphalt Roofing Manufacturers Association)*

proximately 1 inch in from the edge of the roof. Allow approximately ¼ to ⅜ inch of the lower end of each course to project over the rake at each end of the roof. Nails should be spaced 18 inches apart along the ridge and the lap edge on a line about ½ inch in from the edge of the sheet.

Before applying the second course of roll roofing, coat the 2-inch wide lap edge of the underlying sheet with lap cement. Embed the second course in the lap cement so that it overlaps the underlying sheet by exactly 2 inches, and nail it down. Space the nails on 2-inch centers along the lap edge and eave on a line 1 inch in from the edge. The nail spacing along the ridge will be the same as for the first course. If more than one section of roll roofing is required to complete a course, the end laps should be 6 inches wide, embedded in lap cement, and nailed to the roof deck. The nails should be staggered in two rows and spaced on 4-inch centers in each row. Stagger all end laps so that none of them is parallel in adjacent rows.

The procedures used to lay the first two courses of roll roofing also applies to the remaining courses. Roll roofing strips cut 12-inch wide and bent lengthwise along their centers are used to cap the ridge and hips (fig. 11-6).

Details of the concealed nail method of applying smooth

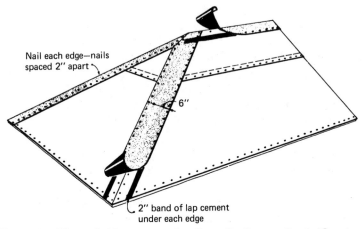

Nail each edge—nails
spaced 2″ apart

6″

2″ band of lap cement
under each edge

Fig. 11-6. Hip and ridge exposed nail application method. *(Courtesy Asphalt Roofing Manufacturers Association)*

surface roll roofing are illustrated in figures 11-7 and 11-8. The concealed nail application method may be outlined as follows:

1. Nail 9-inch wide edge strips along the rakes and eaves and allow approximately ¼ to ⅜ inch of material to overhang the edges of the roof deck. Use two rows of nails spaced on 4-inch centers in lines located ¾ to 1 inch in from each edge of the strip.

2. Apply the first course of smooth surface roll roofing so that its outer edges are flush with the edges of the edge strips at

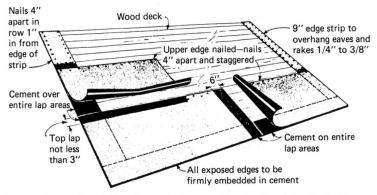

Nails 4″
apart in
row 1″
in from
edge of
strip

Wood deck

9″ edge strip to
overhang eaves and
rakes 1/4″ to 3/8″

Upper edge nailed—nails
4″ apart and staggered

6″

Cement over
entire lap areas

Top lap
not less
than 3″

Cement on entire
lap areas

All exposed edges to be
firmly embedded in cement

Fig. 11-7. Concealed nail application method. *(Courtesy Asphalt Roofing Manufacturers Association)*

the rake and eave. Nail along the top edge of the first course with nails spaced 4 inches apart and slightly staggered. The lower nails in the staggered row should never be more than 2 inches below the top edge of the first course.

3. Cover the edge strip with cement and firmly press the lower edge and rake ends of the first course sheet down until they hold in place.

4. Apply the second course of smooth surface roll roofing so that it overlaps the first course by 3 inches. Nail along the top edge as described in step 2, but do not nail the roll roofing over the edge strip at the rakes until the bottom edge is cemented in place.

5. Cover the 3-inch wide lap area on the first course and the edge strips at the rakes with lap cement, and press the lower edge and rake ends down until they hold in place. Finish nailing along the top edge of the second course sheet at the rakes.

6. Apply the remaining courses by using the nailing and cementing procedures described in steps 2–5.

All end laps should be at least 6 inches wide and staggered so that a lap in one course is not over or adjacent to a lap in the preceding course. The underlying section of the end lap is nailed to the deck with two rows of staggered nails and covered with cement in which the overlapping section of the end lap is embedded.

The ridge and hips are covered with 12 × 36–inch sections cut from the roll roofing strip. As shown in figure 11-8, each section is nailed and covered with cement over that portion which is to be lapped.

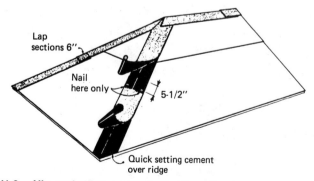

Lap
sections 6″
Nail
here only
5-1/2″
Quick setting cement
over ridge

Fig. 11-8. Hip and ridge concealed nail application method.
(Courtesy Asphalt Roofing Manufacturers Association)

Double Coverage Roll Roofing

Roll roofing can be applied to the roof deck so that it provides double coverage. In other words, there is sufficient overlap between courses to provide a double layer of roll roofing across the surface of the roof deck.

A 19-inch selvage double coverage roll roofing is specially manufactured for this type of roofing application. Some types of double coverage roll roofing are applied with a cold asphalt adhesive, whereas others require the use of hot asphalt. Hot asphalt applications will also differ according to how the roll roofing is saturated and coated. Because of these differences, the manufacturer's instructions should be carefully followed when applying this type of roofing material.

The application of 19-inch selvage double coverage roll roofing to a new roof deck is illustrated in figure 11-9. The procedure may be outlined as follows:

1. Nail a drip edge along the roof eaves and rakes. The drip edge should extend 3 inches onto the roof deck and the nails should be placed about 10 inches apart along its inner edge (fig. 11-10).
2. Cut the 19-inch wide selvage portion from a length of roll roofing to use as a starter strip.
3. Nail the starter strip to the roof deck with about ¼ to ⅜ inch

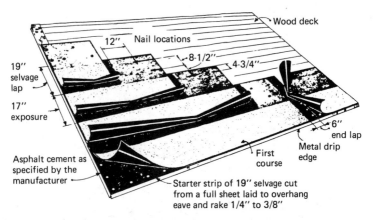

Fig. 11-9. Application of 19-inch selvage double coverage roll roofing. *(Courtesy Asphalt Roofing Manufacturers Association)*

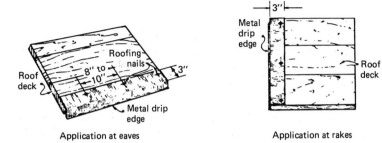

Application at eaves

Application at rakes

Fig. 11-10. Metal drip edge application. *(Courtesy Asphalt Roofing Manufacturers Association)*

of material projecting over the edges of the rake and eave. The nails along the eave and rake should be spaced 6 inches apart in rows 1 inch in from the edge of the roof deck. The nails along the top edge of the starter strip are spaced 12 inches apart in rows 4 inches down from the top edge. A third (middle) row of nails is located exactly halfway between the top and bottom nail rows. As shown in figure 11-9, the nails in the top and middle rows are placed 12 inches apart and staggered.

4. Apply the first course of roll roofing so that its edges are flush with the edges of the starter strip at the rakes and eave. Cement the exposed portion to the starter strip and nail the selvage portion to the roof deck with two staggered rows of nails (fig. 11-9).

5. Apply the remaining courses so that each laps the underlying one the full 19-inch width of the selvage edge.

As shown in figure 11-11, each end lap is 6 inches wide and staggered so that no end lap in one course is adjacent to one in the preceding course. The portion of roll roofing underlying each lap is first nailed to the roof deck and then covered with a 6-inch wide strip of cement. The overlying sheet of roll roofing is then pressed into the cement concealing the nails.

Cut the roll roofing sheets so that they butt at the ridge or hip, and nail them to the roof deck. Both the ridge and hip can be covered with 12 × 36–inch cut strips of roll roofing bent lengthwise along their centers. Allow a 17-inch exposure on both the ridge and hip by overlapping each section a distance equal to the length of the selvage

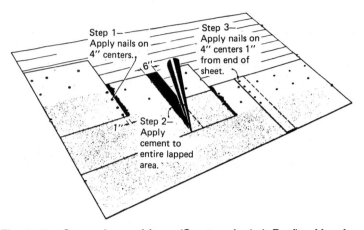

Fig. 11-11. Staggering end laps. *(Courtesy Asphalt Roofing Manufacturers Association)*

portion (fig. 11-12). When covering the hips, begin at the bottom with a starter piece and then lay the first section of roll roofing. Begin at either end of the ridge and work toward the other one.

Vertical application (i.e., application parallel to the roof rakes) is also possible with a 19-inch selvage double coverage roll roofing (fig. 11-13). It follows the same procedures used in the horizontal application method.

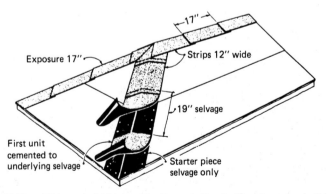

Fig. 11-12. Ridge and hip application method. *(Courtesy Asphalt Roofing Manufacturers Association)*

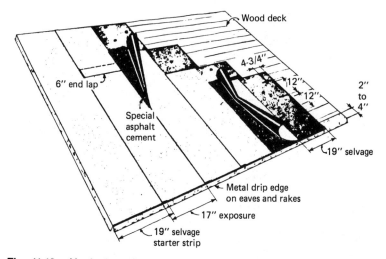

Wood deck

6" end lap

4-3/4"

12"

12"

2" to 4"

Special asphalt cement

19" selvage

Metal drip edge on eaves and rakes

17" exposure

19" selvage starter strip

Fig. 11-13. Vertical application of 19-inch selvage double coverage roll roofing. *(Courtesy Asphalt Roofing Manufacturers Association)*

Other types of roll roofing can also be used in double coverage roofing applications as long as the lapped portion of the sheet is 2 inches wider than the exposed portion. In other words, the exposed portion of a 36-inch roll of roll roofing must be no greater than 17 inches.

Always apply any flashing to the roof deck before the first course of roll roofing is laid (see chapter 8).

Pattern-Edge Roll Roofing

Pattern-edge roll roofing is applied parallel to the roof eaves. The concealed nail application method is used in new construction (fig. 11-14). Both the concealed nail and exposed nail application methods are used for reroofing over existing materials (see the discussion of reroofing).

In new construction, 9-inch wide strips of roll roofing are nailed along the eaves and rakes of the roof deck with two rows of nails spaced 4 inches apart. Each edge strip projects approximately ¼ to ⅜ inch beyond the edge of the eave or rake.

When using the concealed nail application method, the first or starter course is laid with its pattern edge facing up the roof toward the ridge. The straight edge is allowed to extend ¼ to ⅜ inch beyond

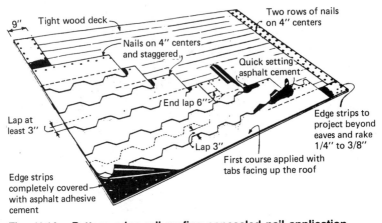

Fig. 11-14. Pattern-edge roll roofing concealed nail application method in new construction. *(Courtesy Asphalt Roofing Manufacturers Association)*

the eaves. Nail the first or starter course along the pattern edge with staggered rows of nails spaced 4 inches apart, but none more than 2 inches below the top of the sheet or within 18 inches of the rake. Thoroughly coat the 9-inch wide edge strip with asphalt roofing cement, press the first or starter sheet down firmly, and nail to the edge of the rake. *Do not nail along the bottom of the sheet.* The bottom of each course will be secured with the roofing cement.

Apply the second and succeeding courses with their pattern edge facing down the roof toward the eaves. Cut each overlying sheet so that its tabs are centered over the cutouts of the course below. Overlap each course so that there is a minimum effective top lap of 3 inches.

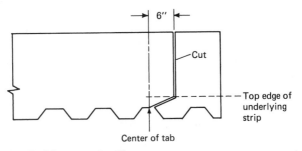

Fig. 11-15. End lap construction.

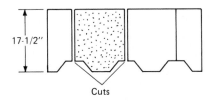

Fig. 11-16. Cutting hip and ridge shingles.

All end laps should be at least 6 inches wide. Form the end lap by cutting diagonally from the center of the cutout on the pattern edge over to a point that will give a 6-inch width and then vertically to the top of the sheet (fig. 11-15).

Nail 2 inches down from the top edge of each sheet with a staggered row of nails spaced 4 inches apart. Nail the end laps to the roof deck with staggered nails spaced 4 inches apart in two rows 1 inch and 5 inches from the end of the sheet. Stagger the end laps so that they are offset between adjacent courses.

Lift the lower edge of each course and apply lap cement in a continuous layer over the full width of the top lap and over 5½ inches of each 6-inch width of end lap. Apply pressure uniformly to the overlying sheet over the entire cemented area.

Hip and ridge shingles are formed by cutting a tab from the pattern-edge roofing sheet. The cuts are made at the center of each cutout on either side of the tab (fig. 11-16). Apply the shingles to the hip or ridge with an overlap of at least one third of the length of the overlying unit. As shown in figure 11-17, this will provide an exposure

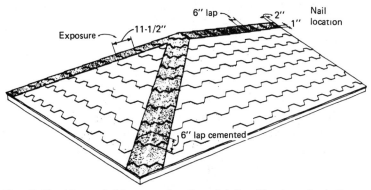

Fig. 11-17. Hip and ridge construction details. *(Courtesy Asphalt Roofing Manufacturers Association)*

of about 11½ inches. Firmly press the first shingle into lap cement applied to either side of the hip or ridge and nail the shingle at the end that will be covered by the overlapping unit. As successive shingles are applied, nail and then coat each portion of the shingle that is to be lapped with lap cement before applying the next one.

After the roof has been completed, check all top and end laps for a good bond. Press down any that may have pulled loose and apply additional lap cement where necessary.

Always apply any flashing to the roof before the first course of roll roofing is laid (see chapter 8).

Roll Roofing Repairs

Small holes, rips, or loose seams that are not too extensive can be repaired by applying roofing cement and pressing the roll roofing back in place.

Larger holes or rips in the roll roofing can be patched. Thoroughly clean the surface around the damaged area and allow it time to day. Cut a piece of roll roofing large enough to cover it. Apply roofing cement to the damaged area, press the patch firmly in place, and nail it around the edges (fig. 11-18). Cover the edges of the patch with cement to prevent it from being lifted by the wind. This will also provide a watertight seal.

Loose, broken, or rusting roofing nails should be removed and replaced with new ones. Do not use the old nail holes when renailing. Cover the old holes with a dab of roofing cement to prevent the possibility of water leaking through the roll roofing and damaging the sheathing.

Reroofing

Pattern-edge roll roofing can be used for reroofing over an existing asphalt shingle roof with a slope of 4-in-12 or more. The concealed nail or exposed nail application method can be used when reroofing.

Pattern-edge roll roofing is always applied with each course parallel to the roof eave. Except for the starter strip, each course must be laid with the diagonal-shaped tabs on the bottom edge and facing the roof eave.

STEP ONE: Cut away roll roofing around damaged area and brush clean the exposed surface.

STEP TWO: Cover exposed area and surrounding roll roofing with roofing cement.

STEP THREE: Nail roll roofing patch over exposed area and cover with roofing cement.

Fig. 11-18. Repairing roll roofing.

The exposed nail method of applying pattern-edge roll roofing to an existing asphalt shingle roof is illustrated in figure 11-19 and may be outlined as follows:

1. Cut back the existing asphalt shingles and nail a 1 × 6–inch board along the rakes and eaves. Allow the boards to extend ³/₈ to ¹/₂ inch beyond the edges of the roof deck and nail them in place.

2. Apply the first or starter course of pattern-edge roll roofing so that its tabs face in the direction of the roof ridge. The lower or straight edge and the ends of the first course must extend ¹/₄ to ³/₈ inch over the rakes and eave.

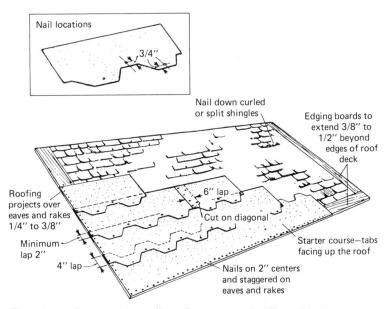

Fig. 11-19. Pattern-edge roll roofing exposed nail application method in reroofing. *(Courtesy Asphalt Roofing Manufacturers Association)*

3. Nail along the bottom edge of the first course with a staggered row of nails spaced 2 inches apart and no more than 2 inches up from the top of the sheet. Nail each sheet along the rake edge.

4. End laps are cut as shown in figure 11-15. The application method is the same as the one described for pattern-edge roll roofing.

5. Apply the remaining courses with the tabs facing the eave and with a minimum horizontal or top lap of 2 inches. Succeeding courses are applied with a nail centered at each tab and at each cutout.

The ridges and hips are formed with 12-inch wide strips bent lengthwise so that 6 inches extends down either side (see fig. 11-6). The ridge and hip sections are laid with a 11½-inch exposure and a 6-inch lap. Nail each ridge or hip strip along its outer edge. Cover the lapped area with a 2-inch band of lap cement before applying the next ridge or hip section.

CHAPTER 12

Built-Up Roofing

- Equipment, Tools, and Materials
- Roof Deck Preparation
- Built-Up Roof Flashing Details
- Hot-Process Roofing Method
- Cold-Process Roofing Method
- Repairing Built-Up Roofs
- Reroofing

Flat roofs on commercial type structures are often covered with a roof membrane consisting of three, four, or five layers of roofing felt, each coated with hot- or cold-mopped asphalt, or with a cap sheet. This type of roof is commonly called a *built-up roof*. It is also customary to refer to a built-up roof as a 10-year, 15-year, or 20-year roof depending on the number of layers or plies in the roof membrane and the method of application.

A built-up roof is generally applied by a roofing contractor who specializes in this type of work. Depending on the job specifications, a reputable contractor will provide a warranty that guarantees the roof membrane for terms of 10, 15, or 20 years. These are limited guarantees subject to the provisions that specified materials be used and that these materials be applied in accordance with the manufacturer's instructions. Furthermore, they must be applied by a local roofing contractor who operates as a field representative of the manufacturer. In general, the roofing contractor is responsible for making certain repairs to the roof membrane during the term of the warranty.

The membrane of a built-up roof is subject to alligatoring, splitting, ridging, and other forms of deterioration caused by its exposure to weather conditions. This deterioration of the roof membrane eventually produces areas on the surface of the roof where leaks can occur. It should be clearly understood from the outset that a built-up roof membrane is as susceptible to leaks as any other type of roof covering material. The condition of the roof should be inspected at regular intervals. If signs of deterioration are detected, the roofing contractor who did the work should be notified. If the roof is not under warranty, however, you will probably have to make the necessary repairs yourself at your own expense. Never sign a roofing contract without first reading the small print. Make sure you understand all the provisions contained in the contract.

Equipment, Tools, and Materials

Many different types of materials are produced for use in built-up roofing. Because of this great variety, care must be taken not to create chemically incompatible combinations. This can sometimes occur when the products of different manufacturers are mixed. Mixing materials will also invalidate any warranty that applies to the roof. The manufacturer of a roofing product will usually recommend other products with which it can be satisfactorily used.

The tools used in built-up roofing will depend on the requirements of the specific job. They will range from simple hand tools used for mopping and spreading the roof coating onto the surface to sophisticated spraying equipment (fig. 12-1). Most of the tools and equipment used in built-up roofing are available through local building supply dealers or rental stores. Some manufacturers of roofing materials will even loan spray equipment free with complete operating instructions in return for a minimum purchase of their purchase. Unfortunately, the minimum purchase is well beyond the means of the average small businessman.

All of the roof covering materials taken together to form a built-up roof, exclusive of the roof deck itself and the flashings, are referred to as the *roof membrane*. Unlike a shingle roof, the roof covering materials of a built-up roof are bonded together to form a single inseparable cover or membrane.

The roofing materials of a built-up roof may be bonded together either by cold or hot applied coatings. These coatings are applied by mopping, brushing, or spraying. When a cold adhesive is used, the application method is referred to as *cold roofing* or *cold-process roofing*, and the completed roof is sometimes called a *cold roof*. Cold-

Fig. 12-1. Asphalt coating being applied over roof deck insulation.

process roofing is most commonly used to maintain or repair an existing built-up roof. Although this roofing method can also be used in new construction, the roof will not last as long as a hot mopped roof.

In *hot-process roofing*, or *hot roofing*, the bonding agent is heated to a high temperature in special kettles before it is applied to the roof. The completed roof is sometimes called a *hot roof*. Because the hot roof process requires special equipment and more workers than the cold roof process, it is generally applied by a professional roofing contractor specializing in this type of work.

Roof Coatings

The roof membrane of a built-up roof consists of one or more plies of roofing felt separated by hot or cold applied roof coatings.

In cold-process roofing, a fibrated cut-back asphalt cement is used as the interply adhesive. Fibrated asphalt or asbestos cut-back coatings or emulsions are used as top coats.

Because hot-process roofing terminology is not standardized, there has been some confusion and misunderstanding between roofing contractors and property owners about job specifications and the types of materials used. Either roofing asphalt or coal-tar pitch can be used for the top coat and interply moppings in hot-process roofing. Although these two bituminous materials are virtually identical in appearance *after* they have been applied to the roof, there are important differences because one is a petroleum by-product and the other is derived from coal. Unfortunately, the terms *roofing asphalt* and *coal-tar pitch* are often incorrectly used as synonyms for a hot-process roof coating. To confuse the issue further, the roof coating on a hot-process roof is also sometimes called *tar*, *pitch*, or *roofing bitumen*.

Roofing asphalt or *asphalt* may be obtained from one of several varieties of naturally occurring bitumen, but it is much more commonly produced as a petroleum by-product. Asphalt obtained from a natural source is sometimes called *bitumen*, *pitch*, or *natural asphalt*.

The three basic types of asphalt produced from petroleum are (1) straight run asphalt, (2) air-blown asphalt, and (3) cracked asphalt. Only air-blown asphalt is used as a roofing material. It reacts with air at approximately 400°–600° F, and has proved less susceptible to temperature change than straight-run asphalt.

Tar is a dark brown or black substance formed from the destructive distillation of such organic materials as wood, coal, and petroleum. *Coal tar* is obtained as a by-product in the manufacture of coke and coke oven gas from soft coal. *Coal tar* (also called *coal-tar pitch* or *roofing pitch*) is one of a number of different coal tar by-products. It is used primarily as a roof covering material.

Strictly speaking, *pitch* is the residue remaining after the distillation of tar. Pitch and tar are chemically identical, but the term *pitch* is used to refer to wood (southern pine) by-products or to pitch obtained from natural deposits. In its natural form, pitch is also called *asphalt*.

There are certain differences between roofing asphalt and coal-tar pitch that will affect their use. For example, the higher melting point of roofing asphalt makes it a more suitable roof coating than coal-tar pitch on steeper slopes. Because coal-tar pitch melts more quickly, it should be protected from the sun with a layer of gravel or slag. One advantage of using coal-tar pitch is that it is a self-sealing roof coating. For this reason, it can be effectively used to repair or recoat an existing roof.

Coal tar chemicals used as roof covering materials are now being largely replaced by petroleum by-products, such as roofing asphalts.

Roofing Asphalts

Roofing asphalts are available in several grades or classes for field application to all types of built-up roof assemblies. The proper grade should be selected for each job, taking into consideration the slope of the roof, weather conditions, and the type of roof assembly.

Roofing asphalt should be heated slowly and the temperature kept relatively stable. This can be accomplished by maintaining a constant fluid level in the kettle and by adding fresh asphalt in small chunks. Kettles must be equipped with a thermostatic device to control asphalt temperatures. For large jobs, asphalt pumps and circulating systems will permit proper mopping temperatures with little heat loss in transporting the hot asphalt from the kettle to the roof surface.

The temperature of the roofing asphalt must be kept within a specific range while being heated in the kettle. The temperature range will depend on the type and grade of roofing asphalt. For example, a *dead level* asphalt has a softening point of 145° F, whereas a *high melt* or *steep asphalt* will begin to soften at 190° F. The kettle temperature will generally be maintained at approximately 400–450° F.

Once again, the temperature maintained in the kettle is very important. Excessive kettle temperatures result in degrading the roofing asphalt, and temperatures below specified minimums result in lack of adhesion. Not only is temperature control important to the quality of the roofing asphalt, it also reduces the amount of smoke pollution.

The recommended temperature range for the roofing asphalt will be indicated by the manufacturer. The type and grade of roofing asphalt selected for the job will depend on the slope of the roof deck and the process used to apply the built-up roof membrane.

Roofing asphalt should be added to the kettle in small chunks. Do not add large chunks. Asphalt is brittle and pieces of a large chunk will sometimes split or break off while it is being handled. If this should happen while it is being added to hot asphalt already in the kettle, the split-off piece may splash hot asphalt onto your face or hands.

Do not apply hot roofing asphalt too thickly, or alligatoring will

Table 12-1. Sieve Requirements According to ASTM Specification D1863–64 (*Courtesy American Society for Testing and Materials*)

Sieve Size	Percentage Total Passing
¾"	100%
½"	90–100%
⅜"	0–70%
No. 4	0–15%
No. 8	0–15%

probably occur within a year. This will result in the development of cracks that will eventually penetrate the roof membrane.

Mineral Surfacing Materials

The surfacing materials used on gravel surfaced built-up roofs are (1) opaque gravel, (2) crushed rock, and (3) crushed blast furnace slag. Each of these surfacing aggregate materials must comply with the provisions and requirements of ASTM Designation D1863–64 (table 12-1).

The size of the surfacing aggregate particles should range from ¼ to ⅝ inch. Furthermore, the aggregate must be dry, hard, and clean before being applied to the roof. Wet aggregate will cause a hot coating, such as asphalt or tar, to foam and form bubbles on its surface. These surface bubbles remain after the hot coating has cooled and hardened.

The surfacing aggregate is embedded in a poured flood coat of hot asphalt. Approximately 300 pounds of slag and 400 pounds of opaque gravel or crushed rock should be used per 100 sqare feet of roof surface.

Roof Deck Preparation

The Built-Up Roof Committee of the Asphalt Roofing Manufacturers Association recommends a minimum of ¼ inch per foot of roof slope to assure adequate drainage. Roofs with slopes less than ¼ inch per foot are highly susceptible to water accumulation. Roof decks with

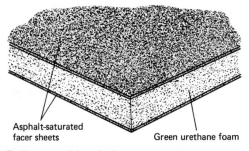

Asphalt-saturated
facer sheets   Green urethane foam

Fig. 12-2. Built-up roof insulation.

slopes less than ³/₄ inch per foot should be designed with interior roof drains and raised gravel stops or parapet walls.

Before a built-up roof can be applied to a deck, the surface of the deck must be thoroughly cleaned. All dirt, debris, and other materials that might prevent the adhesive from adhering to the surface must be removed.

The deck surface must also be dry before the bonding agents and other roofing materials are applied or they will not adhere properly. Any snow, ice, or water on the deck must be removed and sufficient time allowed for the surface to dry before actual roofing begins.

The deck surface should be as smooth as possible. Any warped boards must be removed and replaced. Raised or loose boards or panels must be securely nailed down to underlying purlins, rafters, or beams to prevent deck movement after the roof membrane has been applied to the surface.

If the roof membrane is applied over insulation, the insulation must be completely dry. The insulation must also be suitable as a base for built-up roofs. Urethane, polystyrene, or any other roof insulation material with a soft or spongy composition are generally not recommended as a base for a built-up roof, *unless* the insulation material is sandwiched between asphalt-saturated facer sheets (fig. 12-2).

Built-Up Roof Flashing Details

Most roofs require flashing to protect the deck from leaks at those points on the surface subject to heavy water runoff or water accumulation, or where structural joints are formed. A built-up roof is no exception to this rule, but the flashing details and application methods used are in many ways different from those used on pitched roofs. For that reason, they have been included in this chapter instead of chapter 8.

Metal flashing should always be embedded in plastic asphalt cement before it is applied to the surface of the roof membrane. The top surface of the metal flashing should be coated with a suitable primer to protect it from chemical reactions resulting from contact with bituminous roofing materials.

The principal flashing applications used in built-up roofing are

1. Base flashing
2. Edge flashing
3. Metal gravel stop
4. Vent and stack flashing
5. Projection flashing
6. Interior roof drain flashing
7. Expansion joint flashing

Base Flashing

Base flashing is most commonly used where the roof deck abuts a parapet wall or a vertical wall of the structure. Examples of base flashing applications are illustrated in figures 12-3 and 12-4.

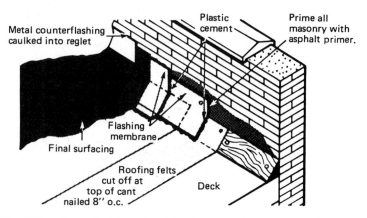

Fig. 12-3. Base flashing of cold-process roofing system. *(Courtesy Flintkote Co.)*

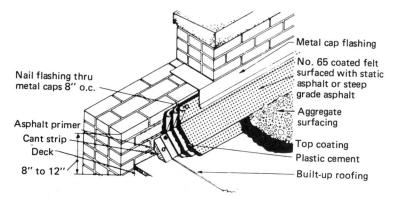

Fig. 12-4. Base flashing of hot-process roofing system. *(Courtesy Flintkote Co.)*

Before the base flashing is applied, a fiberboard *cant strip* with a minimum 4-inch face is installed in the angle formed by the vertical wall and the roof deck. The purpose of the cant strip is to soften the angle formed between the flat and vertical surfaces and to provide support for the flashing. If roof insulation is applied directly to the roof deck, a treated wood nailer the same thickness as the insulation should be installed the length of the joint with one edge abutting the surface of the vertical wall (fig. 12-5). The cant strip is then nailed to the wood nailer. Plastic asphalt cement is applied to the wood nailer before the cant strip is installed to provide additional waterproofing (fig. 12-6).

After the cant strip has been installed, lay the roofing felts of the built-up roof so that they extend up the face and are cut off evenly at the top of the cant. Nail the roofing felts into the top edge of the cant 8 inches on center.

Coat those portions of the masonry wall surface to be covered by the base flashing with an asphalt primer. Install one ply of No. 15 perforated asphalt roofing felt set in plastic asphalt cement extending from the bottom of the cant to approximately 6 inches onto the vertical wall surface above the top of the cant.

Install a second ply of No. 15 perforated asphalt roofing felt in the same manner, but allow it to extend 2 inches beyond the top and bottom of the first ply.

A final ply of roofing felt is applied over the first two to com-

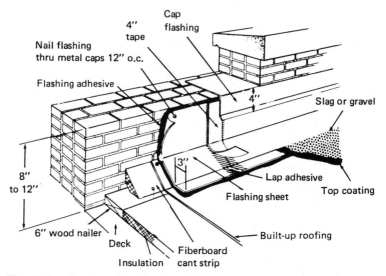

Fig. 12-5. Construction details of insulated built-up roof. *(Courtesy Celotex Corp.)*

plete the flashing membrane. It should be wide enough to extend 2 inches beyond the bottom edge of the second ply. Like the other two plies, it is embedded in plastic asphalt cement, but it is also nailed to the vertical masonry wall with large head roofing nails through tin caps. The nails should be located 8 inches on center along the top edge of the final ply.

After the base flashing membrane has been completed, cover the roof deck surface *and* the flashing with a top coating. Remove 1½ inches of mortar from the first horizontal joint above the base flashing. Caulk the lip or bent edge of a metal counter or cap flashing into

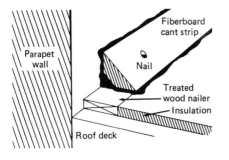

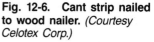

Fig. 12-6. Cant strip nailed to wood nailer. *(Courtesy Celotex Corp.)*

the reglet formed by the removal of the mortar. The metal counter flashing should be large enough to extend downward a minimum of 3 inches over the top edge of the finished flashing (see figs. 12-3 and 12-4).

Edge Flashing

Raised edge flashing should be installed along the eaves and rakes when a building roof has no parapet wall along its edges. Its purpose is to provide a gentle rise in the roofing at the perimeter of the building in order to raise the connection between the bituminous roofing materials and the metal edge trim above any standing water that might accumulate on the flat roof. It also prevents this water from running down the outside face of the building. When raised edge flashing is used, interior drains should be installed on the roof to remove any accumulated water within 24 hours.

Basically a raised edge flashing consists of a tapered edge strip, a wood nailer, and the flashing membrane. A metal fascia strip is sometimes nailed over the edge flashing. A typical raised edge flashing construction is illustrated in figure 12-7.

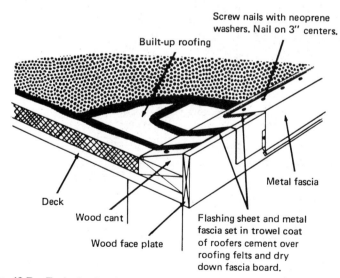

Fig. 12-7. Typical raised edge flashing construction details. *(Courtesy Celotex Corp.)*

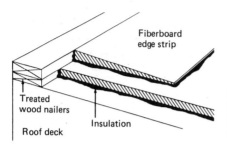

Fiberboard
edge strip

Treated
wood nailers

Insulation

Roof deck

Fig. 12-8. Tapered edge strip cemented to insulation board. *(Courtesy Celotex Corp.)*

The tapered edge or cant strip is either nailed to the wood nailer or secured to the insulation sheet with plastic asphalt cement *before* the roof membrane is applied to the deck. The various plies of roofing felt in the roof membrane should extend across the face of the cant to at least its top edge. If the tapered edge strip is cemented to the insulation and abuts the wood nailer as shown in figure 12-8, then extend the roofing felt to the outer edge of the wood nailer and down its outside vertical edge at least 2 inches where it should be nailed into the nailing strip 6 inches on center. Cover the tapered edge strip with a flashing sheet or roll roofing embedded in plastic asphalt cement. A cross section of this type of construction is shown in figure 12-9. The flashing sheet should be large enough to extend 4 inches beyond the tapered edge strip onto the roof surface and 2 inches down the outside vertical edge of the wood nailer and tapered edge strip. Cover the raised edge flashing with screw nails and neoprene washers on 3-inch centers.

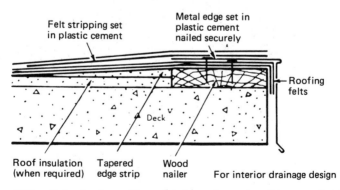

Felt stripping set
in plastic cement

Metal edge set in
plastic cement
nailed securely

Roofing
felts

Deck

Roof insulation
(when required)

Tapered
edge strip

Wood
nailer

For interior drainage design

Fig. 12-9. Raised edge with metal drip edge extending over fascia. *(Courtesy Flintkote Co.)*

Metal Gravel Stop

When the top of the roof is covered with gravel, crushed stone, or slag, a gravel stop should be installed along the edges. As shown in figure 12-10, a gravel stop can also function as a raised flashing edge.

The various layers of roofing felt are applied first, and the flange is stagger-nailed 3 inches on center in lines 3/4 inch from the edge of the flange and 3/4 inch from the lip. The bottom of the flange should be covered with plastic asphalt cement before it is nailed down.

Two plies of No. 15 asphalt roofing felt, one 8 inches wide and one 12 inches wide, are applied over the flange. Coat the metal surface of the flange with a suitable primer. Apply a coating of bitumen or roofing asphalt and the 8-inch felt stripping. Cover the first ply of felt stripping with a bitumen or asphalt coating and apply the 12-inch felt stripping. Apply the top coating to the surface of the roof deck and allow it to cover the 2-ply felt stripping up to the raised edge of the gravel stop. Add the mineral-surfacing aggregate to the top coat while it is still hot.

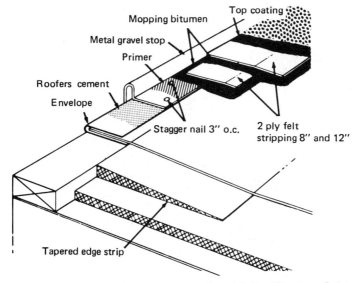

Fig. 12-10. Metal gravel stop construction details. (*Courtesy Celotex Corp.*)

Vent and Stack Flashing

Vents, stacks, and similar types of circular pipes projecting through the roof deck are usually equipped with a metal flange which extends approximately 4 inches onto the surface of the deck (fig. 12-11). The flange should be set in plastic asphalt cement over the roofing felt and nailed 3 inches on center at a distance approximately ³/₄ inch from the edge of the pipe. Cut two collars of roofing felt to fit over the pipe. They should be large enough to overlap the metal flange on all sides by 8 and 12 inches respectively. Embed both roofing felt collars in plastic asphalt cement before applying them around the pipe. By increasing the size of the collars around the pipe, a slight incline is created away from it for better drainage. The final coating is applied to the roof after the flashing around the pipe is completed.

Projection Flashing

Sometimes objects other than circular pipes will project through the roof deck. Usually a vertical wood face plate is constructed around the object and base flashing is applied to the angle formed by the roof deck and the vertical face plate. Counter (cap) flashing is applied over each of the face plates. A typical example of projection flashing is the

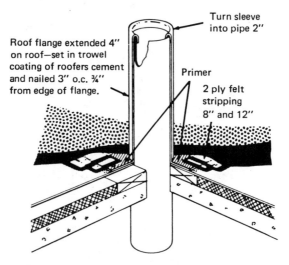

Fig. 12-11. Typical vent pipe flashing. *(Courtesy Celotex Corp.)*

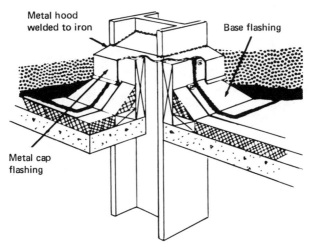

Metal hood
welded to iron

Base flashing

Metal cap
flashing

Fig. 12-12. Projection flashing details. *(Courtesy Celotex Corp.)*

application of wood face plates and metal flashing to an I-beam projecting through the roof deck (fig. 12-12).

Interior Roof Drain Flashing

A flat roof with raised edge flashing will have one or more interior roof drains installed on its surface. The metal flashing flange furnished with the drain should be cemented to the roofing felts with plastic asphalt cement. The metal flange should be covered with two layers of felt stripping embedded in plastic asphalt cement. The top layer of felt should extend 9 inches beyond the flange on all sides, and the bottom layer should extend 6 inches. After the flashing is completed, cover the roof and the flashing with the final coating (fig. 12-13).

Expansion Joint Flashing

Expansion joints are used on large flat roofs to prevent the buildup of destructive stresses caused by the expansion and contraction of structural elements.

As shown in figure 12-14, expansion joint flashing consists of base flashing applied to the two angles formed by the roof surface and the vertical curbs of the joint. The space between the curbs is filled with flexible insulation and covered with a base sheet and metal cap.

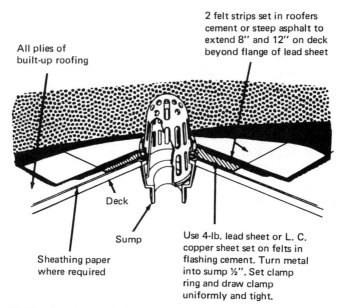

All plies of
built-up roofing

2 felt strips set in roofers
cement or steep asphalt to
extend 8" and 12" on deck
beyond flange of lead sheet

Deck

Sump

Sheathing paper
where required

Use 4-lb. lead sheet or L. C.
copper sheet set on felts in
flashing cement. Turn metal
into sump ½". Set clamp
ring and draw clamp
uniformly and tight.

Fig. 12-13.   Interior roof drain flashing details. *(Courtesy Celotex Corp.)*

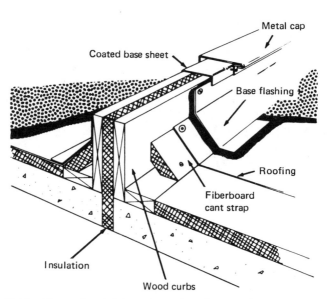

Metal cap

Coated base sheet

Base flashing

Roofing

Fiberboard
cant strap

Insulation

Wood curbs

Fig. 12-14.   Expansion joint construction details. *(Courtesy Celotex Corp.)*

Hot-Process Roofing Method

A hot-process built-up roofing membrane can be installed over either a nailable or a nonnailable roof deck. A nailable deck is any type of surface to which the roofing materials can be nailed. These decks include wood plank decks, plywood decks, and decks made of poured gypsum or structural cement fiber. Nonnailable decks include metal decks and those made from precast, thin shell concrete or reinforced poured concrete.

The procedure used to prepare a nailable deck to receive a built-up roof membrane differs to some extent from the procedure used for a nonnailable deck; however, in either case, the purpose of the preparation is to make certain the roof membrane will adhere properly to the surface of the deck.

Roofing Nonnailable Decks

The first step in preparing a nonnailable deck for a hot roof membrane is to coat it with a suitable asphalt primer. The asphalt primer should be spread over the roof surface at the rate of $1/2$ to $3/4$ gallons per 100 square feet. The primer coating should be kept back at least 4 inches from all slab joints on precast concrete decks. Always solid mop poured concrete decks.

After the asphalt primer has *completely* dried, a base for the roof membrane is added to the surface of the deck. This generally consists of a single ply of a coated base sheet embedded in steep asphalt. These coated base sheets are made of organic felt, saturated and coated on both sides. They are sold under a variety of trade names, including Vaporbar Coated Base Sheet (Celotex) and Empire Base Sheet (Flintkote).

As shown in figure 12-15, each course of the base sheet *must* overlap the underlying one by at least 4 inches. The ends of each course should lap 6 inches, and all laps should be sealed with hot asphalt. The coated base sheet is embedded in the asphalt while it is still hot. Mopping should always start at the low point of the roof. The steep asphalt should be applied at a rate of approximately 23 pounds per 100 square feet.

Three plies of No. 15 perforated asphalt roofing felt (or perforated asphalt-asbestos roofing felt) are applied over the base sheet to complete the roof membrane. Each course of roofing felt is set in a con-

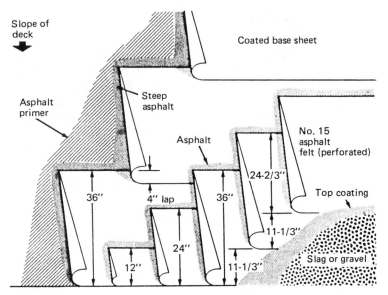

Fig. 12-15. Nonnailable deck with asphalt felt roofing membrane.
(Courtesy Celotex Corp.)

tinuous mopping of steep grade roofing asphalt, which is applied at the average rate of 20 pounds per 100 square feet. The hot asphalt should be applied so that in no place does roofing felt touch.

Begin with starter strips of roofing felt cut 12 and 24 inches wide, followed by full width sheets. Lap each 24³/₄ inches over the preceding ply. Broom each ply to assure complete embedment.

Pour a uniform top coating of steep grade roofing asphalt over the entire surface at the rate of approximately 60 pounds per 100 square feet. While the top coat is still hot, embed not less than 400 pounds of gravel or 300 pounds of slag per 100 square feet.

Roofing Nailable Decks

A nailable roof deck should first be covered with overlapping courses of 36-inch wide 20-pound rosin sheathing paper. As shown in figure 12-16, each course should be nailed to the deck with an overlap of 2 inches. End laps should be at least 6 inches.

Cover the entire surface with overlapping courses of a suitable

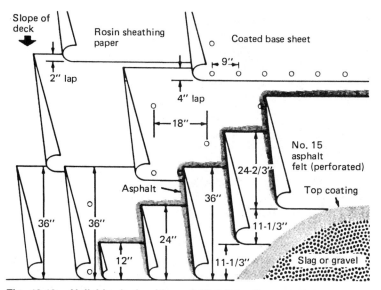

Fig. 12-16. Nailable deck with asphalt felt roofing membrane.
(Courtesy Celotex Corp.)

base sheet, such as 30-pound asphalt-saturated roofing felt. The purpose of the base sheet is to prevent the hot asphalt from penetrating the roof deck sheathing and entering the rafter spaces. Lap each sheet 4 inches over the underlying sheet and lap ends 6 inches. Nail along the lap at 9-inch maximum intervals and stagger nails through the center of each sheet at 18-inch intervals. Use 7/8-inch barbed roofing nails on wood plank decks and 3/4-inch annular threaded-shank nails on plywood decks.

Complete the roof membrane by applying three plies of No. 15 roofing felt, each course set in a continuous mopping of hot steep grade roofing asphalt at the average rate of 20 pounds per 100 square feet so that in no place does felt touch felt.

Use starting strips of roofing felt cut 12 and 24 inches wide, followed by full width sheets. Lap each ply 24 ¾ inches over the preceding one. Broom each ply to assure complete embedment.

Over the entire roof surface pour a uniform top coating of hot steep grade roofing asphalt. While the top coat is still hot, embed not less than 400 pounds of gravel or 300 pounds of slag per 100 square feet.

Cold-Process Roofing Method

A cold-process built-up roofing membrane can be applied over either a nailable or a nonnailable roof deck or over insulation boards. As in hot-process roofing, an asphalt primer must first be applied to the surface of a nonnailable roof deck. No base sheet is required when the roof membrane is applied over a nonnailable roof deck or over insulation boards (figs. 12-17 and 12-18). The application of the remaining three plies of the built-up roof membrane is identical to that described for the hot-process roofing method (see the discussion of roofing nonnailable decks above) except that the sheets of cold-process roofing felt are cemented to the deck with a cold adhesive at the rate of 1½ gallons per 100 square feet. Each ply should be carefully broomed to assure complete embedment.

A cold-process emulsion is applied to the entire surface as a top coat at the rate of 5 gallons per 100 square feet. Two coats are brushed onto the surface of the roof deck to complete the roof membrane.

On nailable roof decks, a layer of 20-pound rosin sheathing paper is nailed to the sheathing before the three plies of the built-up roof membrane are applied (fig. 12-19). The remaining details of applying a cold-process roof to a nailable roof deck are described in the preceding sections.

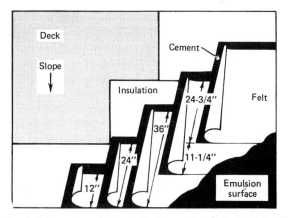

Fig. 12-17. Cold-process roof applied over insulated deck. *(Courtesy Flintkote Co.)*

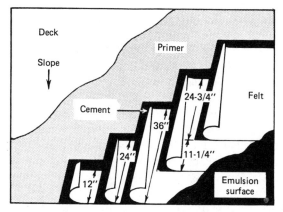

Fig. 12-18. Cold-process roof applied over nonnailable deck. *(Courtesy Flintkote Co.)*

Repairing Built-Up Roofs

A properly maintained built-up roof will have a service life of from 10 to 20 years. Sunlight is an important factor in the deterioration of this type of roof because the ultraviolet rays oxidize and shrink the coatings. The sun also bakes out the roofing oils, which causes a pliable

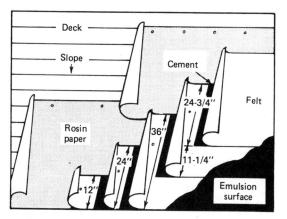

Fig. 12-19. Cold-process roof applied over nailable deck. *(Courtesy Flintkote Co.)*

roof to turn hard and brittle. As the coatings shrink and pull back from the edges of the roof, the underlying roofing felt is exposed and begins to rot. Foot traffic, vibration, and shaking also contribute to roof deterioration.

Carefully inspect the roof for blisters, breaks, large cracks, and other surface irregularities. Circle them with chalk and remove gravel, nails, and other loose debris from the marked area. If the roof membranes are still pliable and in good condition except for minor surface irregularities or holes in the flashing, only minor repairs are necessary. Blisters can be repaired by first making an x-type cut and folding back the flaps. The exposed cavity is then filled with a plastic roofing cement, the flaps are pressed down in their original position, and roofing cement is troweled over the flaps. All buckles, ridges, folds, and other surface irregularities can be repaired in the same way except that only a single lengthwise cut is required.

If the roofing felts are just beginning to dry out but are still in a soft, pliable condition, they can be restored by brushing or spraying the roof surface with a cold-applied mineral rubber resurfacer and sealant. If the roofing felts are dried out and brittle, a primer should be applied before the resurfacing top coat. The primer is a penetrating oil that restores the roofing felts to a pliable condition and seals cracks. *Remove all gravel and debris from the roof surface before applying a primer or resurfacing top coat.*

Cold-process resurfacing coatings can be applied without having to hire professional roofers to do the work. The manufacturers of these top coatings will often supply all the necessary equipment to apply them if the order is large enough.

Reroofing

Reroofing over an existing built-up roof is not generally recommended because defects in the old roofing material are easily transferred to the new covering, especially if the defects are severe or of long duration.

CHAPTER 13

Roofing with Wood Shingles and Shakes

- Tools and Equipment
- Roof Deck Preparation
- Wood Shingle Roofs
- Wood Shake Roofs
- Shingle and Shake Panel Roofs
- Shingle and Shake Finishes
- Reroofing

Western red cedar is the commercial wood most commonly used in the production of roofing shingles and shakes (fig. 13-1). It has an extremely fine and even grain, considerable strength, and a low expansion and contraction rate. It also sheds water well and provides a high degree of thermal insulation for the structure. As a roofing material, western red cedar shingles and shakes are twice as resistant to heat transmission as asphalt shingles, and three times as resistant as built-up roofing.

Wood fiber roofing is a relatively new product available in the form of rigid hardboard panels that are produced with surfaces designed to resemble individual cedar shingles (fig. 13-2). After a period of time, the wood fiber material weathers to a light gray color in much the same way that cedar does. Installing wood fiber shingle panels is easier, quicker, and less expensive than installing individual cedar shingles. Roofing with individual cedar shingles and shakes, however, results in a less uniform and more natural and rustic appearance.

Fig. 13-1. Rough-textured western red cedar shake roof. *(Courtesy Red Cedar Shingle & Handsplit Shake Bureau)*

Fig. 13-2. Roof covered with wood fiber roofing panels. *(Courtesy Masonite Corp.)*

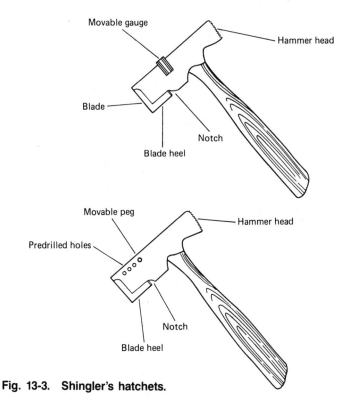

Fig. 13-3. Shingler's hatchets.

Tools and Equipment

Most of the tools and equipment used in other types of roofing are also used when applying wood shingles or shakes. These tools and various types of equipment are described in chapter 2. Many roofers also use a shingler's or roofer's hatchet similar to those illustrated in figure 13-3. This is a multiple-purpose tool that is used to align, nail, or split the shingles or shakes. Each hatchet is equipped with an adjustable weather exposure gauge for cutting or fitting the roofing pieces.

Roof Deck Preparation

Wood shingles will absorb potentially damaging moisture unless there is adequate air circulation beneath them. For this reason, they are nor-

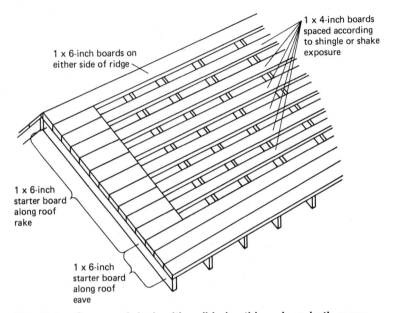

1 x 6-inch boards on
either side of ridge

1 x 4-inch boards
spaced according
to shingle or shake
exposure

1 x 6-inch
starter board
along roof
rake

1 x 6-inch
starter board
along roof
eave

Fig. 13-4. Open roof deck with solid sheathing along both eaves and rakes.

mally applied over spaced sheathing with no underlayment. Examples of spaced sheathing are illustrated in figures 13-4 and 13-5. Closed sheathing is often used along the eaves and rakes if the undersides of the rafters are left open.

If the local building code requires closed (solid) sheathing for a wood shingle roof, the roof deck must first be covered with an underlayment of roofing felt and then built up with 1 × 4–inch wood furring strips to provide adequate air circulation beneath the shingles (fig. 13-6). Leave ⅛-inch gaps in the furring strips to provide for roof drainage.

The 1 × 4's used in spaced sheathing or to build up a closed roof deck should be spaced on center the same distance as the shingle or shake exposure selected for the roof.

Because wood shakes are produced with rough, irregular surfaces, enough air can circulate around them to prevent moisture accumulation and absorption. As a result, they can be applied over either spaced (open) or closed sheathing. Closed sheathing is recommended for areas where heavy snows or strong winds are common.

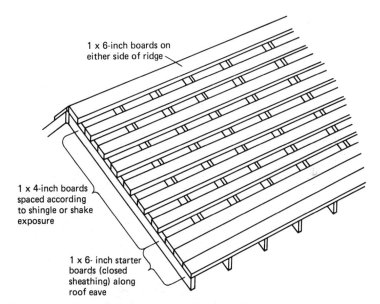

Fig. 13-5. Open roof deck with solid sheathing along eaves.

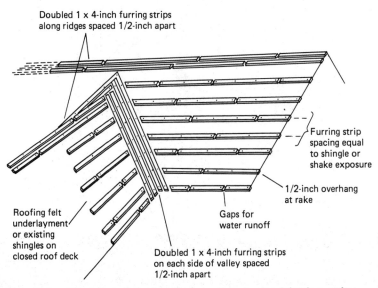

Fig. 13-6. Closed roof deck with 1 × 4–inch wood furring strips.

Underlayment

CLOSED ROOF DECKS

Sweep the surface clean and cover any knot holes with small pieces of tin. Use mastic tape to secure the tin to the sheathing. Cover the roof deck sheathing with an underlayment of 36-inch wide 15-pound asphalt-saturated roofing felt. Lap the roofing felt 4 inches along the horizontal joint and 6 inches along the vertical joint. Nail 1 × 4–inch wood furring strips over the underlayment and space them according to the shingle or shake exposure.

OPEN ROOF DECKS

An underlayment is not used over spaced sheathing if wood shingles are applied to the roof. If shakes are applied, lay a 36-inch wide strip

Fig. 13-7. Construction details of shake roof with roofing felt interply between each shake course. *(Courtesy Red Cedar Shingle & Handsplit Shake Bureau)*

Fig. 13-8. Step flashing along side of chimney. *(Courtesy Red Cedar Shingle & Handsplit Shake Bureau)*

of 30-pound asphalt-saturated roofing felt along the roof edge and 18-inch wide strips between each shake course (fig. 13-7).

Flashing

Metal flashing is preferred for wood shingle or shake roofs (figs. 13-8 to 13-10). Copper flashing is often recommended, but other types of less expensive, rust and corrosion resistant metals can also be used just as effectively. Flashing application procedures are described in chapter 8.

Preformed metal drip edges are not used along the eaves and rakes of wood shingle roofs. On wood shake roofs, a drip edge should be nailed along the eave before the roofing felt is laid and along the rake afterward.

Fig. 13-9. Flashing details along bottom of chimney. *(Courtesy Red Cedar Shingle & Handsplit Shake Bureau)*

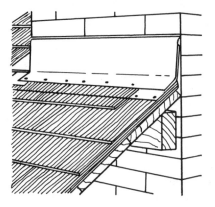

Fig. 13-10. Base and counter flashing used along chimney and vertical masonry walls. *(Courtesy Red Cedar Shingle & Handsplit Shake Bureau)*

Wood Shingle Roofs

Wood shingles are available as individual shingles or roof panels. Individual wood shingles are commonly produced in No. 1 Blue Label, No. 2 Red Label, and No. 3 Black Label roofing grades.

Wood shingles have two smooth sides that are produced by sawing both faces of a cedar block (fig. 13-11). Each of the three roofing grade shingles is cut in 16-, 18-, and 24-inch lengths with thicknesses at the butt of .40-, .45-, and .50-inch respectively.

Shingle grades are determined by the Red Cedar Shingle & Handsplit Shake Bureau. No. 1 Blue Label is the premium grade of wood shingles used in roofing. These shingles are cut entirely from the heartwood, and are 100 percent clear and 100 percent edge-grained. A flatgrain and a limited amount of sapwood are found in the No. 2 Red Label grade of wood shingles. No. 3 Black Label is the lowest grade of wood shingle used in roofing. It is strictly a utility grade shingle used primarily for inexpensive applications on secondary buildings.

To accurately estimate the number of shingles or roof panels required to cover a roof, the roof pitch or slope must be determined first (see chapter 2). After the roof pitch or slope has been determined, find the best maximum exposure to use. Exposure is the amount of uncovered shingle surface exposed to the weather. A well constructed shingle roof is usually three shingle layers thick through each exposed section of shingle (fig. 13-12). Once the roof pitch or slope and shingle exposure are known, the wood shingle type and size can be selected by using the information listed in table 13-1.

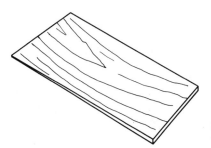

Fig. 13-11. Tapered, smooth-faced wood shingle.

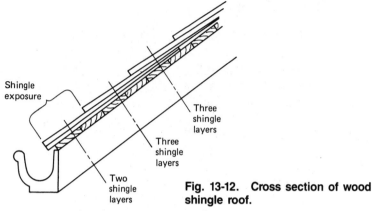

Shingle exposure

Three shingle layers

Three shingle layers

Two shingle layers

Fig. 13-12. Cross section of wood shingle roof.

Applying Wood Shingles

Proper weather exposure is important and depends largely on the pitch or slope of the roof. On roof slopes of 4-in-12 and steeper, the standard exposures for No. 1 grade shingles are 5 inches for 16-inch shingles, 5½ inches for 18-inch shingles, and 7½ inches for 24-inch shingles. On roof slopes less than 4-in-12, but not less than 3-in-12, reduced exposures of 3¾ inches, 4¼ inches, and 5¾ inches, respectively, are required. Reduced exposures are also recommended for No. 2 and No. 3 shingles on all 4-in-12 or steeper roof slopes (see table 13-1).

Use only rust-resistant nails when applying wood shingles. Rusting nails lose their holding power and create unsightly stains down the roof surface. Galvanized steel, aluminum, or copper roofing nails may be used to apply wood shingles. Use 3d nails for 16-inch and 18-inch

Table 13-1. Wood Shingle Types, Sizes, and Exposures (*Courtesy Red Cedar Shingle & Handsplit Shake Bureau*)

	Maximum exposure recommended for roofs								
Pitch or Slope	No. 1 Blue Label			No. 2 Red Label			No. 3 Black Label		
	16"	18"	24"	16"	18"	24"	16"	18"	24"
3-in-12 to 4-in-12	3¾"	4¼"	5¾"	3½"	4"	5½"	3"	3½"	5"
4-in-12 and steeper	5"	5½"	7½"	4"	4½"	6½"	3½"	4"	5½"

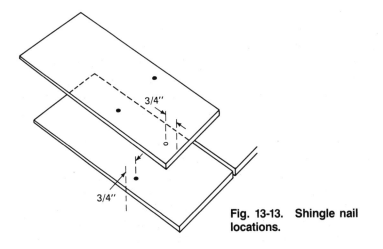

Fig. 13-13. Shingle nail locations.

shingles and 4d nails for 24-inch shingles *in new construction*. A threaded, ring-shank nail is sometimes recommended for plywood roof sheathing less than ½ inch thick.

Only two nails should be used in each shingle. The *exact* location of the nails will depend on the amount of shingle exposure that is dictated by roof pitch, the type of coverage, and the shingle manufacturer's recommendations. To avoid splitting the wood, nails should be placed approximately ¾ inch from each edge of the shingle. Make certain that the nails are also located at least ¾ inch to 1½ inches above the butt line of the next course so that they will be completely covered by those shingles (fig. 13-13). Drive the head of the roofing nail flush against the surface of the shingle. This will provide sufficient holding power. *Never* drive the nail so hard that its head crushes the wood (fig. 13-14).

Lay the shingles with a ⅛-inch to ¼-inch space between them to allow for expansion. A 1½-inch side lap is recommended between the joints in successive shingle courses. Never allow two joints to be aligned when separated by a single course of shingles.

The bottom edge of a shingle course can be kept in a straight line by butting the shingles against a 1 × 4–inch board nailed temporarily to the roof (fig. 13-15). Thatch, weave, and other application styles are possible by varying the shingle butt line.

Begin shingling by laying a double or triple thick starter course along the roof eave (fig. 13-16). The best quality shingle will be on

Correct

Incorrect

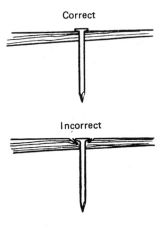

Fig. 13-14. Correct and incorrect methods of driving nails. *(Courtesy Red Cedar Shingle & Handsplit Shake Bureau)*

top. Allow the starter course to extend approximately 1½ inches beyond the eave or the end of the roof sheathing to insure proper drainage into the gutters (fig. 13-17). The starter course and all other shingle courses should also extend 1½ inches beyond the rake (gable) edge at each end of the roof.

Lay successive shingle courses up the roof to the ridge line with each course overlapped to provide the required exposure. Remember to maintain a separation of at least 1½ inches between joints in adjacent courses. Do not allow direct alignment between joints in alternate courses (fig. 13-18).

Laying a 36-inch wide strip of roofing felt along the roof eaves will provide some protection against ice dam formation during the

Board temporarily nailed to shingles

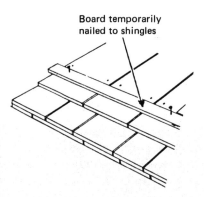

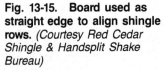

Fig. 13-15. Board used as straight edge to align shingle rows. *(Courtesy Red Cedar Shingle & Handsplit Shake Bureau)*

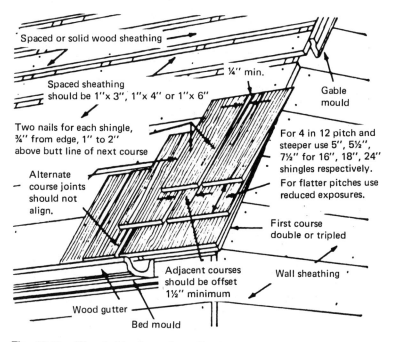

Spaced or solid wood sheathing

¼" min.

Spaced sheathing should be 1"x 3", 1"x 4" or 1"x 6"

Gable mould

Two nails for each shingle, ¾" from edge, 1" to 2" above butt line of next course

For 4 in 12 pitch and steeper use 5", 5½", 7½" for 16", 18", 24" shingles respectively. For flatter pitches use reduced exposures.

Alternate course joints should not align.

First course double or tripled

Adjacent courses should be offset 1½" minimum

Wall sheathing

Wood gutter

Bed mould

Fig. 13-16. Wood shingle roof application details. *(Courtesy Red Cedar Shingle & Handsplit Shake Bureau)*

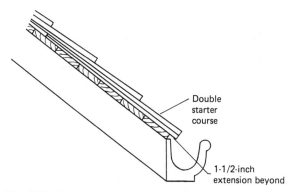

Double starter course

1-1/2-inch extension beyond

Fig. 13-17. Shingle extension beyond sheathing along roof eave. *(Courtesy Red Cedar Shingle & Handsplit Shake Bureau)*

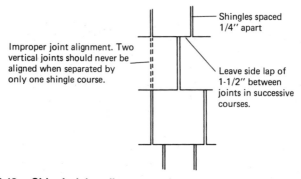

Improper joint alignment. Two vertical joints should never be aligned when separated by only one shingle course.

Shingles spaced 1/4" apart

Leave side lap of 1-1/2" between joints in successive courses.

Fig. 13-18. Shingle joint alignment, side lap, and spacing. *(Courtesy Red Cedar Shingle & Handsplit Shake Bureau)*

cold winter months. Apply the roofing felt up the roof deck far enough to extend just beyond the inside surface of the exterior walls (fig. 13-19).

Hip, Ridge, and Rake Details for Wood Shingle Roofs

The hips and ridges should be of the Boston-type construction with covered (protected) nailing. Hip and ridge details are illustrated in figure 13-20. Either the 6d- or 10-d length nail is recommended for applying hip and ridge shingles. The nail must be long enough to penetrate the underlying sheathing.

Factory assembled hip and ridge units are available from the shingle manufacturer. Hip and ridge shingles can also be cut to size at the site, but extreme care must be used to maintain a uniform width.

The same amount of exposure should be used when applying hip and ridge shingles as was used for the shingles on the rest of the roof. Overlap the hip or ridge shingles so that the laps alternate (see fig. 13-20). Cut and plane the overlapping shingles to form an even edge with the lapped ones. Ridge flashing is often installed under the shingles of a Boston-type ridge to provide additional protection to the joint (fig. 13-21).

As shown in figure 13-20, hip shingling should begin at the eave with a double starter course. Two 6d or 10d nails are used with each shingle and placed where they will be covered by the next hip shingle. Continue up the hip with *single* hip shingles overlapped to provide the required exposure. Cut the edges of the last shingle to provide a snug fit with the ridge or another hip.

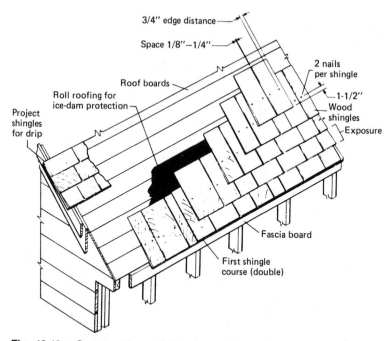

Fig. 13-19. Construction details of wood shingle roof.

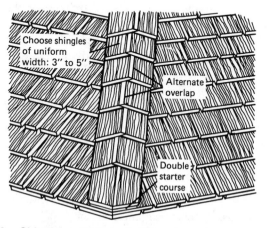

Fig. 13-20. Shingle roof hip and ridge construction details.
(Courtesy Red Cedar Shingle & Handsplit Shake Bureau)

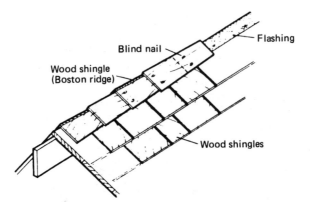

Fig. 13-21. Flashing installed under wood shingles of Boston-type ridge.

The ridge is shingled by starting at one end of the roof and working toward the other or at both ends and working toward the center. In either case, the first shingles are laid doubled just as they are when shingling a hip (see fig. 13-20). The two ridge shingles are laid with the joints alternately facing opposite directions and secured to the roof deck with 10d nails. The nails should be placed where the next overlapping shingle will cover them. Continue the ridge by laying *single* overlapping shingles along the roof ridge line. The amount of overlap will depend on the exposure required. The nails must be placed where they will be covered by the overlap of the next shingle. At the opposite end of the roof, nail a second shingle over the last shingle in the ridge course. The last shingle will be laid with its tip or tail facing outward from the end of the roof. The shingle applied over this one will be positioned with its butt end facing outward.

The shingles ending at each rake must be cut and planed to form an even, straight edge. Allow the shingles to extend (overhang) approximately 1½ inches beyond the edge of the roof at the rakes.

Valley Details for Wood Shingle Roofs

Cover the valley with a suitable metal flashing before applying the shingles. Leading wood shingle manufacturers recommend 26-gauge, center-crimped, painted galvanized steel or aluminum for valley flashing (fig. 13-22).

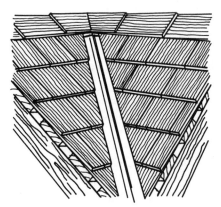

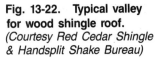

Fig. 13-22. Typical valley for wood shingle roof. *(Courtesy Red Cedar Shingle & Handsplit Shake Bureau)*

The width of the flashing will depend on the pitch of the roof. The flashing should extend a minimum of 7 inches from both sides of the valley center line on roofs with one-half pitch or steeper. Roofs with less than one-half pitch should have valley flashing that extends at least 10 inches from both sides of the valley center line (fig.13-23).

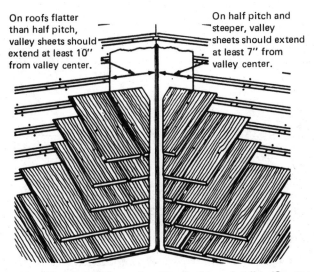

Fig. 13-23. Wood shingle roof valley flashing details. *(Courtesy Red Cedar Shingle & Handsplit Shake Bureau)*

Shingles that extend into the valley should be cut to the proper miter and their edges should form a straight line parallel to the valley center line. Never break joints into the valley, lay shingles with their grain parallel with the center line of the valley, or nail through the metal flashing.

Swept Eave Shingling

Some roofs have eaves that curve or sweep upward (fig. 13-24). Wood shingles can be applied to this type of roof if the curve is no greater than one inch per foot. The amount of exposure is determined both by roof pitch and the type of wood shingle used. When in doubt, always follow the shingle manufacturer's recommendations.

Figure 13-25 illustrates the construction details of a typical swept eave roof. The shingles must be soaked in water overnight and installed the next day. Soaking the shingles makes them more pliable and insures that they will conform to the curve of the roof without splitting. Begin by laying a double course of shingles along the eave with approximately 1½-inch overhang. The remaining courses are applied in the conventional manner.

Fig. 13-24. Swept eave roof. *(Courtesy Masonite Corp.)*

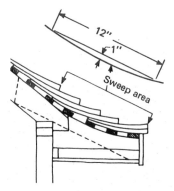

Fig. 13-25. Construction details of swept eave roof. *(Courtesy Red Cedar Shingle & Handsplit Shake Bureau)*

Shingling Convex, Concave, and Apex Roof Junctures

CONVEX JUNCTURE

The convex juncture is commonly found on barn roofs (fig. 13-26). The section of the roof above the juncture usually has a normal pitch. At the juncture line, however, the pitch of the lower section of roof becomes much steeper. Construction details of a typical convex juncture are shown in figure 13-27.

When shingling a convex juncture, the lower portion or slope of the roof is shingled first. Shingling begins at the eaves with a double starter course. After the final shingle course has been nailed in place, a metal flashing strip is installed along the juncture. The flashing strip should be about 8 inches wide. It must be wide enough to cover the

Fig. 13-26. Roof with convex juncture.

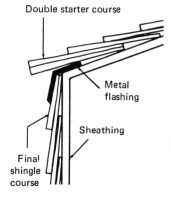

Double starter course

Metal flashing

Sheathing

Final shingle course

Fig. 13-27. Construction details of typical convex roof juncture. *(Courtesy Red Cedar Shingle & Handsplit Shake Bureau)*

nails in the final course of shingles on the lower slope. Bend the metal flashing strip so that it conforms to the angle formed at the juncture, allowing approximately 4 inches to overlap each slope. Bend the metal carefully to avoid fracturing it.

After you have installed the flashing strip, apply a double starter course along the edge of the upper roof slope. Approximately 1 to 1½ inches of overhang may be allowed if so desired. Apply the remaining shingle courses on the upper slope in the usual manner.

Fig. 13-28. Roof with concave juncture.

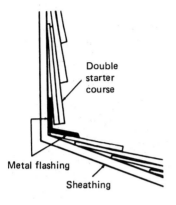

Double
starter
course

Metal flashing

Sheathing

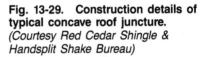

Fig. 13-29. Construction details of typical concave roof juncture. *(Courtesy Red Cedar Shingle & Handsplit Shake Bureau)*

CONCAVE JUNCTURE

A concave roof juncture is formed when the upper portion of a roof assumes a steeper pitch than the lower one (fig. 13-28). The construction details of a typical concave roof juncture are shown in figure 13-29.

Begin shingling a concave roof juncture by laying a double starter course along the eave or lower slope. After applying the final course of shingles on the lower roof slope, install a metal flashing strip conforming to the angle formed at the concave juncture. The flashing strip should be wide enough to cover the nails in the final course of shingles on the lower slope.

Shingling of the upper slope should begin with a double starter course applied along its lower edge. The remaining courses are applied in the usual manner.

APEX JUNCTURE

Some roofs have angles which form apex junctures. Construction details of a typical apex roof juncture are shown in figure 13-30. Note that the apex juncture shown is formed by a vertical wall and roof slope.

Shingling must be completed to the juncture. Before the juncture itself is covered, apply a 12-inch wide metal flashing strip over the juncture. Bend the flashing strip to conform to the angle formed by the juncture, but allow 8 inches of the strip to cover the roof slope and 4 inches to cover the top of the wall.

Complete the roof slope by laying a course of shingles over the metal flashing. Allow the shingle tips of this course to extend slightly

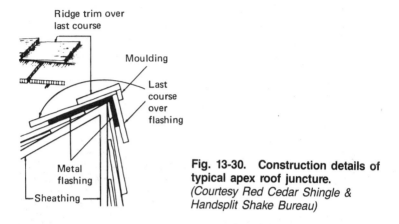

Fig. 13-30. Construction details of typical apex roof juncture. *(Courtesy Red Cedar Shingle & Handsplit Shake Bureau)*

beyond the juncture line. Complete the wall by laying shingles up to the juncture line. Trim the tip from each shingle in the last wall course to provide a snug fit with the overlapping roof shingles.

Apply a molding strip along the top edge of the last course of wall shingles. As shown in figure 13-30, the top edge of the molding strip is flush with the top edge of the last course of roof shingles. Apply a single wood strip along the roof parallel with the juncture and overlapping the molding strip.

Wood Shingle Repairs

The most frequent problems encountered with a wood shingle roof are cracked or splintered shingles, loose roofing nails, and shingles that have been lifted by the wind.

If the roof deck has an underlayment, check its condition before attempting repairs. The roofing felt should be black and flexible. If the roofing felt is crumbly, dry, and hard, the existing roof should be removed and the roof deck reroofed.

If both sections of a cracked shingle remain, nail them to the roof and cover both the nail heads and joint with a suitable roofing cement (fig. 13-31). The nail holes should be drilled before nailing to prevent splitting the shingle. Where the damage is more extensive and a replacement shingle is not available, a temporary repair can be made by inserting a piece of galvanized steel or aluminum under the damaged shingle and nailing through the shingle and metal with two nails (fig. 13-32). The piece of galvanized steel should be wide

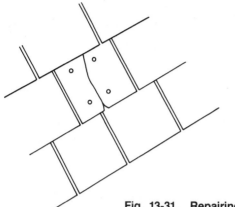

Fig. 13-31. Repairing cracked shingle.

enough to extend 2 inches beyond both edges of the shingle and four inches under the butt line of the overlying shingle in the next course.

Replace badly curled shingles, those with extensive splintering or other signs of deterioration, and cracked or split shingles with missing pieces. Individual shingles can be removed by first splitting them and removing the pieces, and then cutting the roofing nails with a shingle ripper or hacksaw (fig. 13-33). It may be necessary to lift the

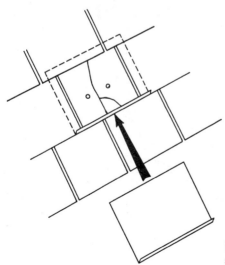

Fig. 13-32. Temporary shingle repair.

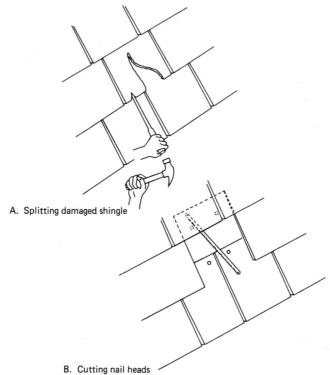

A. Splitting damaged shingle

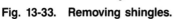

B. Cutting nail heads

Fig. 13-33. Removing shingles.

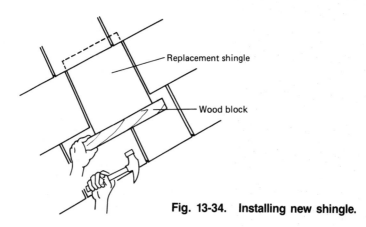

Replacement shingle

Wood block

Fig. 13-34. Installing new shingle.

overlying shingle slightly to provide enough room to insert the shingle ripper or hacksaw. After the shingle has been removed, make certain that the old nails are cut off flush with the roof deck sheathing or underlayment. Take care when cutting them not to damage the sheathing or underlayment.

Trim the replacement shingle to the required width, slide the new shingle into place, and tap it gently with a hammer and wood block to align its butt with the other shingles in the course (fig. 13-34). Nail the shingle to the roof deck and cover the nail heads with roofing cement to protect them from corrosion or leakage through the nail holes.

Lifted shingles can often be nailed back down in place. Use two to four nails and cover the nail heads with roofing cement. If the shingle is bowed, split it down the center and remove about ¼ inch of wood from the inside edge of one section to form a joint for roofing cement. Nail the two sections on either side of the joint. Cover the nails and fill the joint with roofing cement (fig. 13-35).

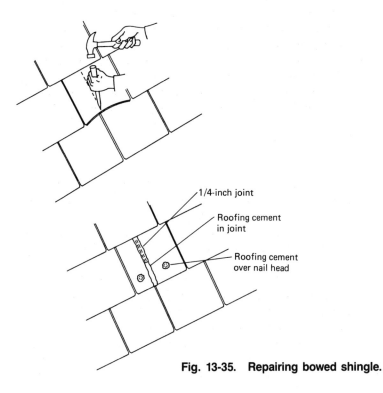

1/4-inch joint

Roofing cement in joint

Roofing cement over nail head

Fig. 13-35. Repairing bowed shingle.

Wood Shake Roofs

Wood shingles have a relatively smooth surface, whereas shakes have at least one highly textured, natural-grained split surface. The difference results from the methods used to produce them. Wood shingles are sawed. Shakes are split (rived) from cedar blocks (fig. 13-36).

Wood shakes are available as individual cedar shakes or roof panels. There are three types of wood shakes:

1. Straight-split
2. Tapersplit
3. Handsplit

STRAIGHT-SPLIT SHAKES

Straight-split shakes are cut (handsplit) from the same end of the cedar block with a heavy steel blade called a froe and a wooden mallet (see figure 13-36). They are hand split without reversing the block which

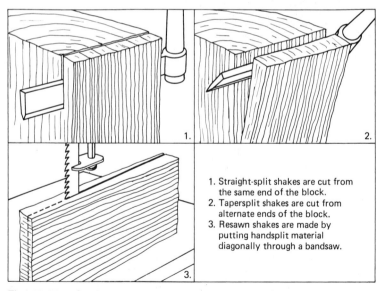

1. Straight-split shakes are cut from the same end of the block.
2. Tapersplit shakes are cut from alternate ends of the block.
3. Resawn shakes are made by putting handsplit material diagonally through a bandsaw.

Fig. 13-36. Cutting straight-split, tapersplit, and handsplit shakes.
(Courtesy Red Cedar Shingle & Handsplit Shake Bureau)

Fig. 13-37. Straight-split shake.

results in a medium-textured shake of generally uniform thickness (fig. 13-37). Each shake is 3/8 inch thick and available in either 18-inch or 24-inch lengths. Straight-split shakes are sometimes called barn shakes and are applied only when triple coverage is desired.

TAPERSPLIT SHAKES

Tapersplit shakes are cut from alternate ends of the cedar block with a froe and mallet. To obtain the tapered thickness for these shakes, the block is turned end for end after each split is made (fig. 13-38). These medium-textured shakes are approximately ½ inch thick at their butt end and 24 inches long. They may be applied to provide either double or triple coverage.

Fig. 13-38. Tapersplit shake.

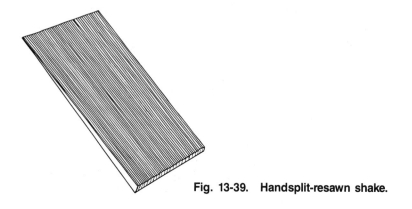

Fig. 13-39.   Handsplit-resawn shake.

HANDSPLIT SHAKES

Handsplit (or resawn) shakes have one heavily textured side and one smooth side. They are produced by splitting a cedar block into boards of the desired thickness with a froe and mallet and then passing the board through a thin band-saw to form two tapered shakes (fig. 13-39). These shakes are thicker and heavier through the butt than other shakes, ranging in thickness from ½ to ¾ inch. They are produced in 18-, 24-, and 32-inch lengths, and may be applied to provide either double or triple coverage.

A typical wood shake roof panel is shown in figure 13-40. It consists of sixteen 24-inch cedar shakes bonded to ½ inch thick exterior sheathing grade plywood. Each panel is 8 feet long. Only 15 panels are required to cover 100 square feet of roofing area at a 10-inch exposure. Each of the 24-inch shakes on the roof panel has a handsplit face 10 inches to 12 inches from the butt. They are sawn the rest of the way to form a uniform extraheavy tip.

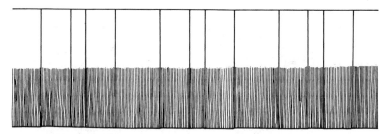

Fig. 13-40.   Typical wood shake roof panel.

Applying Wood Shakes

Begin roofing by laying a 36-inch wide strip of 30-pound roofing felt over the sheathing boards along the roof eave. Lay a double or triple layer starter course at the eave line. The bottom course or courses can be either 15-inch or 18-inch shakes. The 15-inch shake is designed specifically for this purpose.

Secure each handsplit shake to the sheathing with two galvanized steel, aluminum, or copper roofing nails. Drive the nails into the shake about one inch from each edge, and one or two inches above the butt line of the overlapping shakes of the next course. The nails must be long enough to penetrate at least ½ inch into the sheathing boards. The two-inch length of the sixpenny (6d) nail is usually long enough, but longer nails may have to be used for special shake thicknesses or exposures. The nails should be driven in flush with the shake surface. Avoid driving the nails so hard that they damage the wood.

After the starter course has been laid, lay an 18-inch wide strip of 30-pound asphalt-saturated roofing felt over the top portion of the shakes in each course. Lay the roofing felt interply strip with its bottom edge located above the butt at a distance equal to twice the exposure length of the shakes. This allows the top edge of the roofing felt to extend onto the sheathing.

Figure 13-7 illustrates the construction details of a shake roof with a roofing felt interlay inserted between each shake course. The 24-inch shakes used in this example are laid with a 10-inch exposure. This exposure requires that the bottom edge of the roofing felt be applied 20 inches above the shake butts (twice the exposure distance). As a result, 4 inches of the roofing felt cover the tops of the shakes and the remaining 14 inches extend onto the sheathing boards. Note that the use of a roofing felt interply is not necessary when tapersplit or straight-split shakes are applied in snow-free areas at weather exposures less than one third the shake length.

The space between shakes should be approximately ¼ to ⅜ inch to allow for expansion when moisture is absorbed. These spaces or joints should be offset by at least ½ inch adjacent courses. In 3-ply roof construction, the joints in alternate rows should not be in direct alignment.

Straight-split shakes differ from taper-split shakes by having the same thickness throughout their length. When applying straight-split shakes, the smooth end should be laid uppermost. Doing so will produce a tighter and more weather-resistant roof.

Ridges are constructed in much the same way as hips. On un-broken ridges that terminate in a gable at each end, however, begin by laying a double starter course at each gable and then work from both gables to the center of the ridge. When the center is reached, splice the two courses with a small saddle of shake butts.

Hip, Ridge, and Rake Details for Wood Shake Roofs

Prefabricated shakes are available from manufacturers for use on hips and ridges. Shakes can also be cut to size at the site for this purpose. In either case, an 8-inch wide strip of 15-pound roofing felt must first be placed over the ridge and hip crowns before any shakes are laid.

Select shakes approximately 6 inches wide if the hips and ridges are to be fabricated at the site. Nail wood guide strips to the roof on both sides of the hips and ridges. Each guide must be positioned exactly 5 inches from the hip and ridge center lines.

Use nails long enough to penetrate at least ½ inch into the underlying sheathing boards. Two 8d nails on each side of the shake should be adequate for this purpose.

Nail the first hip shake in place with its long edge resting against the wood guide strip and its bottom edge flush with the edge of the

Fig. 13-41. Shake hip roof construction details. *(Courtesy Red Cedar Shingle & Handsplit Shake Bureau)*

roof. Cut back the edge of the shake projecting over the center of the hip to form a bevel. Apply the shake on the opposite side and cut back the projecting edge to fit the bevel cut of the first shake. Double the starting course and then proceed up the hip by laying a single shingle on either side of the hip line. The shakes in the successive courses are applied alternately in reverse order (fig. 13-41). Use the same exposure for the hip shakes as was given for the shakes on the roof.

Valley Details for Wood Shake Roofs

Either an open or a closed valley can be used on a wood shake roof, but the former is the more common of the two. The construction details of a typical open valley are shown in figure 13-42.

Both types of valleys are lined with roofing felt and metal flashing.

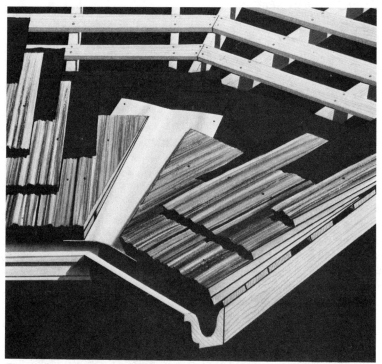

Fig. 13-42. Construction details of typical open valley. *(Courtesy Red Cedar Shingle & Handsplit Shake Bureau)*

Construction of an open valley begins by laying a strip of 15-pound roofing felt over the sheathing the entire length of the valley. The roofing felt is overlaid with metal flashing at least 20 inches wide with a 4- to 6-inch head lap. Galvanized metal flashing should be painted on both sides with a good grade of metal paint to provide protection against corrosion. Copper or aluminum flashing do not require painting. Shakes adjoining the valley must be trimmed parallel to the center line.

Begin the construction of a closed valley by nailing a 1 × 6-inch wood strip into the saddle. Cover the wood strip with 15-pound roofing felt and then lay metal flashing over it. The metal flashing should be 20 inches wide with a 2 inch head lap. Lay the shakes into the valley and trim their edges to form a joint extending along the valley center line.

Shake Roof Variations

Distinctive shake roof designs can be created by using special application methods. The one shown in figure 13-43 produces a perspective or dimensional effect. It is obtained by gradually reducing the exposure of each shake course from the eaves to the ridge line. It requires the use of 24-inch shakes on the lower half of the roof and 18-inch shakes on the upper half. Application of the shakes begins at the eaves with 24-inch shakes laid with a 10-inch exposure. The exposure is gradually reduced to 8½ inches as successive courses are laid over the bottom half of the roof. The courses of 18-inch shakes used to cover the upper half of the roof begin with an 8-inch exposure that is gradually reduced to 5 inches. The gradual reduction of exposure is illustrated in figure 13-44.

A textured appearance can be created by laying the shakes with their butts placed slightly above or below the horizontal line of each course (fig. 13-45). This application method produces an irregular and random roof pattern. Still greater irregularity can be obtained by interspersing longer shakes at random points along each course. A more rugged appearance is possible by mixing rough and relatively smooth surface shakes.

The gable line of a shake roof can be tilted upward slightly by inserting a strip of cedar bevel siding the full length of each gable end (fig. 13-46). The bevel siding strip is installed with its thick edge flush with the edge of the roof deck sheathing. Tilting the gable line results in an inward pitch of the roof surface. This inward pitch

Fig. 13-43. Shake roof with perspective or dimensional effect. *(Courtesy Red Cedar Shingle & Handsplit Shake Bureau)*

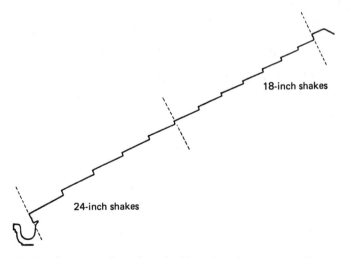

18-inch shakes

24-inch shakes

Fig. 13-44. Cross section of roof with reduced exposure. *(Courtesy Red Cedar Shingle & Handsplit Shake Bureau)*

Fig. 13-45. Shake roof with textured appearance. *(Courtesy Red Cedar Shingle & Handsplit Shake Bureau)*

accentuates the gable line and channels moisture away from the gable edge.

Wood Shake Repairs

The procedures used to repair or replace wood shingles also apply to wood shakes (see the discussion of wood shingle repairs). When removing wood shakes, however, special care should be taken to avoid damaging the roofing felt overlapping each shake course.

Fig. 13-46. Shake roof with tilted gable line. *(Courtesy Red Cedar Shingle & Handsplit Shake Bureau)*

Shingle and Shake Panel Roofs

Panelized shingles and shakes are available from several manufacturers for use in roofing and reroofing. These roof panels eliminate the need to handle the many small pieces normally found in traditional shingle and shake roofing applications. As a result, the work progresses much

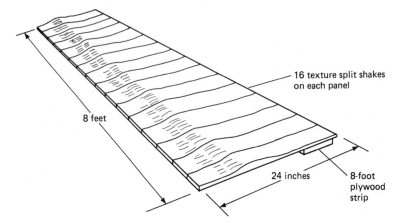

Fig. 13-47. Construction details of Shakertown roof panel.

more quickly and with less waste, clutter, and clean-up. The panels are easy to handle, usually self-aligning, and do not require workers with special skills or training to apply them.

Both cedar and wood fiber roof panels are available for use in new construction and reroofing. Cedar panels commonly consist of up to 16 shingles or shakes bonded to an 8-foot long plywood base (see fig. 13-40). Construction details of typical cedar panels are shown in figures 13-47 and 13-48. Each panel combines both the shingles or shakes and the sheathing in a single unit. As a result, they may be applied over both open and closed (solid) roof decks.

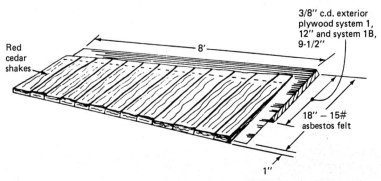

Fig. 13-48. Construction details of Foremost-McKesson roof panel.
(Courtesy Foremost-McKesson Building Products, Inc.)

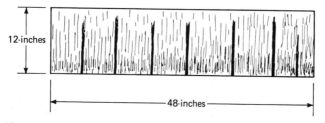

Fig. 13-49.   Masonite wood fiber roof panel.

Wood fiber (hardboard) roof panels are manufactured to resemble the rough textured surface of handsplit shakes (fig. 13-49). Each panel is a nominal 12 × 48 inches and must be applied over a closed roof deck with sheathing a minimum ½ inch thick. Because wood fiber roof panels lack the structural strength of cedar shingle or shake panels, they cannot be applied over spaced sheathing. In reroofing, wood fiber panels can be applied over a single layer of old roofing in relatively good condition. More than one layer of old roofing or roofing in poor condition should be stripped down to the bare roof deck.

Applying Cedar Panels

Cedar shingle or shake roof panels are normally used on roofs with slopes of 4-in-12 or more. They should be nailed to the sheathing (closed roof decks) or directly to the rafters (open roof decks) with eightpenny (8d) galvanized box head nails.

Roofing felt is not required with cedar shingle roof panels or shake panels having shakes with less than an 8-inch exposure. Cedar shake roof panels with exposures greater than 8 inches require an 18-inch wide strip of 30-pound roofing felt applied as an interply between the panel courses.

To apply the roof panels, begin by snapping a chalk line or nailing a 1 × 6–inch straight edge near the ends of the rafters. Position the chalk line or straight edge (starter strip) so that the starter roof panel will carry water runoff to the center of the gutter. Make sure the chalk line or straight edge is exactly parallel to the ridge line because it will be used to align the starter panels (fig. 13-50).

Starting at the gable edge, nail the starter roof panels at each rafter with two nails. Drive the first nail into the panel 1 inch down

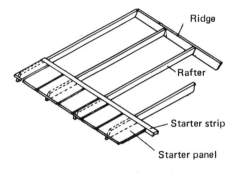

Fig. 13-50. Straight
edge (starter strip) used
to align starter panel.
*(Courtesy Shakertown
Corp.)*

from the top of the plywood strip and the second nail 2 inches up
from the bottom of the strip. Make certain that the ends of the starter
panels break on rafter centers. Double the starter panels by laying
full size shingle panels directly over them (fig. 13-51). Although the
panels are designed to center on 16-inch or 24-inch rafter spacing,
they may meet between them. When this happens, make certain that
the panels on adjacent courses do not end between the same two
rafters. Plywood chips may be used between panel joints to provide
a stronger roof underlayment.

Cut each panel at the rakes so that its outside edge is flush with
the outside edge of the rafter. Remove the last shingle and apply the
rake molding over the plywood edge of the sheathing so that the top
edge of the molding is flush with the top edge of the sheathing (fig.
13-52). Apply a shingle or shingles to obtain the desired overlap and
lay the next course of panels with a minimum 1½-inch offset. Vertical joints in alternate courses should be varied enough to avoid a
straight line.

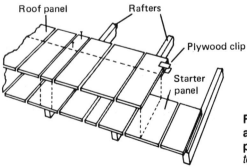

Fig. 13-51. Roof panel
applied over starter
panel. *(Courtesy Shaker-
town Corp.)*

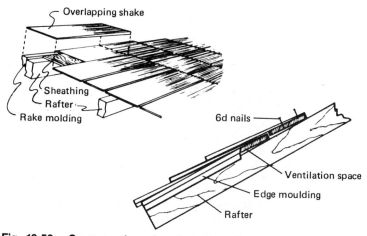

Fig. 13-52. Cross section of roof at rake. *(Courtesy Shakertown Corp.)*

Use a saber saw to cut the panels around vents and other roof projections. Install the metal flashing between the top and bottom panels with a small portion extending beyond the bottom edge of the two panels to minimize ice-dam formation (fig. 13-53).

A saber saw should also be used to cut the panels around the chimney if the flashing is installed prior to roofing. Flashing methods are described in chapter 8. If the panels are applied first, remove some of the shingles around the chimney and weave individual shingles into the flashing.

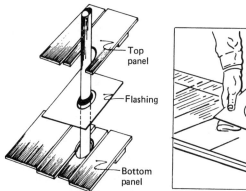

Fig. 13-53. Vent pipe details.

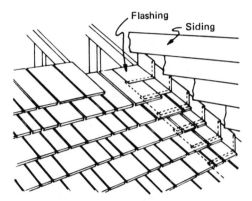

Fig. 13-54. Vertical wall flashing. *(Courtesy Shakertown Corp.)*

Flashing at the juncture of a vertical wall with the roof should be applied after each panel course is laid (fig. 13-54). Nail the flashing above the exposure line and near the outside edge of the flashing.

Boards measuring 1 × 6 inches should be recessed on both sides of the valley rafter to provide a strong support base for the valley flashing. As shown in figure 13-55, the boards should be installed so that their top surface is flush with the top of the valley rafter.

Roofs of ½ pitch or steeper should have at least 9 inches of flashing metal on either side of the valley center line. Roofs of lesser pitch should have wider valley metal. If the slope of the roof on one side of the valley is steeper than the opposite side, extend the flashing further up the side of lesser pitch to accommodate a faster flow of water runoff.

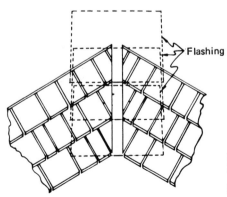

Fig. 13-55. Recessed 1 × 6–inch boards. *(Courtesy Shakertown Corp.)*

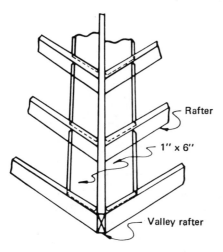

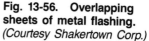

Fig. 13-56. Overlapping sheets of metal flashing. (Courtesy Shakertown Corp.)

Use the same flashing metal recommended by the Red Cedar Shingle & Handsplit Shake Bureau for conventional shingle roofs. Apply precut flashing sheets after each panel course is laid (fig. 13-56). Nail the flashing near its outside edge and well above the panel exposure line. The service life of the flashing can be extended by painting the sheets on both sides with a suitable metal paint before they are installed.

Ridge units are available from the panel manufacturer for use on both hips and ridges. To construct a hip, cut the panels at the proper angle to fit the hip line and nail them to both sides of the hip. Then, cover the length of the hip crown with a narrow strip of roofing felt and apply the shingle ridge units over the felt with the same exposure as the rest of the roof (fig. 13-57). The ridge is constructed in a similar manner.

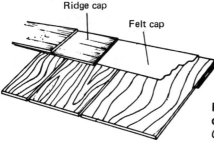

Fig. 13-57. Ridge and hip details. (Courtesy Shakertown Corp.)

Applying Wood Fiber Panels

Wood fiber panels must be installed over closed roof decks of ½-inch to ⅜-inch thick plywood or wood sheathing boards (maximum 6-inch nominal width) on roofs with slopes of 4-in-12 or greater. The thickness of the plywood will depend on the type of wood fiber panel used. For example, Masonite Woodruf Traditional panels require ½-inch thick CDX plywood sheathing, whereas Masonite Woodruf Rustic panels are installed over ⅜-inch thick exterior plywood conforming to APA specifications. Check the roofing manufacturer's installation instructions or ask for information at a local building and supply outlet before installing the panels. Note that wood fiber roof panels should not be installed directly over roof insulation such as wood fiber, foam, fiberglass, or perlite.

Attach each panel to the roof deck with large-head galvanized steel roofing nails. The nails should be no less than 12-gauge metal with ⅜-inch diameter heads. Each nail should be long enough to penetrate through the panel and plywood deck or at least ¾ inch into solid wood sheathing. If staples are used as fasteners, they must be at least 16-gauge metal and long enough to penetrate 1 inch into the wood deck. The staple crown must be a minimum of ⁷⁄₁₆ inch and must bear tightly against the surface without penetrating it. *Do not countersink or drive nails or staples at an angle or leave them exposed to the weather.*

Nail a 1 × 3–inch wood starter strip along the roof eave and nail a metal drip edge over the starter strip (fig. 13-58). If the roof has a slope of 4-in-12 or greater, cover the roof deck sheathing with an underlayment of nonperforated No. 15 (or heavier) asphalt-saturated roofing felt. (The South Florida Building Code requires an underlayment of No. 30 felt.) On roofs with low slopes, install a double layer of No. 15 felt or a single layer of No. 30 felt. Lap the 36-inch wide felt 3 inches along the top, 6 inches along the sides, and carry the felt 12 inches over ridges, hips, and valleys from both sides. Stagger the side laps 7 inches apart to avoid forming a continuous vertical seam. Nail a metal drip edge to each rake *after* the felt has been laid.

Install the first course of panels along the roof eave beginning at either the left or right rake and working across the roof to the opposite rake. Install the panels with an overhang of approximately 1½ inches at the eave and not less than 1 inch at the rake. Use a chalk line to insure that each course is parallel to the ridge line. Fasten each

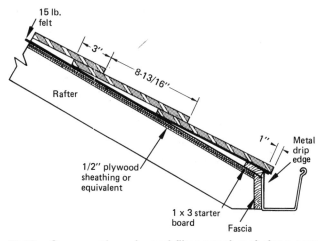

Fig. 13-58.   Cross section of wood fiber panel roof along eave.
(Courtesy Masonite Corp.)

panel to the roof deck with nails or fasteners spaced 8 inches apart
in a row 2¼ inches from the top edge (fig. 13-59). The number of
fasteners will vary from 5 to 8, depending on the type of panel and
local building code requirements. Insert a metal joint tab under each
butt joint as the panels are installed across the deck. The first fastener
is driven into the panel 2½ inches from the butt joint after the metal
joint tab has been inserted. Stagger butt joints at least 12 inches in

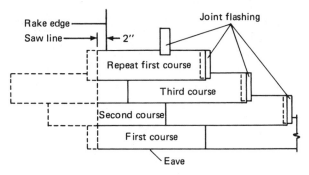

Fig. 13-59.   Installing wood fiber roof panels. *(Courtesy Masonite
Corp.)*

adjoining courses to avoid the possibility of forming a continuous vertical joint. Leave a 1/16-inch space between the panels at the butt joint. The head lap and exposure will vary, depending on the type of panel.

Valleys may be closed or open. Both require the same type of flashing. In either case, the roof panel can be cut with standard woodworking tools to conform to the valley angle.

The construction details of a typical open valley are illustrated in figure 13-60. To construct an open valley, begin by centering a 36-inch wide strip of 30-pound asphalt-saturated roofing felt in the valley and nailing it along its outside edges with enough nails to hold it firmly and smoothly in place. Then, install metal flashing of galvanized steel, aluminum, or copper so that it is centered over the roofing felt and nail it to the roof deck along its outside edges. If more than one section of flashing is used, lap each section a minimum of 6 inches and seal the joint with roofing cement.

Ridges and hips should be flashed with 6-inch wide metal or 30-pound asphalt-saturated roofing felt. Preformed hip and ridge caps, or shingles cut to a 6-inch width, are then applied over the flashing with the same head lap and exposure as the roof panels (figs. 13-61 and 13-62). Use two nails or staples for each side of the ridge or hip cap.

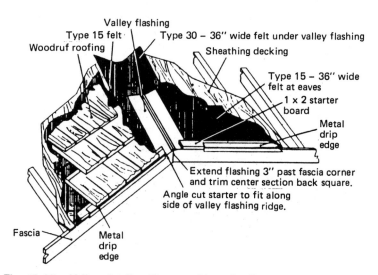

Fig. 13-60. Valley details. *(Courtesy Masonite Corp.)*

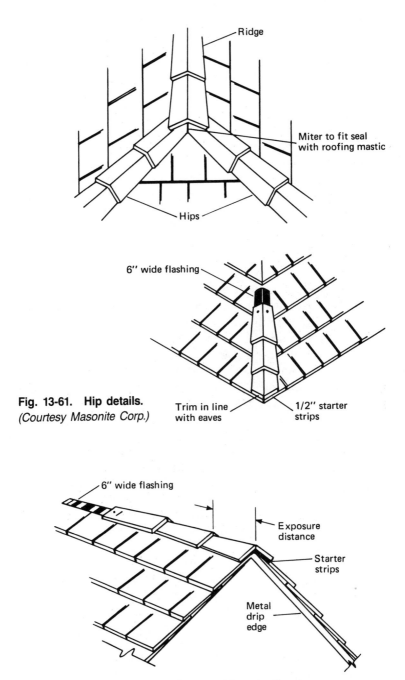

Fig. 13-61. Hip details. *(Courtesy Masonite Corp.)*

Ridge

Miter to fit seal
with roofing mastic

Hips

6" wide flashing

Trim in line
with eaves

1/2" starter
strips

6" wide flashing

Exposure
distance

Starter
strips

Metal
drip
edge

Fig. 13-62. Ridge details. *(Courtesy Masonite Corp.)*

Shingle and Shake Finishes

Cedar shingles and shakes, and wood fiber (hardboard) roof panels, are often applied to the roof unfinished. These roofing materials weather by stages, changing from their original color to an attractive medium gray or dark gray. In their natural state, they can last for years without any special attention. Cedar shingles, cedar shakes, and wood fiber (hardboard) roof panels can be painted or stained in solid colors or semitransparent tones to coordinate with other colors on the structure.

Reroofing

Cedar shingles can be applied over any roofing material except tile, slate, and mineral fiber (asbestos-cement) shingles. The shingles may be nailed directly over the old roofing material, to battens, or directly to the roof sheathing after the old roof covering has been removed.

Use fivepenny (5d) galvanized steel or aluminum roofing nails if the old roof covering is not removed. The nails must be long enough to penetrate the old shingles and the sheathing underneath. If the old roofing is removed, shorter nails may be used. The nails recommended for reroofing are rust-resistant and will therefore not stain the wood shingles.

If the old roof covering material is removed, check the roof deck sheathing for damage and make any necessary repairs. Check the condition of the flashing on closed decks and replace damaged sections. Apply the shingles as in open-deck new construction (see the discussion of applying wood shingles in this chapter). On a closed roof deck, battens will have to be nailed to the sheathing to provide proper ventilation for the shingles.

If the old roof covering material is to remain and the new shingles are nailed directly over the existing roofing, begin by cutting away a 6-inch strip of old roofing along the eaves and gables (figs. 13-63 and 13-64). Fill this space with 1 × 6-inch boards (fig. 13-65). Remove the old ridge covering and replace it with bevel siding boards with their butt edges overlapped at the peak (fig. 13-66). The thin edge of the ridge boards should point downward toward the roof eaves.

Cover the old valley flashing with 1 × 6-inch boards nailed on each side of the valley center line (fig. 13-67). Cover the boards with

Fig. 13-63. Cutting back old shingles along eave.

new metal flashing. Nail a drip edge along each roof eave and rake, and begin shingling (fig. 13-68). The shingling procedure will be the same as that used in new construction (see the discussion of applying wood shingles in this chapter).

Fig. 13-64. Cutting back old shingles about six inches from gable edges. *(Courtesy Red Cedar Shingle & Handsplit Shake Bureau)*

Fig. 13-65. Nailing 1 × 6–inch boards along gable edges and eaves.

In wet climates wood shingles should not be nailed directly to the old roof covering material or the roof deck sheathing. Unless the shingles are properly ventilated they will absorb excessive amounts of moisture. This can be avoided by nailing 1 × 2–inch or 1 × 3–inch battens to the roof before applying the shingles. The 1 × 3–inch battens are used for shingle exposures of 5½ inches or more.

Fig. 13-66. Nailing bevel siding along ridge.

Fig. 13-67. Nailing 1 × 6–inch boards in valley.

Allow the ends of the battens to extend about ½ to ¾ inch beyond the edge of the roof rake. Provide drainage gaps at regular intervals along the battens and leave the space between parallel battens open at the rake to allow ventilating air to enter and circulate beneath the shingles. Nail a metal drip edge along the ends of the battens parallel to the rake.

Fig. 13-68. Nailing red cedar shingles over old roofing.

Fig. 13-69. Reroofing with wood shakes. *(Courtesy Red Cedar Shingle & Handsplit Shake Bureau)*

Shakes can be applied over the same types of roofing materials as wood shingles. The roof preparation procedures for reroofing with either wood shingles or shakes are also essentially the same. After the roof deck or existing roof has been properly prepared, apply the shakes and a roofing felt interply between each shake course as in new construction (fig. 13-69).

Wood-fiber roof panels may be used in the same reroofing applications as individual wood shingles and shakes. The existing roof must be applied over solid sheathing, and its surface must be smooth, flat, and free of protruding nails. An underlayment should be installed if the existing roof leaked. If necessary, a metal drip edge should be installed along the eaves. Apply molding along the rakes to conceal the old roofing materials. If the eaves or rakes are in poor condition, cut back the old roofing far enough to receive a 1 × 6–inch edging strip. Remove old hip and ridge shingles and apply 6-inch wide metal flashing or No. 30 roofing felt flashing centered over the crown. Worn, deteriorated, or damaged valley flashing should be replaced. If the existing valley is an open one and deep, install two wood strips of sufficient size in the valley to bring its level up to the flat plane of the roof and then install the flashing. Apply the roof panels over the existing roof as in new construction.

CHAPTER 14

Slate Roofing

- Roofing Slate
- Slate Roofing Nails
- Tools and Equipment
- Roof Construction Details

- Slating
- Repairing Slate Roofs
- Reroofing with Slate

Slate is one of the most expensive roofing materials, but it more than compensates for its initial high installation cost by providing an attractive roof that will last four or five times as long as one covered with asphalt shingles (fig. 14-1). If the owner of a house or building is interested in low first cost, a slate roof is certainly not the answer to his roofing problem. On the other hand, a slate roof is definitely the answer if a strong roof with a long service life is desired. The better grades of slate have been known to last 100 years. Lower quality grades are expected to last at least 50 years or more.

Roofing Slate

Slate is a stone that requires no admixture of materials, heat treatment, or special processes to convert it to a roofing material. It is simply extracted from the ground in blocks, which are then split and trimmed to the desired size and thickness.

In addition to being a particularly tough and durable roofing material, slate is completely fireproof and waterproof. It is also highly resistant to climatic changes and will not disintegrate as other roofing materials have a tendency to do. However, it may become brittle

Fig. 14-1. Rough-textured slate roof. *(Courtesy Evergreen Slate Co., Inc.)*

with age and will crack or split when struck by a falling branch or other hard object.

A principal disadvantage of using slate is its weight (900 to 1,000 pounds per roofing square), which requires reinforced framing to support it. Slate is also difficult to install. For this reason, slating is usually done by trained and experienced professional roofers.

Slate is available in a variety of colors, such as gray (neutral), green, blue, purple, black, or red; or in mottled colors, such as gray and green, blue and gray, blue and black, or green and purple. Some slate colors are permanent and nonfading; whereas others (so-called weathering slate) will fade to softer tones after extended exposure to the weather. Ribbon slate has one or more dark stripes crossing the unexposed portion. The useful life of ribbon slate is 60 to 75 years, which makes it cheaper than "one-hundred-year" slate. A slate roof may be composed of slates of a single color, or a harmonious grouping of two or more colors.

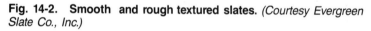

Fig. 14-2. **Smooth** and **rough textured slates.** *(Courtesy Evergreen Slate Co., Inc.)*

Slates are cut to at least 30 or 40 different sizes and a number of different thicknesses and weights. Some of these different slate sizes and weights are listed in table 14-1. Slate is also cut with either a smooth or rough textured surface (fig. 14-2). Many of the different slate sizes are identified by special names. The most commonly used slate sizes are called "large ladies" (16 × 8 inches), "countess" (20 × 10 inches), and "duchess" (24 × 12 inches).

The standard thickness of smooth and rough textured slate is commonly ¼ to ⅜ inch, but it is also available in textural thicknesses of ½ and ¾ inch. A graduated roof appearance is possible by laying ¾-inch thick slate at the eaves and graduating to ¼-inch slate at the ridge. Several thicknesses of slate may be intermingled in each course for additional roughness and texture.

A common method of applying slate to a sloping roof is to use slate of one uniform standard length and width with all slates laid to a line (fig. 14-3). If desired, this pattern may be varied by laying two or more sizes of slate.

A textured slate roof can be formed by using a textured or rough grade of slate (fig. 14-4). Another method is to use slates of varying

Table 14-1. Standard Slate Sizes and Weights
(Courtesy Evergreen Slate Co., Inc.)

Sizes

Length	Standard Widths	Exposure with standard 3" head lap
24"	16"–14"–12"	10½"
22"	14"–12"–11"	9½"
20"	14"–12"–11"–10"	8½"
18"	14"–12"–11"–10"–9"	7½"
16"	14"–12"–11"–10"–9"–8"	6½"
14"	12"–11"–10"–9"–8"–7"	5½"
12"	12"–10"–9"–8"–7"–6"	4½"
10"	10"–9"–8"–7"–6"	3½"

To relieve uniformity of shadow line, architects have continued to specify one length in its random widths.

Weights

Standard Smooth Texture		700–800 lbs. per Sq.
Standard Rough Texture		800–900 lbs. " "
¼"	" "	900 lbs. " "
⅜"	" "	1200 lbs. " "
½"	" "	1800 lbs. " "
¾"	" "	2500 lbs. " "

Fig. 14-3. Uniform courses of smooth surface slate. *(Courtesy Evergreen Slate Co., Inc.)*

Fig. 14-4. Textured slate courses. *(Courtesy Evergreen Slate Co., Inc.)*

thickness, size, and color laid so that the bottom line of each course is slightly uneven (see fig. 14-1).

Slates of any thickness laid tile fashion may be used as a surfacing material for flat or low-pitched roofs instead of slag or gravel. In most cases, however, the minimum recommended slope for a slate roof is 4-in-12 or greater.

Slate can also be used on mansard roofs. Because of the short rafters on this type of roof, a small size slate with a 5- or 6-inch exposure is used. Some slate companies produce a special slate for mansard roofs.

Slate Roofing Nails

Always use nails that will resist rust and corrosion. Solid copper, copper weld, or zinc-coated nails are recommended for slate roofs. Special slating nails are designed with large thick heads that will fit into countersunk holes in the slate.

Use 3d (1¼-inch) nails with slates of standard thickness and 18 inches or less in length, and 4d (1½-inch) nails with slates 20 inches or longer. The general rule of thumb is to add 1 inch to twice the thickness of the slate to determine the length of nail that should be used. For example, slate ¼ inch thick should be nailed down with nails at least 1½ inches long (¼ × 2 plus 1 = 1½ inches).

A 6d (2-inch) nail is recommended for nailing slates to hips and ridges, because greater penetration and holding power is required at these points in the roof structure.

Nail holes are machine-punched in each slate in accordance with federal specifications. Extra holes can be made by using a hammer and center punch to mark their location and then drilling them with an electric drill. When making these extra holes, make certain that the slate is placed on a flat and smooth surface.

Slates are generally produced with two nail holes machine-punched in the upper end or head of each slate. The advantage of nailing the slate at its upper end is the additional protection provided the nail heads by two layers of slate. In areas of the country where strong winds are common, greater roof strength can be obtained by nailing each slate at the middle. Longer nails are required, particularly if open battens are used to support the slate, but the leverage effect of strong winds is greatly reduced when the center nailing method is used.

Tools and Equipment

Two special tools used in slating are illustrated in figures 14-5 and 14-6. The slate nail ripper is inserted under the slate until a nail is hooked

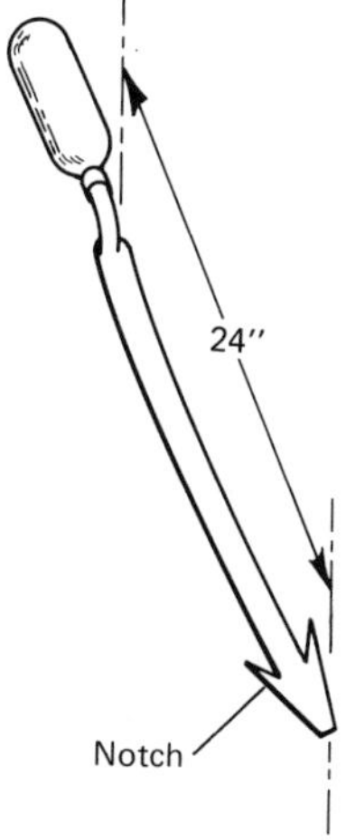

24"

Notch

Fig. 14-5. Slate nail ripper.

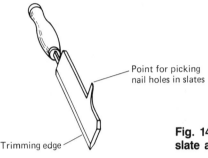

Point for picking
nail holes in slates

Trimming edge

**Fig. 14-6. Slating tool for trimming
slate and picking nail holes.**

by one of the notches. When the ripper is pulled out, it cuts the nail off level with the surface. If a slate nail ripper is unavailable, a hacksaw blade can be used to cut the nails. Some roofers still use the cutting tool shown in figure 14-6 for trimming slate and picking nail holes. Slate can also be trimmed with a cold chisel and hammer. The slate is first scored along a straight edge with the chisel and the unwanted section is knocked off. In addition to nailing the slates to the roof, a claw hammer can be used to mark extra nail holes with a center punch. The holes are then drilled with an electric drill. Another useful tool for working with slate is a putty knife, which can be used to coat the bottom of a replacement slate with asphalt roofing cement or clear butyl cement.

Ladders, scaffolds, safety harnesses, and similar types of equipment are described in chapter 2. The work safety measures described in the second chapter also apply to slating and should be carefully read and followed.

Roof Construction Details

Slate producers generally consider the weight of a slate roof to be relatively insignificant when compared with the combined weights of rafters, sheathing, wind pressure, and the water saturation of certain other types of roofing materials, and do not recommend any special roof construction details.

Regardless of what a slate company states in its sales literature, the local building code should be consulted before applying a slate roof. Many of these local codes will require special roof construction

or reinforcement of existing roofs. Some will specify the use of 1 ×
6–inch tongue-and-groove sheathing board on the roof deck.

Roof Deck Preparation

Cover the sheathing of the roof deck with horizontal, overlapping layers
of 30-pound asphalt roofing felt. The joints in each course of roofing
felt should be lapped toward the roof eaves and ends by at least 4
inches.

Lap the roofing felt 4 inches over all hips and ridges, and 2 inches
over the flashing metal of any valleys or built-in gutters. Secure the
roofing felt along laps and at ends with roofing nails to hold it in place.

Flashing

Rust and corrosion resistant copper flashing should be used when roofing with slate. Application procedures are described in chapter 8. Other
types of metal flashing can be used if they have the same rust and
corrosion resistant characteristics as copper flashing. Copper flashing
is generally preferred, however, because the flashing should have the
same long service life as the slate.

Slating

The application of a slate roof is called *slating* or *slate roofing*. This
is a type of roofing that generally requires experienced workers.

For years the popular method of slating called for the delivery
of a mixed assortment of slate to the job where it was sorted and graded
into a variety of different sizes and thicknesses. The courses along the
roof eave contain the heaviest and longest slates. Medium-sized slates
are used in the center courses, and the smallest slates are used along
the roof ridge. This slating method is sometimes called *random* (or
texture) *slating*.

Standard slating (or *standard commercial slating*) is a method
that uses slates that have already been graded at the quarries before
shipment to the job site. The slates are graded by length, width, and
thickness. Standard slating is a modern application method used
primarily in commercial roof construction. It is also used on less expensive house roofs.

Special precautions must be taken when working on a slate roof, because slate is a brittle roofing material and will crack or break if too much weight is placed on it. Consequently, a footing scaffold or a similar device should be used to more evenly distribute the worker's weight when working on a slate roof.

When laying the slate, allow a 2-inch projection at the eaves and 1 inch at all gable ends. The slate should be laid in horizontal courses with the standard 3-inch head lap, and each course should break joints with the preceding one. Slates at the eaves or cornice line are doubled and canted 1/4 inch by nailing a wood cant strip along the edge of the roof.

Slates overlapping flashing should be laid so that the nails do not penetrate the metal. Any exposed nail heads must be covered with a suitable roofing cement.

All hip and ridge slates must be laid in roofing cement spread thickly over the unexposed surfaces of the underlying courses of slate. These slates should be nailed securely in place and pointed with roofing cement.

Slate should be laid so that all joints are broken and there is a minimum lap of 3 inches over the slate in the course below it. It is customary to provide a 3-inch lap for sloping roofs with a rise of 8 to 12 inches per foot. A 4-inch lap is used if the rise is 4 to 6 inches per foot.

The amount of exposure for each slate is equal to one half the length of the slate minus the lap dimension:

$$\text{Slate Exposure} = \frac{\text{Length of Slate} - \text{Lap}}{2}$$

A typical method for laying slate can be outlined as follows:

1. Inspect and reinforce the roof framing and sheathing as required.
2. Cover the roof sheathing with a suitable 30-pound asphalt roofing felt (see the discussion of roof deck preparation).
3. Nail a wood cant strip along the eave or cornice line (fig. 14-7). The cant strip along the eave will tip the bottom end of the slate upward slightly. A similar strip (usually lath) nailed along both sides of the roof ridge provides additional support for the slates, but it is not always used.

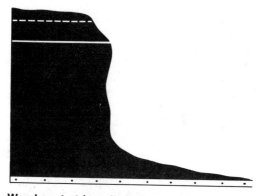

Fig. 14-7. Wood cant strip nailed along roof eave.

4. Lay a starter course of slate along the roof eave with each slate laid lengthwise approximately 1/16 inch apart and projecting 1 inch beyond the edge of the roof (fig. 14-8). Begin and end the starter course with the slates extending 1 inch beyond the rakes. Nail the slates through the machine-punched nail holes. Do not drive the nails too tightly against the slate because you may crack or break it. The nail head should be driven in just far enough to barely touch the surface of the

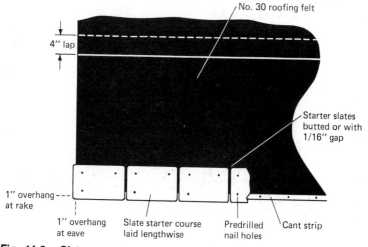

No. 30 roofing felt

4" lap

Starter slates butted or with 1/16" gap

1" overhang at rake

1" overhang at eave

Slate starter course laid lengthwise

Predrilled nail holes

Cant strip

Fig. 14-8. Slate starter course.

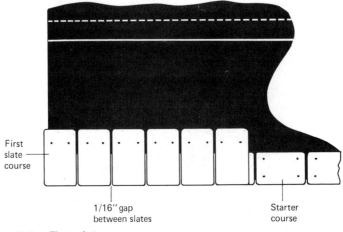

First slate course

1/16" gap between slates

Starter course

Fig. 14-9. First slate course.

slate. Use a counter punch and hammer to make a third hole in the upper-right corner of each slate.

5. Begin the first course of slate at the left roof rake with a full slate that extends 1 inch beyond the rake line and 1 inch beyond the eave line. Lay the slates of the first course vertically (at a right angle to the starter course). Complete the first course across the roof with the same projection (1 inch) at the right roof rake (fig. 14-9). Leave approximately a

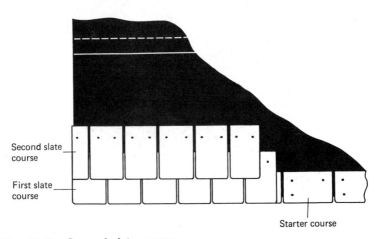

Second slate course

First slate course

Starter course

Fig. 14-10. Second slate course.

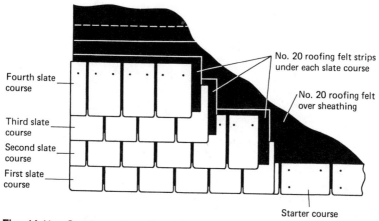

Fourth slate course

Third slate course

Second slate course

First slate course

No. 20 roofing felt strips under each slate course

No. 20 roofing felt over sheathing

Starter course

Fig. 14-11. Overlapped roofing felt.

$1/16$-inch space between each slate. The bottom edge of each slate in the first course should be even with the bottom edge of each slate in the starter course.

6. Begin the second course with a half slate and allow the same projection beyond the left roof rake as was provided for the first course. Complete the second course to the right roof rake and cut the slate (if necessary) to maintain the required projection beyond the rake (fig. 14-10).

The remaining slate courses are laid according to the instructions outlined in steps 3–7 above. The following points are important to remember:

1. Maintain the same projection beyond the rakes on gable roofs for each slate course. The roof line at the rake must be even.
2. Lay the slate so that the vertical joints on adjoining courses break and are at least 3 inches apart.

Some slate roofs are laid with a strip of 30-pound asphalt roofing felt covering the unexposed portion of each course. The roofing felt strip provides a cushion for the next slate course (fig. 14-11).

Valley Details

Either open or closed valleys can be used in slate roofing. The application procedure is similar to the one used when applying wood shingles or shakes to a valley (see chapter 13). Each slate must be cut to conform to the angle of the valley. Slate can be cut by scoring it deeply along a straight edge with a chisel. The unwanted portion can then be removed by tapping it lightly with a hammer or hitting it with the hand.

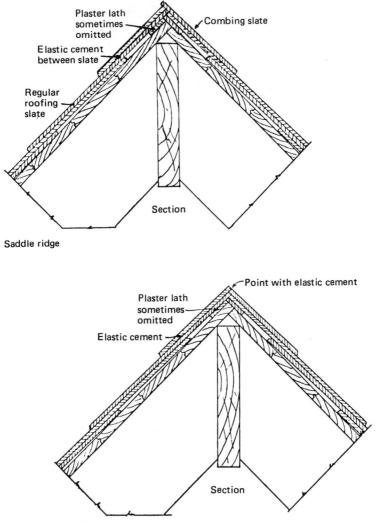

Fig. 14-12. Sectional views of saddle and strip saddle ridges.
(Courtesy National Slate Association)

Ridge Details

Slate can be laid to form the following four types of ridges:

1. Saddle ridge
2. Strip saddle ridge
3. Comb ridge
4. Coxcomb ridge

Most slate roofing instructions will specify a saddle ridge or a strip saddle ridge. Sectional views of both types of ridges are shown in figure 14-12. An overlapping plaster lath or beveled strip is sometimes used along both sides of the ridge line for additional reinforcement, but it may be omitted. Each ridge slate is first laid in elastic cement spread thickly over the unexposed surface of the undercourse of slate and then nailed securely in position with two roofing nails. The nails are covered with a spot of slater's elastic roofing cement. The cement is available in colors to match as nearly as possible the general color of the slate. Point the joint formed by the slate at the ridge line with elastic cement.

The principal difference between a saddle ridge and a strip saddle ridge is that the slates on the latter form butt joints, whereas on a saddle ridge they overlap approximately 3 inches (figs. 14-13 and 14-14).

A sectional view of a comb ridge is shown in figure 14-15. In this type of ridge application, the combing slate projects about ⅛ inch beyond the ridge line. The combing slate can be laid with the grain vertical or horizontal (figs. 14-16 and 14-17).

A coxcomb ridge is similar in detail to a comb ridge except that

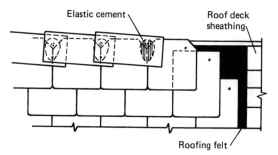

Fig. 14-13. Saddle ridge construction details. *(Courtesy National Slate Association)*

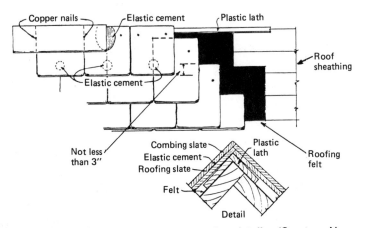

Fig. 14-14. Strip saddle ridge construction details. *(Courtesy National Slate Association)*

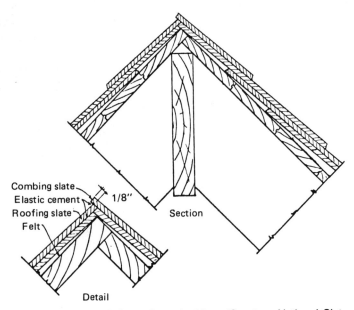

Fig. 14-15. Sectional view of comb ridge. *(Courtesy National Slate Association)*

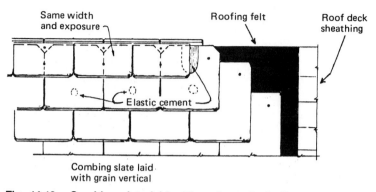

Fig. 14-16. Combing slate laid with grain vertical. *(Courtesy National Slate Association)*

the combing slates are laid to alternately project on either side of the ridge.

The slate course applied along the ridge should break joints with the preceding one. In other words, never allow a joint in the ridge course to line up with a joint in the preceding (underlying) course on the roof deck.

The ridge course should be given the same exposure as the other roof courses. This may mean cutting the slate used in the ridge course or using smaller slate of a proper size.

The projection of the ridge course at the roof rake must be exactly the same as the other courses on the roof.

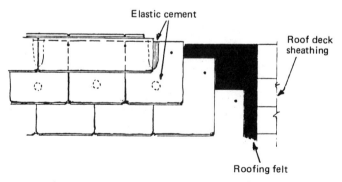

Fig. 14-17. Combing slate laid with grain horizontal. *(Courtesy National Slate Association)*

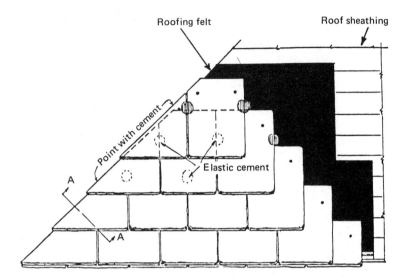

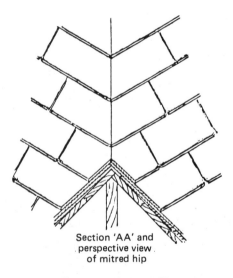

Section 'AA' and
perspective view
of mitred hip

Fig. 14-18. Mitred hip construction. *(Courtesy National Slate Association)*

Hip Roofs

There are four principal types of hip construction used in slate roofing:

1. Saddle hip
2. Mitered hip
3. Boston hip
4. Fantail hip

Beveled strips, or one or two strips of plaster lath, are sometimes nailed along the hip line to provide additional reinforcement and a slightly elevated nailing base.

Both the mitered and fantail hips are constructed by extending each roof course to the roof rake and trimming the last slate to conform to the angle of the hip (figs. 14-18 and 14-19). Fantail hip construction differs by using slates with their slips trimmed at a right angle to the hip line. Point with elastic roofing cement along the line at which the hip slates meet and secure them with two or three roofing nails placed in the unexposed portion of each slate. The nails should be embedded in a spot or circle of roofing cement for additional holding power.

The saddle hip is formed by laying overlapping slates up the hip

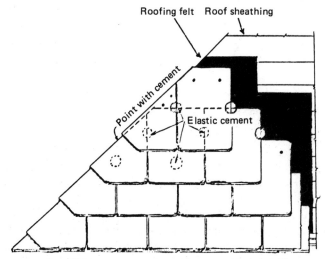

Fig. 14-19. Fantail hip construction. *(Courtesy National Slate Association)*

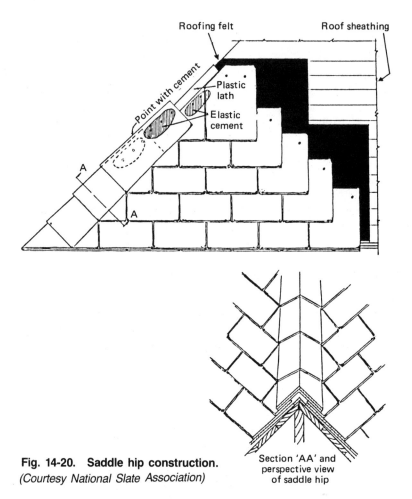

Fig. 14-20. Saddle hip construction.
(Courtesy National Slate Association)

Section 'AA' and
perspective view
of saddle hip

ridge after each course of slates on the roof has been extended to the
rake and trimmed to conform to the angle of the hip (fig. 14-20). The
nailing procedure and the use of roofing cement is the same as that
described for constructing mitered and fantail hips.

The Boston hip is constructed by applying overlapping slates, as
is the case with a saddle hip, but a narrower slate (marked *B* in fig.
14-21) is used to complete each course of roof slates. The nailing and
cement application procedure is the same as that described for constructing mitered and fantail hips.

Hip slates should always be laid from the bottom of the hip

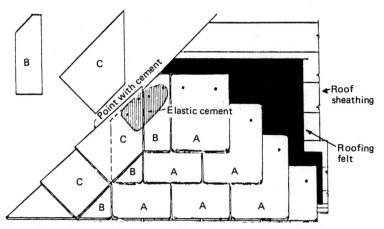

Fig. 14-21. Boston hip construction. *(Courtesy National Slate Association)*

toward the ridge line of the roof if the hip construction requires that the slates cap the hip. The nail holes in each hip slate are made with a counter punch after the amount of exposure has been marked on the slate with a piece of chalk or a pencil. The hip exposure should match that of the main slate courses used on the roof.

Repairing Slate Roofs

Individual slates will sometimes break or crack (fig. 14-22). When this occurs, the broken or cracked slate should be repaired or replaced immediately to avoid the possibility of roof leaks. A cracked slate can be repaired by filling the crack with asphalt roofing cement, putty,

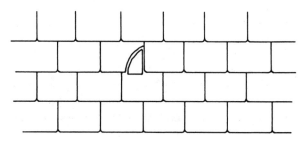

Fig. 14-22. Cracked slate.

Fig. 14-23. Sealing hairline crack with putty.

or a suitable synthetic sealer (fig. 14-23). These materials can also be used to reattach a loose slate. A slate too damaged to repair should be replaced with a new one.

Do not attempt to remove a damaged slate by breaking it into smaller pieces with a hammer because you will run the risk of breaking the slates in the lower course. The nails should be cut first; the damaged slate can then be pulled out. A shingle ripper or a long chisel blade without its handle should be inserted under the damaged slate. Several hammer blows against the nail cutter or chisel blade should be enough to cut the nails and free the slate (figs. 14-24 and 14-25).

Fig. 14-24. Removing damaged slate with shingle ripper and hammer.

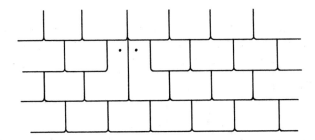

Fig. 14-25. Slate removed.

After you have removed the damaged slate, cut a 5-inch wide strip of copper long enough to extend 6 inches under the slate course above the damaged one and still bend around the bottom edge of the replacement slate (fig. 14-26). Nail the copper strip in position with two roofing nails and cover the new nails and the tops of the old nails with roofing cement. Place the slate on a flat surface and mark the locations of the new nail holes with a center punch. Make certain the new holes are *not* aligned with any existing ones on the roof. Use an electric drill to make new holes in the slate. Coat the bottom of the replacement slate with asphalt roofing cement. The cement can be applied with a putty knife. Insert the new slate, press it down, and bend the metal strip up around its bottom edge to give it support (fig. 14-27). Trim off any excess metal from the copper strip.

The individual slates will generally outlast the nails used to fasten them to the roof deck. This condition is common on very old slate roofs. If you inspect the roof and find that many of the nails have

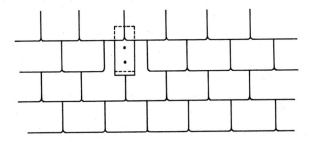

Fig. 14-26. Nailing metal strip.

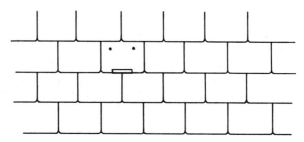

Fig. 14-27. **Bending metal strip around bottom of replacement slate.**

rusted or are loose, the *entire* slate roof must be removed and replaced with a new one. Remove the slates *and* the roofing felt underlayment and examine the condition of both the flashing and the roof deck sheathing. Replace any damaged flashing and broken or warped sheathing boards. Protruding nails should be nailed flush with the surface and the roof deck swept clean of any dirt or debris. Apply a new roofing felt underlayment and lay the new courses of slate.

When removing slate from an existing roof, try to save slates that are still in good condition. New slates should be the same size and thickness as the old ones and should closely match them in color. Do not be concerned about a difference in color tone between the old and new slates because this adds to the attractiveness of a slate roof. Make certain that the variation in color tone is uniformly distributed across the roof deck.

Reroofing with Slate

Slate may be used for reroofing. It can be applied over old material, but better results are assured when the old roof covering is removed. Inspect the roof rafters and sheathing to make certain they are strong enough to support the weight of a slate roof. It may be necessary to reinforce them. Consult the local building code before applying a slate roof. The procedures used in reroofing with slate are generally the same as those used in new construction. If the old roofing material is not removed, trim the shingles back to the eave and rake edges before laying the slate.

CHAPTER 15

Tile Roofing

- Tools and Equipment
- Roof Deck Preparation
- Roofing Tile
- Laying Tile
- Application Procedures
- Types of Tile Roofs

- Tile Roof Repair and Maintenance
- Reroofing with Tile

Tile has been used as a roofing material for centuries. Some of the earliest and finest examples of tile roofs date from the time of ancient Greece and Rome where they were used on both public buildings and private residences. Its popularity as a roofing material is still widespread in regions with warm climates, such as the Mediterranean area, Latin America, Florida, the southwestern United States, and California. Elsewhere it has been replaced to a great extent by cheaper roofing materials, particularly asphalt-based products.

Tools and Equipment

Tile roofing is applied with the same basic tools and equipment used in the application of other roofing materials. These roofing tools and equipment are described in chapter 2. Roofing tiles should be cut with a circular saw equipped with a masonry (carborundum) blade. *Safety goggles must be worn when cutting tile with a circular saw.* Flying particles of clay or concrete tile may cause serious injury to unprotected eyes.

Roof Deck Preparation

Always consult the local building code before laying tile. Because of the extra weight of a tile roof, special roof construction may be necessary. Inspect the roof framing and reinforce the roof with braces and extra rafters if the framework needs strengthening.

Either a closed or open roof deck can be used to support a tile roof. A closed roof deck is generally constructed of 1 × 6-inch tongue-and-groove sheathing board or plywood sheathing panels of suitable thickness with 1 × 2-inch battens mounted between the roof eaves and ridge. The use of battens is always recommended and is required for roofs with slopes of 7-in-12 or more. The battens should be made from strips of douglas fir, redwood, or cedar. Both redwood and cedar battens offer the best resistance to moisture. The battens should be nailed to the sheathing with nails placed approximately every 2 feet along the strip (fig. 15-1). Adequate means of drainage must be provided at the battens to prevent the accumulation of moisture on the

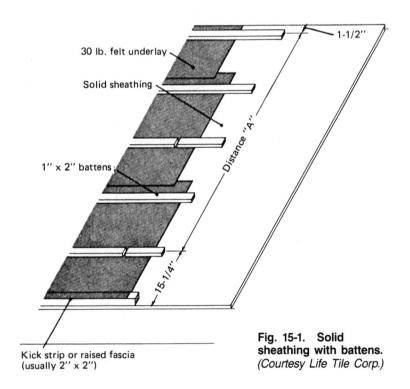

Fig. 15-1. Solid sheathing with battens. *(Courtesy Life Tile Corp.)*

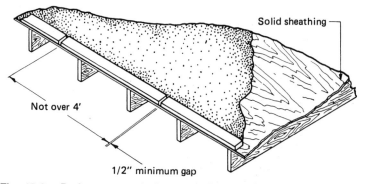

Fig. 15-2. Drainage gaps in battens. *(Courtesy Monier Co.)*

roof. This can be done by allowing ½-inch gaps every 4 feet, as in figure 15-2, or by slightly raising the battens off the roof deck with shims cut from asphalt shingles.

As shown in figure 15-1, a 2 × 2–inch kick strip is nailed along the edge of the sheathing directly above the fascia board. The kick strip slightly elevates the bottom edge of the first course of tile to give it the same cant as succeeding tile courses. In new construction, a raised fascia board is sometimes used instead of a kick strip to raise the first tile course. In either case, a 28-gauge galvanized metal strip should be nailed along the eave to support the underlayment over the raised fascia board or kick strip (fig. 15-3). This will prevent the accumulation of water along the edge of the roof.

An open roof deck is constructed with spaced sheathing boards. The spacing of the sheathing boards will depend on the size of the tile. There are two basic types of open roof decks used to support a flat tile roof: (1) spaced sheathing and (2) combination solid and spaced sheathing.

A typical open roof deck with spaced sheathing is shown in figure 15-4. The spacing of the 1 × 6–inch battens depends on the size of the tile. Either a kick strip nailed directly to the top of the fascia board or a raised fascia board may also be used on an open roof deck in new construction to elevate the bottom edge of the first course of tile.

Sometimes a transition from solid to spaced sheathing is used to support a tile roof (fig. 15-5). The solid sheathing may consist of either ½-inch or ¾-inch thick plywood or 2-inch thick tongue-and-groove sheathing board.

The spaced battens must be mounted exactly parallel to the roof

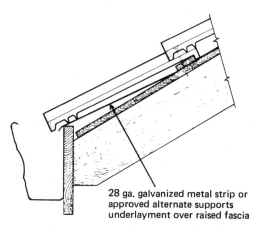

28 ga. galvanized metal strip or
approved alternate supports
underlayment over raised fascia

Fig. 15-3. Underlayment along roof eave supported by galvanized metal strip to prevent ponding of water. *(Courtesy Monier Co.)*

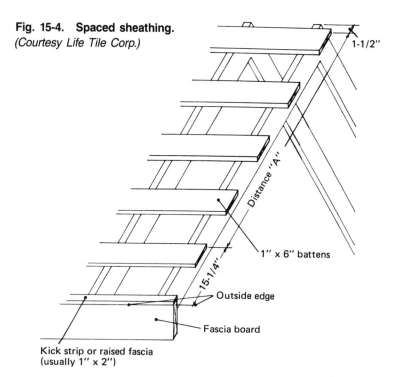

Fig. 15-4. Spaced sheathing.
(Courtesy Life Tile Corp.)

1-1/2"

Distance "A"

1" x 6" battens

15-1/4"

Outside edge

Fascia board

Kick strip or raised fascia
(usually 1" x 2")

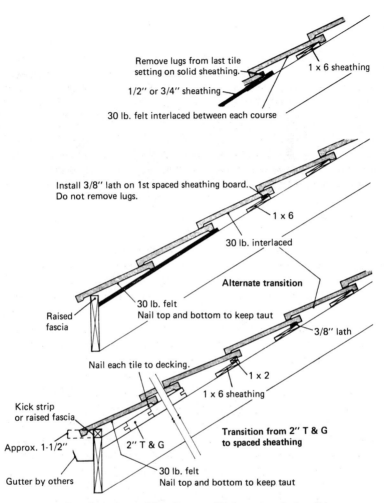

Remove lugs from last tile setting on solid sheathing.

1 x 6 sheathing

1/2" or 3/4" sheathing

30 lb. felt interlaced between each course

Install 3/8" lath on 1st spaced sheathing board. Do not remove lugs.

1 x 6

30 lb. interlaced

Alternate transition

Raised fascia

30 lb. felt
Nail top and bottom to keep taut

3/8" lath

Nail each tile to decking.

1 x 2

1 x 6 sheathing

Kick strip or raised fascia

Transition from 2" T & G to spaced sheathing

Approx. 1-1/2"

2" T & G

Gutter by others

30 lb. felt
Nail top and bottom to keep taut

Fig. 15-5. Overhang transition from solid to spaced sheathing. *(Courtesy Life Tile Corp.)*

eaves and as near the proper spacing as possible. The overall appearance of the finished roof will depend on this. Suggested layout calculation procedures are provided by the tile manufacturers. The layout method for spacing the battens on the roofs shown in figures 15-1 and 15-4 may be outlined as follows:

1. Install a kick strip directly above the fascia board. The kick strip is not required if a raised fascia board is used.
2. Measure from the outside edge of the kick strip or raised fascia board 15¼ inches up the roof deck. This point will represent the top edge of the first batten. Snap a chalk line across the roof deck sheathing (closed roof) or rafters (open roof) and nail the batten in place.
3. Measure down from the roof ridge 1½ inches and snap a chalk line. This line will mark the position of the top edge of the uppermost batten. Nail the batten in place.
4. Measure from the top edge of the lowest batten to the top edge of the highest batten. This measurement represents distance A in figures 15-1 and 15-4. Divide this measurement into equal spaces not to exceed 14 inches.
5. Consult the layout calculation chart shown in table 15-1. The calculations in this table are based on the first batten being installed at 15¼ inches from the outside edge of the kick strip or raised fascia board. The table provides information on equal batten spacing and the total number of battens or courses. Example: If distance A is 84 inches, it can be divided into six equal spaces ($84 \div 14 = 6$). Six spaces will require a total of seven battens.

The ridge and hip tiles on many roofs are supported by wood stringers or nailers. The size used will depend on the size of the tile and the slope or pitch. As a general rule, 2 × 6's standing on edge and toenailed to the ridge board are used to support large curved tiles. Flat tiles, on the other hand, can be fully supported by 2 × 2's or 2 × 4's nailed flat against the ridge board.

The cover tiles on some Spanish, mission, and custom tile roofs are supported by wood nailing strips running perpendicular to the roof eaves. The size of the nailing strips (usually 1 × 2–inches or 1 × 4–inches) will depend on the type and size of the tile and the tile manufacturer's application instructions.

Underlayment

Follow the tile manufacturer's instructions for applying the underlayment. If an underlayment is required, sweep the roof deck clean with a broom and cover any knot holes in the sheathing with small pieces

Table 15-1.   Layout Calculation Chart *(Courtesy Life Tile Corp.)*

Distance "A"	Equal Spacing	Total Courses	Distance "A"	Equal Spacing	Total Courses
84"	14"	7	162"	13½"	13
87"	12⁷⁄₁₆"	8	165"	13¾"	13
90"	12⅞"	8	168"	14"	13
93"	13⁵⁄₁₆"	8	171"	13⅛"	14
96"	13¾"	8	174"	13⅜"	14
99"	14" *	8	177"	13⅝"	14
102"	12¾"	9	180"	13¹³⁄₁₆"	14
105"	13⅛"	9	183"	13⅙"	14
108"	13½"	9	186"	13¼"	15
111"	13⅞"	9	189"	13½"	15
114"	12⅝"	10	192"	13¹¹⁄₁₆"	15
117"	13"	10	195"	13¹⁵⁄₁₆"	15
120"	13⁵⁄₁₆"	10	198"	13³⁄₁₆"	16
123"	13⅝"	10	201"	13⅜"	16
126"	14"	10	204"	13⁹⁄₁₆	16
129"	12⅞"	11	207"	13¾"	16
132"	13³⁄₁₆"	11	210"	14"	16
135"	13½"	11	213"	13⁵⁄₁₆"	17
138"	13¹³⁄₁₆"	11	216"	13½"	17
141"	14" *	11	219"	13¹¹⁄₁₆"	17
144"	13¹⁄₁₆"	12	222"	13⅞"	17
147"	13⅜"	12	225"	13¼"	18
150"	13⅝"	12	228"	13⅜"	18
153"	13⅞"	12	231"	13⁹⁄₁₆"	18
156"	13"	13	234"	13¾"	18
159"	13¼"	13	237"	13¹⁵⁄₁₆"	18

*Slight adjustment must be made for these distances.

of tin before laying the roofing felt. Use mastic tape to secure the tin patches to the sheathing.

The underlayment on a closed roof deck will generally consist of overlapping layers of 36-inch wide asphalt-saturated roofing felt. Use either a double layer of 15-pound or a single layer of 30-pound felt. Lap the courses of roofing felt 4 inches along the horizontal edge and 6 inches along the vertical. Secure the roofing felt with nails driven along the horizontal laps and at the ends to hold it in place. On open roof decks, the underlayment may consist of 16-inch wide strips of 30-pound asphalt-saturated roofing felt laid between each course of tile.

Flashing

Metal flashing is preferred for tile roofs. Copper flashing is often recommended, but other types of less expensive, rust and corrosion resistant metals can also be used just as effectively. Flashing application procedures are described in chapter 8. Flashing specifically required for tile roofs includes the following:

1. Metal drip edges applied along the roof eaves *before* the underlayment is laid.
2. Metal drip edges applied along the roof rakes *after* the underlayment is laid.
3. Minimum 24-inch wide crimped metal flashing applied over 90-pound mineral-surfaced roll roofing in the valleys.

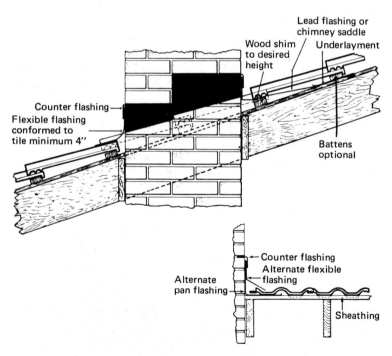

Fig. 15-6. Chimney and vertical wall flashing details. *(Courtesy Monier Co.)*

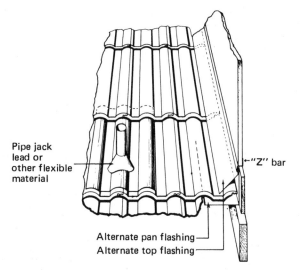

Fig. 15-7. Vent pipe flashing and flashing where roof abutts.
(Courtesy Monier Co.)

Some examples of different types of tile roof flashing applications are illustrated in figures 15-6 and 15-7. Additional flashing details for tile roofs are described in the appropriate sections of this chapter.

Roofing Tile

Modern roofing tile is made from baked clay, concrete, or metal. Clay tile is the most expensive and is used in quality residential construction. Metal tile is generally used on commercial buildings (see chapter 16). This chapter covers the application of both clay and concrete tile.

Clay Roofing Tile

Clay roofing tiles are made from a specially prepared clay that is baked in a kiln maintained at the proper temperature. Baking the clay at a properly maintained temperature is important because it affects the quality of the tile. If the baking temperature in the kiln is too low, the tile will be weak and porous. As a result of its higher porosity, the tile will absorb more moisture and eventually decay. Warping and color variations generally result when the kiln temperature is too high.

In every respect, the methods used in manufacturing clay roofing tile are similar to those used in manufacturing brick.

Clay roofing tiles are available in a variety of colors, ranging from various shades of red to dark brown or blue. Exposure to the elements over an extended period of time will cause clay roofing tiles to fade to lighter shades. This color fading occurs only with clay roofing tiles, not those made of concrete or metal, and the fading occurs in individual tiles on the roof at different rates. Eventually the roof will be covered by tiles of many different shades and tones of the same color, and it is this characteristic of clay roofing tile that provides the roof with such a distinctive and attractive appearance.

Concrete Roofing Tile

Concrete roofing tiles are available in a variety of different designs and colors. They are less expensive than clay roofing tiles and their colors will not fade after extended exposure to weather conditions. Lightweight concrete roofing tiles weigh approximately 100 pounds per roofing square less than the lightest weight of clay roofing tiles.

Laying Tile

Laying tile can be made easier by snapping vertical and horizontal chalk lines across the roof to serve as a guide for the horizontal courses and vertical rows of tiles (fig. 15-8). For all types of tiles except interlocking tiles, it is possible to adjust the horizontal guide lines to accommodate tiles that may be smaller than the manufacturer's advertised dimensions (some reduction in size may occur during the baking process for clay tiles) and to insure that full-size, uncut tiles are laid along the roof ridge, hips, and dormers. Adjustments of the horizontal guide lines should always be made in favor of increasing the lap and reducing the exposure.

When loading tile on the roof, space the piles to insure an equal distribution of weight. This precaution is particularly important for gable roofs. Suggested loading distributions for gable and hip roofs are illustrated in figures 15-9 and 15-10.

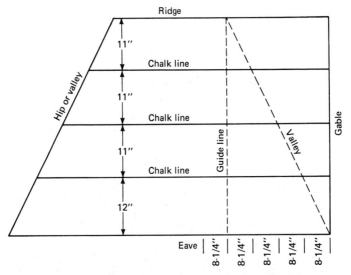

Fig. 15-8. Chalk line locations for tile roof having tile average exposure of 11 inches and average width exposure of 8¼ inches.

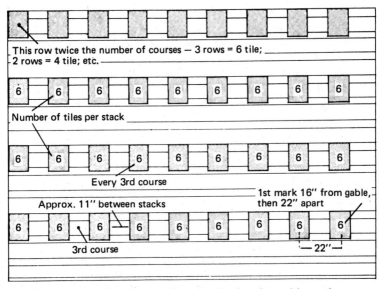

Fig. 15-9. Suggested tile loading distribution for gable roof.
(Courtesy Life Tile Corp.)

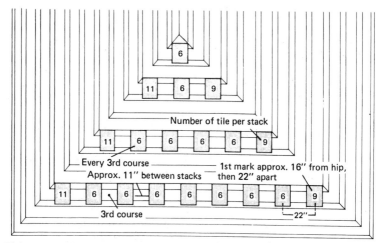

Fig. 15-10. Suggested tile loading distribution for hip roof. *(Courtesy Life Tile Corp.)*

Application Procedures

The application procedures illustrated in figures 15-11 to 15-19 apply to all the Monray tile profiles from the Monier Company, a leading manufacturer of concrete roofing tiles. With some modifications, these procedures can also be used to apply clay and concrete tiles from other tile manufacturers. Whenever possible, however, read and carefully follow the tile manufacturer's application instructions.

FIELD TILES

Field tiles with interlocking devices or nailing flanges along their left side are applied by beginning at the right roof rake and working across the roof to the left roof rake. Mission tiles, English or shingle tiles, and similar types of tiles produced without interlocking devices or nailing flanges may be applied by starting at either rake and working across the roof to the opposite rake.

RIDGE DETAILS

The top courses of field tiles are brought to within one inch of each other on each side of the ridge and the space is capped with a ridge

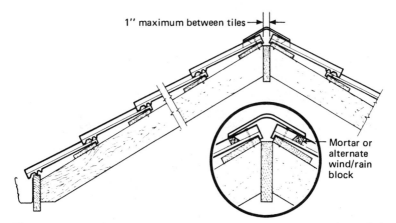

Fig. 15-11. Roof cross section with ridge details. *(Courtesy Monier Co.)*

tile (fig. 15-11). All ridge tiles are fastened to the roof with nails or wires (fig. 15-12). If there is no ridge board, and the ridge tiles are fastened with wire, anchor the wire to the nails in the adjacent field tile courses. Lay all ridge tiles in the same direction. Do not use mortar between tile joints. Embed the ridge tiles in a thin layer of mortar

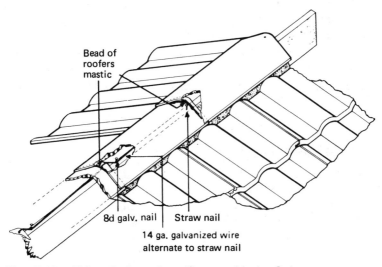

Fig. 15-12. Ridge tile fastening. *(Courtesy Monier Co.)*

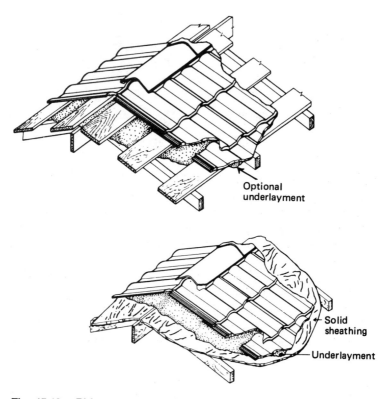

Optional
underlayment

Solid
sheathing

Underlayment

Fig. 15-13. Ridge construction details. *(Courtesy Monier Co.)*

in the water course only. Ridge details for roofs with either spaced or solid sheathing are shown in figure 15-13.

GABLE JUNCTURE DETAILS

Two recommended methods of finishing a ridge where it meets the gable are illustrated in figure 15-14. One method calls for mitering the rake and ridge tiles to fit the gable juncture. The other method requires cutting a cover piece from a rake tile to fit the juncture. In either case, all cracks should be filled with mortar to prevent moisture penetration.

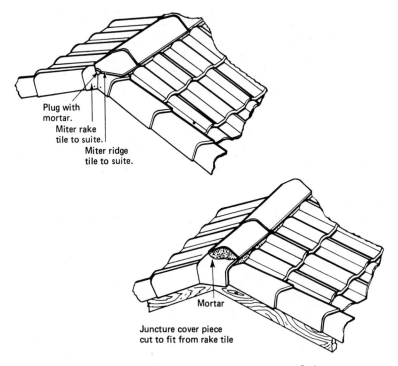

Plug with mortar.
Miter rake tile to suite.
Miter ridge tile to suite.

Mortar

Juncture cover piece cut to fit from rake tile

Fig. 15-14. Gable juncture details. *(Courtesy Monier Co.)*

HIP DETAILS

Hips are formed by bringing cut field tile close to one another on either side of the hip and covering the gap between them with a hip tile. A specially shaped hip starter tile is used at the foot or base of the hip. The hip tiles are embedded in mortar, but no mortar is used in the overlapping tile joints. Hip construction details for roofs with either spaced or solid sheathing are shown in figures 15-15 and 15-16.

VALLEY DETAILS

The construction details of the open and closed valleys used on tile roofs are illustrated in figures 15-17 and 15-18. The open valleys are

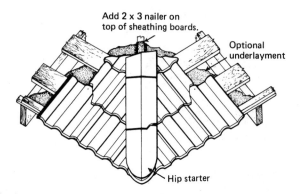

Fig. 15-15. Hip construction on roofs with spaced sheathing. *(Courtesy Monier Co.)*

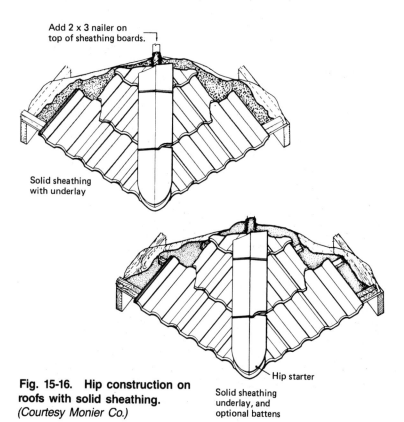

Fig. 15-16. Hip construction on roofs with solid sheathing. *(Courtesy Monier Co.)*

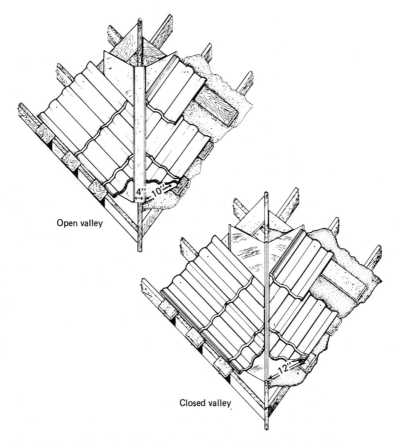

Fig. 15-17. Valley construction on roofs with spaced sheathing.
(Courtesy Monier Co.)

recommended for areas where falling leaves or pine needles are a common problem. Note that a W-shaped valley metal is used in open valleys.

GABLE END DETAILS

Two frequently used methods of finishing the roof at the gable ends are to install rake tiles or a barge board and flashing (fig. 15-19). The methods are the same for both spaced and solid roof sheathing.

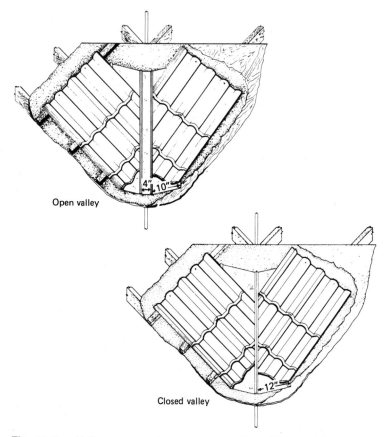

Fig. 15-18. Valley construction on roofs with solid sheathing. *(Courtesy Monier Co.)*

Nails and Fasteners

Large-head noncorrosive copper, aluminum, or galvanized steel nails are used to fasten tiles to a roof deck. For best results, always use the type nail specified by the tile manufacturer. Tile roofing nails should be long enough to penetrate 3/4 inch into or through the sheathing or battens.

The nails are driven through one or more machine-punched holes in the tile. They should be driven in far enough to just barely touch the surface of the tile. Driving them further may cause the tile to crack or break.

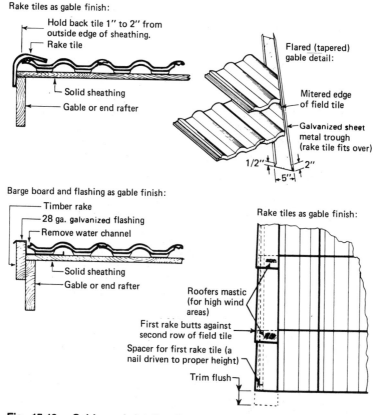

Rake tiles as gable finish:

Hold back tile 1" to 2" from outside edge of sheathing.

Rake tile

Solid sheathing

Gable or end rafter

Flared (tapered) gable detail:

Mitered edge of field tile

Galvanized sheet metal trough (rake tile fits over)

1/2"   2"

5"

Barge board and flashing as gable finish:

Timber rake

28 ga. galvanized flashing

Remove water channel

Solid sheathing

Gable or end rafter

Rake tiles as gable finish:

Roofers mastic (for high wind areas)

First rake butts against second row of field tile

Spacer for first rake tile (a nail driven to proper height)

Trim flush

Fig. 15-19. Gable end details. *(Courtesy Monier Co.)*

Use noncorrosive 14-gauge copper wire anchored to copper nails when wire fastening tiles to the roof. Tile tie systems consisting of copper, galvanized wire, or brass strips are also available from some manufacturers for securing tiles.

Cutting and Trimming Tile

Tiles must be cut or trimmed to fit along rakes, hips, ridges, and valleys, or around vent pipes, stacks, and other objects that protrude through the roof surface.

Care must be taken when cutting or trimming tile because it is a hard, brittle material that is easy to crack or chip. Mark the cutting line with a straight edge and place the tile on a flat, wood surface. Cut the tile with a circular saw fitted with a masonry (carborundum) blade.

Mortar

Mortar is sometimes used to provide additional holding power for the tiles, particularly along hips and ridges, and to protect the roof from moisture penetration.

Use a cement mortar capable of establishing a strong bond and tight seal under the tile. When mixing mortar use 3 parts sand to 1 part plastic cement. Color pigments are available to match or approximate the color of the tile.

Tile should be immersed in water for at least two minutes before being applied to the roof deck. Dampening the tile avoids the problem of the mortar drying out before it sets and cures.

Types of Tile Roofs

Roofing tiles are produced in the form of field tiles and a variety of different accessory tiles. The field tiles are used to cover the flat expanse of the roof deck and may consist of either individual flat (butted or interlocking) tiles or overlapping pans and covers. Hip and ridge tiles are designed to complement the field tiles for hip and ridge applications. Ridge closures or cat faces are individual units used to close the ridge at gable ends where the rake and ridge units meet. Hip starter units are applied at the bottom of each roof hip. The hip is completed with standard hip tiles and a special terminal tile where the hip meets the ridge. Rake and eave tiles or eave closures are used to close the roof rakes and eaves.

The traditional tile roofs are the flat, Spanish, and mission styles. Spanish tile roofs consist of overlapping S-shaped tiles. Mission or barrel-type roofs are formed from overlapping curved pans and covers. Many custom tile roofs can be created by combining flat pans with cover tiles of different designs. Roman and Greek tile roofs fall within this category. Tile manufacturers also create their own variations of traditional tile roof styles.

Flat Tile Roofs

A flat tile roof bears a strong resemblance to a slate roof. The principal differences are that flat roofing tiles are slightly thicker than slate and have greater uniformity. Examples of field and accessory tiles used on flat tile roofs are shown in figures 15-20 and 15-21.

Concrete flat field tiles are manufactured with small nibs or lugs at the head (top) by which they can be hung from horizontal roof

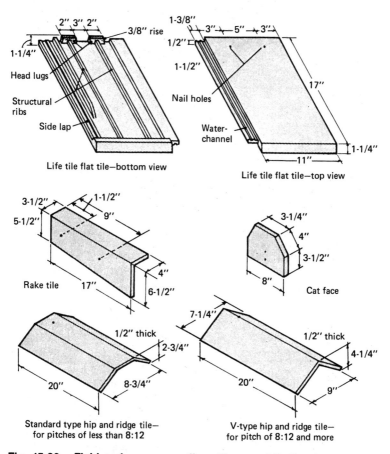

Fig. 15-20. Field and accessory tiles. *(Courtesy Life Tile Corp.)*

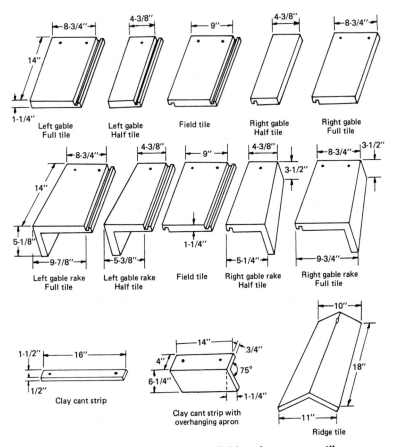

Fig. 15-21. Lincoln interlocking flat field and accessory tiles. *(Courtesy Gladding, McBean & Co.)*

battens without nailing (fig. 15-22). Every third or fourth course of tiles is nailed to provide additional holding power. The lower ends of these tiles are sometimes curved downward to provide a close, tight fit against the underlying course. Flat tiles with nibs on both the bottom and top ends are also available.

Some flat tiles are designed to interlock along the sides (see fig. 15-23). The interlocking of each tile course results in a more watertight roof surface. The water channel on the side of the tile collects rain water and directs it down the roof to the gutters. The tiles shown in figure 15-23 overlap to provide a watertight roof surface.

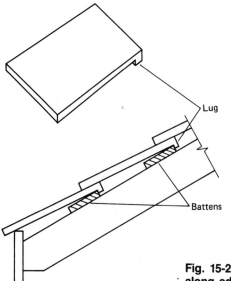

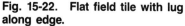

Fig. 15-22. Flat field tile with lug along edge.

Flat or English shingle tiles are attached to the roof with two large-head roofing nails through machine-punched holes near the top end of each tile (fig. 15-24). These flat tiles resemble slate in both appearance and method of application.

Flat tiles can be applied to a closed (solid sheathing) roof deck with a pitch under 3-in-12, but the surface must be covered with a watertight membrane before the tiles are laid (fig. 15-25). The common practice is to cover the roof deck with two layers of 30-pound roofing felt mopped together with hot asphalt. A 12-inch wide strip of 90-pound roofing felt inserted along the eaves of the roof provides additional reinforcement to the angle at the kick strip.

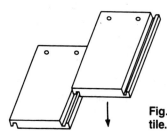

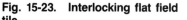

Fig. 15-23. Interlocking flat field tile.

Fig. 15-24. English shingle tile (noninterlocking flat field tile).

Redwood lath strips 2-inches wide are nailed to the roof perpendicular to the eaves. These strips should be nailed over the rafters and spaced no farther apart than 24 inches on center. They are embedded in the hot asphalt when the roof is mopped.

Top mop the entire roof deck (including the lath strips) with hot asphalt and then nail the 1 × 2–inch battens to the lath strips with 8d nails. When flat tiles are applied to a closed (solid sheathing) roof deck with a pitch of 3-in-12 or more, it is not necessary to hot mop the underlayment (fig. 15-26).

When an open roof deck with a pitch of 4-in-12 or more is used to support the tiles, 16-inch wide strips of roofing felt are interlaced between each course of tile (fig. 15-27). The top edge of the strip is

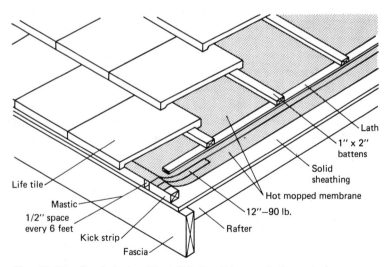

Fig. 15-25. Roof deck with solid sheathing and slope under 3-in-12. *(Courtesy Life Tile Corp.)*

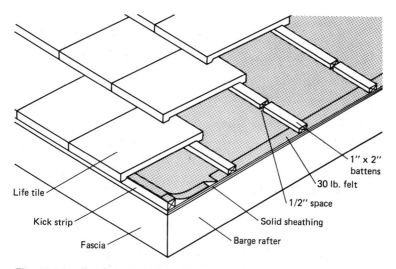

Fig. 15-26. Roof deck with solid sheathing and slope of 3-in-12 or steeper. *(Courtesy Life Tile Corp.)*

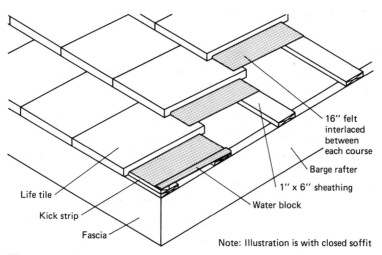

Fig. 15-27. Roof deck with spaced sheathing and slope of 4-in-12 or steeper. *(Courtesy Life Tile Corp.)*

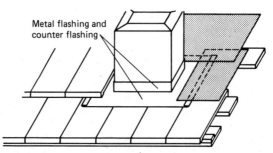

Metal flashing and counter flashing

Fig. 15-28.   Chimney flashing details. *(Courtesy Life Tile Corp.)*

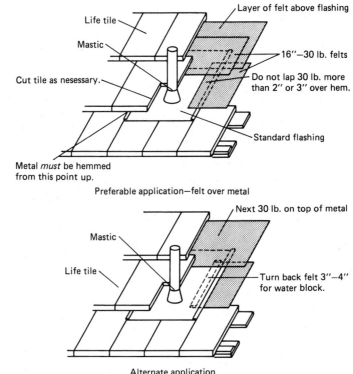

Life tile

Mastic

Cut tile as nesessary.

Metal *must* be hemmed from this point up.

Layer of felt above flashing

16″–30 lb. felts

Do not lap 30 lb. more than 2″ or 3″ over hem.

Standard flashing

Preferable application—felt over metal

Next 30 lb. on top of metal

Mastic

Life tile

Turn back felt 3″–4″ for water block.

Alternate application

Fig. 15-29.   Recommended pipe and vent flashing methods.
(Courtesy Life Tile Corp.)

nailed to the sheathing boards and the bottom edge overlaps the top surface of the underlying course of tiles.

Always lay flat field tiles so that vertical joints in one course break with those in adjoining courses. This can be accomplished by starting every other course with a narrower tile provided by the tile manufacturer for this purpose or cutting one from a standard size field tile.

Chimney, pipe, and vent flashing details for flat tile roofs are illustrated in figures 15-28 and 15-29. The tile should be cut back far enough to allow a free flow of water. Do not use mortar or obstruct the flow of water between the tile and any projections.

Valley details for a flat tile roof are shown in figures 15-30 and 15-31. The valleys should be cut as close to the proper angle as possible. Leave approximately a 4-inch width (2 inches on each side) for an open valley. Butt the tiles for a closed valley.

As shown in figure 15-31, the valley flashing metal must be cut in a V shape at the roof eave so that side hems are to the outside of the fascia or kick strips. Remove lugs when laying tile over the flashing metal. Do not nail through the valley flashing metal. Mastic may be used to mount small tile pieces. Miscellaneous construction details for flat tile roofs are shown in figures 15-32 to 15-35.

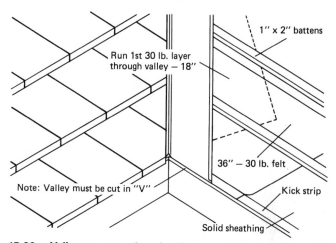

Fig. 15-30. Valley construction details for roof decks with solid sheathing. *(Courtesy Life Tile Corp.)*

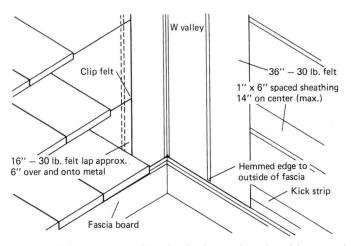

Fig. 15-31. Valley construction details for roof decks with spaced sheathing. *(Courtesy Life Tile Corp.)*

French Tile Roofs

The French tile roof is a variation of the flat tile roof. It is constructed by covering the roof deck with interlocking tiles having fluted surfaces (fig. 15-36). Grooves or channels on the exposed surface of the

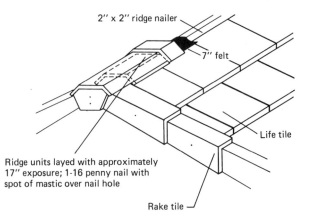

Fig. 15-32. Flat tile roof ridge construction details. *(Courtesy Life Tile Corp.)*

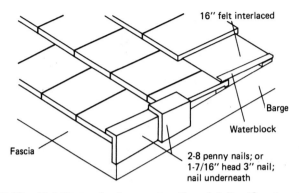

Fig. 15-33. Flat tile roof rake construction details. *(Courtesy Life Tile Corp.)*

tile direct the water runoff away from the joints. Construction details of a typical French tile roof are illustrated in figure 15-37. Note the use of a cant strip along the edge of the eave. The cant strip replaces the starter strip and provides the necessary elevation for the first tile course (fig. 15-38).

Hip cover tiles are nailed to a wood strip or stringer which is attached to the hip ridge (fig. 15-39). The space between the stringer and the rows of tiles is sealed with roofing cement. As shown in figure

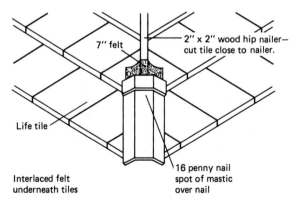

Fig. 15-34. Flat tile roof hip construction details. *(Courtesy Life Tile Corp.)*

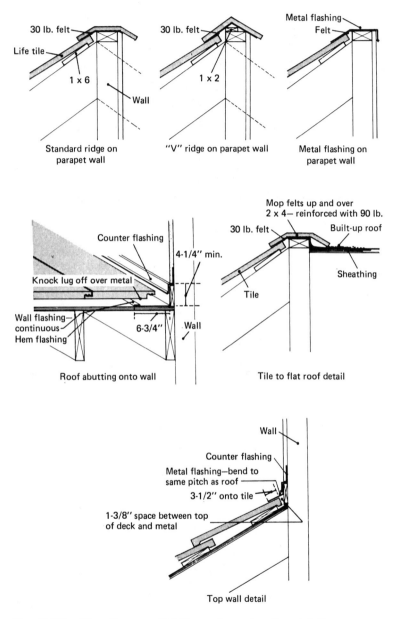

Fig. 15-35. Miscellaneous flat tile roof construction details. *(Courtesy Life Tile Corp.)*

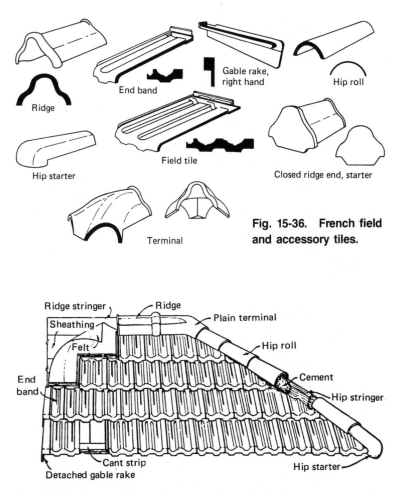

Ridge

End band

Gable rake, right hand

Hip roll

Hip starter

Field tile

Closed ridge end, starter

Terminal

Fig. 15-36. French field and accessory tiles.

Ridge stringer

Ridge

Plain terminal

Sheathing

Hip roll

Felt

End band

Cement

Hip stringer

Cant strip

Detached gable rake

Hip starter

Fig. 15-37. French tile roof construction details. *(Courtesy Ludowicki Tile Co.)*

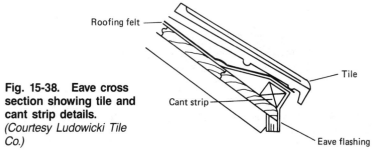

Roofing felt

Tile

Fig. 15-38. Eave cross section showing tile and cant strip details. *(Courtesy Ludowicki Tile Co.)*

Cant strip

Eave flashing

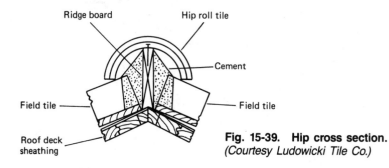

Ridge board

Hip roll tile

Cement

Field tile

Field tile

Roof deck
sheathing

Fig. 15-39. Hip cross section.
(Courtesy Ludowicki Tile Co.)

15-36, the main ridge cover tiles differ in design from those used to cover a hip ridge. Other construction details of French tile roofs are illustrated in figure 15-40.

Spanish Tile Roofs

Spanish tiles are semicircular convex tiles manufactured with a nailing flange along one edge of the field tile pieces (fig. 15-41). The tiles

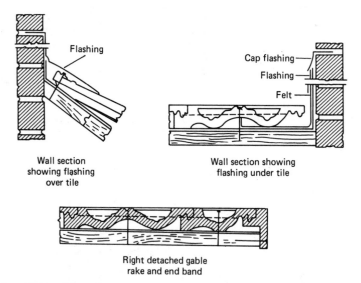

Flashing

Cap flashing

Flashing

Felt

Wall section
showing flashing
over tile

Wall section showing
flashing under tile

Right detached gable
rake and end band

Fig. 15-40. Miscellaneous construction details for French tile roofs.
(Courtesy Ludowicki Tile Co.)

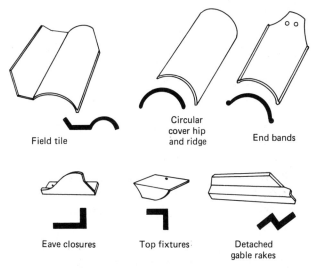

Field tile

Circular
cover hip
and ridge

End bands

Eave closures

Top fixtures

Detached
gable rakes

Fig. 15-41. Spanish field and accessory tiles.

used to cover the hips and ridges resemble mission cover tiles. Construction details of a typical Spanish tile roof are illustrated in figure 15-42.

The type of underlayment used under a Spanish tile roof and the method used to apply it to the roof deck are described in the sec-

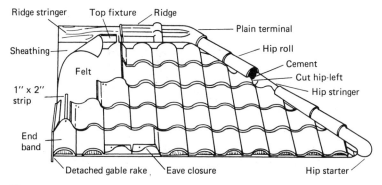

Fig. 15-42. Spanish tile roof construction details. *(Courtesy Ludowicki Tile Co.)*

Fig. 15-43. Nail locations for Spanish tiles.

tion on mission tiles (see the discussion of mission tile roofs in this chapter).

The cover portion of each Spanish field tile is laid to overlap the flange of the tile in the adjoining vertical row to form a continuous S pattern across the roof. Each field tile is nailed directly to the roof deck by nails driven through machine-punched holes in the tile flange (fig. 15-43).

Each vertical row of tiles is closed at the eaves with a specially designed eave closure tile (fig. 15-44). End bands are used to complete tile courses at the rakes where they are fastened to a wood nailing strip that is parallel to the edge of the roof. When installed, the end bands overlap the gable rake tile on one side and the nailing flange of the last field tile in each horizontal course (fig. 15-45). Other construction details are illustrated in figure 15-46. Flashing details common to all types of roofs (including Spanish tile roofs) are described in chapter 8.

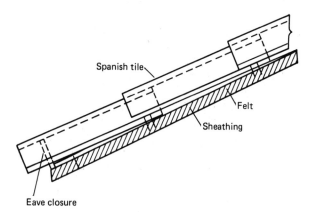

Spanish tile

Felt

Sheathing

Eave closure

Fig. 15-44. Roof section at eave. *(Courtesy Ludowicki Tile Co.)*

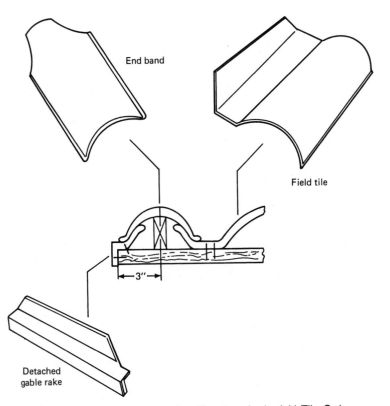

Fig. 15-45. Roof section at rake. *(Courtesy Ludowicki Tile Co.)*

Begin roofing by mounting the eave closure tiles over the drip edge along each roof eave. *Always* lay the eave closure tiles from right to left along the roof eave (*never* left to right). This procedure is necessary with Spanish field tiles because the nailing flange is on the left side of each tile. This flange will be covered by the curved or barrel section of the field tile in the adjacent vertical row. Make certain that the spacing of each closure tile produces the desired width exposure in the vertical rows of field tiles.

Lay the first course of field tiles from right to left along the roof eave. The rake tiles can be applied as each course of field tiles is completed or after all the field tiles on the roof are in place. The joint

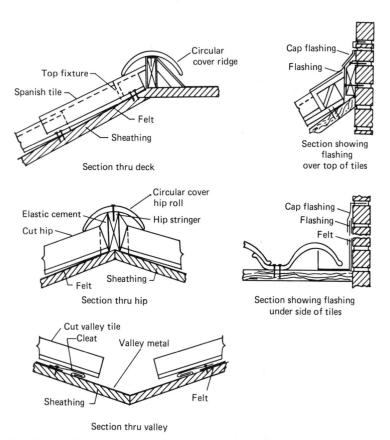

Circular cover ridge

Top fixture

Spanish tile

Felt

Sheathing

Section thru deck

Cap flashing

Flashing

Section showing flashing over top of tiles

Circular cover hip roll

Elastic cement

Cut hip

Hip stringer

Felt

Sheathing

Section thru hip

Cap flashing

Flashing

Felt

Section showing flashing under side of tiles

Cut valley tile

Cleat

Valley metal

Sheathing

Felt

Section thru valley

Fig. 15-46. Miscellaneous construction details for Spanish tile roof. *(Courtesy Ludowicki Tile Co.)*

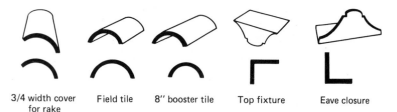

3/4 width cover for rake

Field tile

8" booster tile

Top fixture

Eave closure

Fig. 15-47. Mission field and accessory tiles. *(Courtesy Ludowicki Tile Co.)*

between each field tile and each gable rake tile should be sealed with mortar.

Each successive course of field tiles should overlap the underlying one by 3 inches. Use stringers under the ridge and hip tiles. Fill the space around the stringers with mortar. Mortar should also be used to seal any cracks or breaks where the last course of field tiles meet the ridge and hip tiles.

Mission Tile Roofs

Mission tile roofs are constructed by laying semicircular barrel-shaped tiles in vertical rows across the roof deck (fig. 15-47). The concave surface (or pan) of the tile in one row is overlapped by the convex (or cover) surface in an adjoining row to form a continuous S-shaped pattern across the roof. Construction details of a typical mission tile roof are illustrated in figure 15-48. Because of the type of tile used, this roof is also sometimes called a *straight barrel* tile roof.

The underlayment for roofs with pitches ranging from 3-in-12 to 5-in-12 consists of 30-pound roofing felt laid parallel to the roof eaves. The roofing felt is laid with a 4-inch side lap and 6-inch end lap. It is blind nailed with large-headed roofing nails. After the underlayment is laid, the entire surface should be hot mopped.

Each concave tile surface or pan is nailed directly to the roof

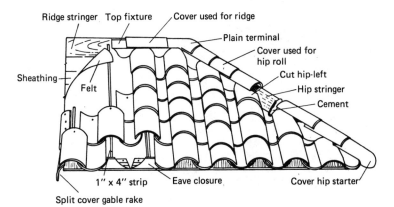

Fig. 15-48. Mission tile roof construction details. *(Courtesy Ludowicki Tile Co.)*

deck with a single nail inserted through a machine-punched hole in the upper end of the tile. The lap of the tiles in either the concave (pan) or convex (cover) rows will depend on the exposure desired. It will usually be at least 3 inches.

Each convex tile surface or cover is nailed to a vertical 1 × 4–inch wood strip that is fastened to the roof deck. The nail is inserted through a machine-punched hole in the upper end of the tile. Each convex tile must be laid to overlap the edges of the concave tile surfaces or pans in the adjoining parallel rows. Overlapping the edges of the concave tiles with the convex tiles provides a more waterproof roof surface.

Another method of securing the cover tile to the roof deck is to wire it to a nail driven into the sheathing (figs. 15-49 to 15-51). This application method creates a slight irregularity in vertical tile alignment, producing an effect very similar to that of Mediterranean tile roof construction. Wiring the cover tiles to the roof deck eliminates the need for the 1 × 4–inch wood nailing strips. The double row of eave tiles can be eliminated if a 1 × 2–inch kick strip or raised fascia board is used along the roof edge.

A special three-quarter width cover is provided by some tile manufacturers for closing the end of the roof at the rake. The end

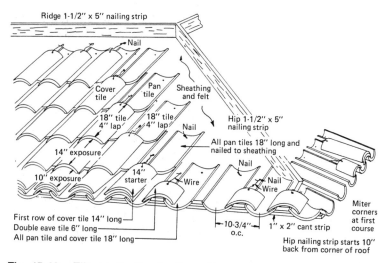

Fig. 15-49. Tile application by the wiring method. *(Courtesy Gladding, McBean & Co.)*

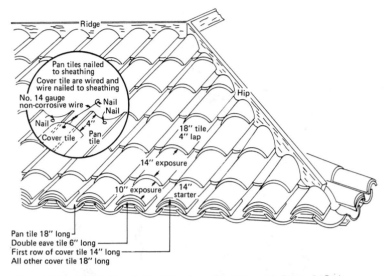

Fig. 15-50. Field tiles laid. *(Courtesy Gladding, McBean & Co.)*

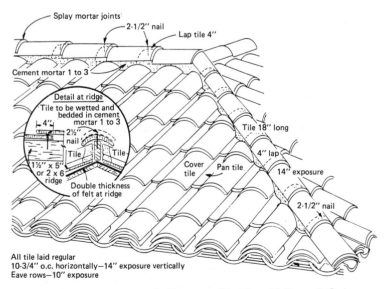

Fig. 15-51. Completed roof. *(Courtesy Gladding, McBean & Co.)*

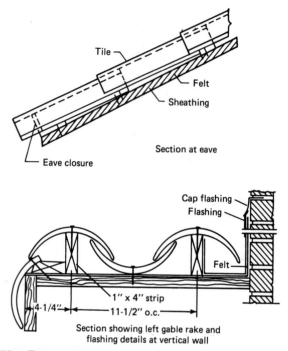

Section at eave

Section showing left gable rake and
flashing details at vertical wall

Fig. 15-52. Eave and rake construction details. *(Courtesy Ludowicki Tile Co.)*

of each vertical row of convex tiles is closed at the eave with a specially designed eave closure tile section that is generally provided by the tile manufacturer. The eave closure tile is nailed directly to the roof deck. Construction details for eave and rakes are illustrated in figure 15-52.

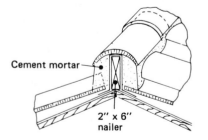

Fig. 15-53. Hip and ridge construction details. *(Courtesy Gladding, McBean & Co.)*

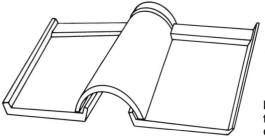

Fig. 15-54. Roman
tile roof pan and
cover combination.

Ridge cover tiles are nailed to a vertical wood strip (or stringer) attached to the roof ridge (fig. 15-53). Roofing cement is applied between the wood strip and each vertical course of tile to form a waterproof seal.

The space surrounding the nailing strips under the gable, hip, and ridge tiles should be filled with cement mortar. Mixing instructions for the cement mortar are described in the section on Spanish tile roofs (see the discussion of Spanish tile roofs).

Mission tile roofs use the same types of flashing as other types of tile roofs. Flashing details are described in this chapter and in chapter 8.

Roman and Greek Tile Roofs

Roman and Greek tile roofs reproduce the distinctive tile roofing patterns used throughout the Mediterranean area since the days of early Greece and Rome. Roman tile roofs use flat pans with curved cover tiles (fig. 15-54). An angular-shaped cover tile is used with the flat pans on a Greek tile roof (fig. 15-55).

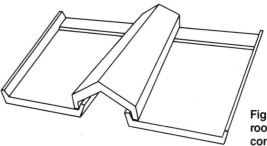

Fig. 15-55. Greek tile
roof pan and cover
combination.

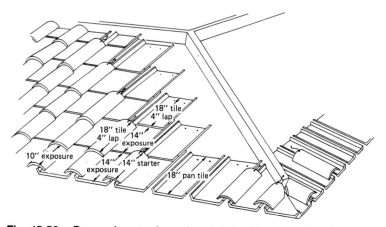

Fig. 15-56. Pan and cover fastening details. *(Courtesy Gladding, McBean & Co.)*

The construction details of a Roman tile roof on which the cover tiles are fastened to the roof with 14-gauge copper wire and nails are illustrated in figures 15-56 to 15-58. The pans (flat tiles with raised edges) are first laid in overlapping courses across the roof with the

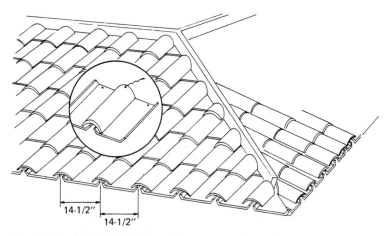

Fig. 15-57. Field tiles laid and roof ready to receive ridge and hip tiles. *(Courtesy Gladding, McBean & Co.)*

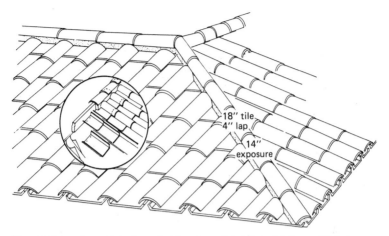

Fig. 15-58. Completed roof. *(Courtesy Gladding, McBean & Co.)*

required spacing between each vertical row. Cover tiles are then used to overlap the vertical spaces and connect adjacent pans. Some tile manufacturers require that the cover tiles used on the roof plane be nailed to vertical nailing strips (fig. 15-59). The nail is driven through a machine-punched hole near the upper end of the cover tile.

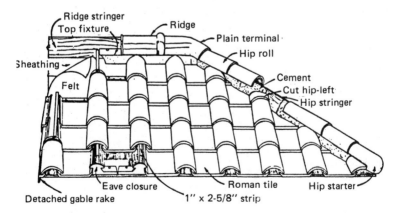

Fig. 15-59. Roof with cover tiles nailed to wood nailing strips. *(Courtesy Ludowicki Tile Co.)*

Regardless of which method is used to attach the cover tile on the roof plane, hip and ridge cover tiles are nailed directly to a wood board or stringer.

Flashing details and the points at which mortar is applied are the same as for Spanish or mission-style tile roofs.

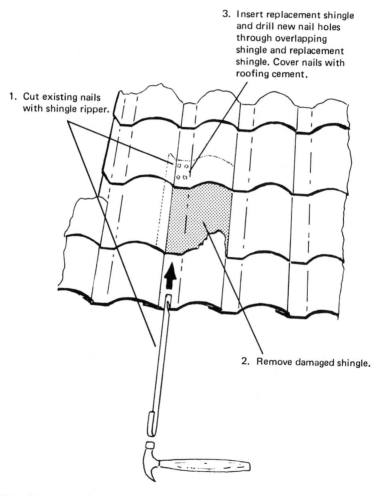

3. Insert replacement shingle and drill new nail holes through overlapping shingle and replacement shingle. Cover nails with roofing cement.

1. Cut existing nails with shingle ripper.

2. Remove damaged shingle.

Fig. 15-60. Replacing damaged tile.

Tile Roof Repair and Maintenance

A properly laid tile roof generally requires little maintenance. Occasionally an individual tile will crack, break, or tear loose, but damage is usually the result of a tree limb striking the roof or someone walking on the roof instead of a defect in the tile. Damaged tiles must be removed and replaced with new ones or a leak may develop.

A cracked tile can be repaired with a number of different synthetic sealers. Ask a local building supply dealer for advice. If the tile is too damaged to repair, remove it and replace it with a new one. Remove the damaged tile by first cutting the nails fastening it to the roof with a shingle ripper (also called a *nail ripper*) or a hacksaw blade. Pull the damaged tile loose and sweep out any dirt or debris. Examine the surface of the roofing felt underlayment under the tile for rips or holes. These should be repaired with a patch and roofing cement before installing the new tile.

Select a new tile of the same size, shape, and color. Slide the replacement tile in place and drill two new nail holes with an electric drill (fig. 15-60). Drill through both the overlapping tile and the underlying replacement tile at a point approximately 1 inch below the old nail holes. Nail the tiles in place with roofing nails and cover the nail holes with roofing cement.

Cracked or replacement flat-shingle tiles are sometimes reinforced with a piece of copper or aluminum flashing long enough to fit under the entire length of the tile and wide enough to extend under adjoining tiles. The flashing should be about 1/4 inch longer than the length of the tile when installed. The extra 1/4 inch is bent up to hold the tile in place (fig. 15-61). Cut off the existing nails with a hacksaw blade before inserting the flashing, drill two new holes, and cover the nail heads with roofing cement.

Reroofing with Tile

Tile has been used in reroofing but it is not a common practice. Reroofing is a practical, low-cost solution to roofing problems when a relatively inexpensive roof needs to be upgraded. As a result, an inexpensive roof covering material is generally recommended for reroofing; more expensive tile is reserved for new construction.

A.

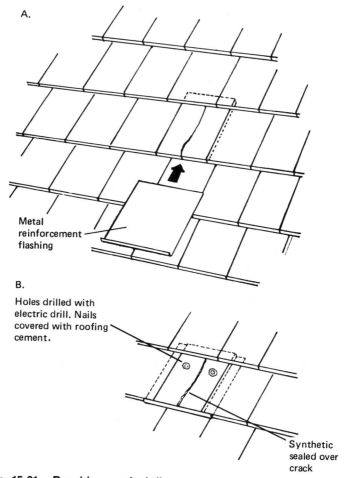

Metal
reinforcement
flashing

B.

Holes drilled with
electric drill. Nails
covered with roofing
cement.

Synthetic
sealed over
crack

Fig. 15-61. Repairing cracked tile.

If the decision is to reroof with tile, be sure to inspect the roof framing to make certain it is capable of supporting the extra weight of a tile roof. If the roof framing does not appear to be strong enough, reinforce it with additional bracing and rafters.

Prepare the existing roof so that it provides a flat, smooth surface for the tile courses. Repair or replace damaged sections of the old roofing, and make certain that the materials are in relatively good condition and do not show signs of rot or advanced stages of deteriora-

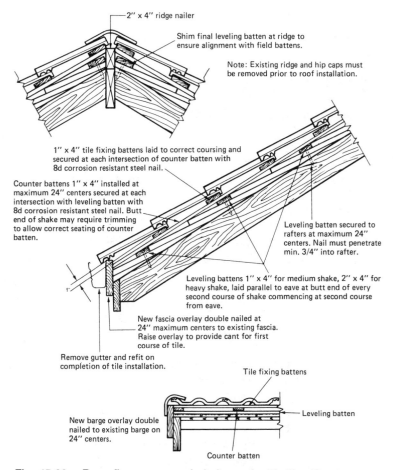

2" x 4" ridge nailer

Shim final leveling batten at ridge to ensure alignment with field battens.

Note: Existing ridge and hip caps must be removed prior to roof installation.

1" x 4" tile fixing battens laid to correct coursing and secured at each intersection of counter batten with 8d corrosion resistant steel nail.

Counter battens 1" x 4" installed at maximum 24" centers secured at each intersection with leveling batten with 8d corrosion resistant steel nail. Butt end of shake may require trimming to allow correct seating of counter batten.

Leveling batten secured to rafters at maximum 24" centers. Nail must penetrate min. 3/4" into rafter.

Leveling battens 1" x 4" for medium shake, 2" x 4" for heavy shake, laid parallel to eave at butt end of every second course of shake commencing at second course from eave.

New fascia overlay double nailed at 24" maximum centers to existing fascia. Raise overlay to provide cant for first course of tile.

Remove gutter and refit on completion of tile installation.

Tile fixing battens

Leveling batten

New barge overlay double nailed to existing barge on 24" centers.

Counter batten

Fig. 15-62. Reroofing over wood shake roof with tile. *(Courtesy Monier Co.)*

tion. Regardless of the condition of the old roofing, however, these materials will have to be removed if any of the sheathing boards are warped and need to be replaced. After the existing roof has been properly prepared, battens and tile courses can be applied as in new construction. Construction details of a wood shake roof reroofed with tile are shown in figure 15-62.

CHAPTER 16

Installing Gutters and Downspouts

- Gutter-and-Downspout Drainage System
- Designing a Roof Drainage System
- Vinyl Gutters and Downspouts
- Aluminum Gutters and Downspouts
- Downspout Drainage Connections
- Repair and Maintenance

The roof drainage system is designed to carry water away from the exterior walls of the structure. This is done to protect the edges of the roof and the walls from water damage. If water is allowed to enter the wall cavities, it will soak the insulation and reduce its effectiveness. Water inside the walls will also cause wood framing studs to rot, create stains on inner wall surfaces, and damage the sheathing and siding materials. Carrying the water away from the walls prevents seepage into the basement. A major cause of moisture in the basement is the collection of water along the foundation walls. One way to prevent this problem is to ensure that water runoff from the roof is carried a suitable distance away from the wall. Finally, an efficient roof drainage system will considerably reduce winter snow and ice buildup along the roof eaves (fig. 16-1).

The gutter-and-downspout drainage system is the most commonly used method of draining water runoff from the roofs of residential structures. This type of drainage system is used with sloping roofs. The water is collected in gutters hung from the edge of the roof or attached to the fascia board or rafter ends. The gutters empty the

Fig. 16-1. Winter snow and ice buildup. *(Courtesy Genova, Inc.)*

water into vertical downspouts that are used to carry the water away from the exterior walls of the structure (fig. 16-2). On flat roofs the water is often drained from one or more openings located on the roof and then conducted through an inside wall conduit to an underground drain pipe. A gutter system is not used with this type of roof.

Gutter-and-Downspout Drainage System

A typical gutter-and-downspout drainage system contains the following components and parts:

1. Gutters
2. Downspouts
3. Elbows
4. Inside and outside miters

Fig. 16-2. Aluminum gutter and downspout. *(Courtesy Alcoa Building Products, Inc.)*

    5. Gutter ends or end caps
    6. Slip joint connectors
    7. Gutter screens
    8. Hangers, hooks, and brackets
    9. Pipe bands, straps, and fasteners

Most of these components and parts are found in the aluminum gutter-and-downspout roof drainage system illustrated in figure 16-3.

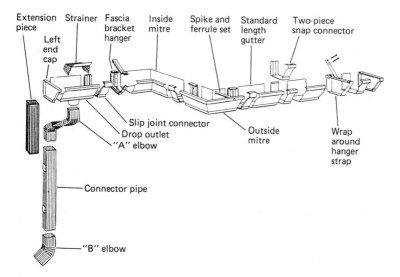

Fig. 16-3. Typical gutter and downspout system. *(Courtesy Howmet Aluminum Corp.)*

The *gutter* (also called an *eave trough* or *trough*) receives the water runoff from the roof and channels it to a downspout. The *downspout* (sometimes called a *conductor pipe* or *leader*) carries the water from the gutter to the ground. Gutters are available in half-round or formed types (fig. 16-4). The downspouts may be either round or rectangular in shape. The downspouts are corrugated to provide greater strength against bursting.

A *conductor elbow* should be attached to the bottom of the downspout to direct the flow of water away from the exterior wall.

 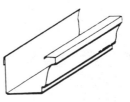

Half-round gutter   Formed gutter

Fig. 16-4. Typical gutter cross sections.

Elbows are also used at each offset gutter and downspout connection. These are the inside and outside miters shown in figure 16-3.

Roof drainage systems are made from a variety of materials, including aluminum, galvanized steel, copper and bronze, vinyl, and wood. The most popular type is made of aluminum. Construction details of a typical aluminum gutter and downspout are shown in figure 16-2. Galvanized steel gutters and downspouts are also widely used, but they do not provide the same protection against rust and corrosion, and maintenance costs are correspondingly higher.

Vinyl gutters and downspouts are designed for easy do-it-yourself installation (fig. 16-5). Only a handsaw and a screwdriver are required. Most components and parts can be easily snapped together and installed by one person. The gutters, downspouts, and accessories are made of solid PVC vinyl that will not chip, peel, rust, or rot. The material is resistant to temperature extremes as well as the scratching and denting common to metal gutters and downspouts.

Copper and bronze gutters are more expensive than either aluminum or galvanized steel systems, but they resist rust and corrosion better. Although copper and bronze gutters do not require a protective coating, they may stain the exterior wall surface after an ex-

Fig. 16-5. Vinyl gutter and downspout. *(Courtesy Plastmo Vinyl Raingutters)*

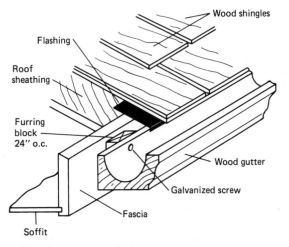

Fig. 16-6. Wood gutter installation.

tended period of time. This problem can be avoided by coating them with paint or a suitable varnish.

Wood roof drainage systems are rarely used. They are usually used to drain wood shingle or shake roofs. Construction details of a typical installation are shown in figure 16-6. The gutters are mounted on the fascia board that runs along the roof edge (eave). They are screwed to the fascia with a 1 × 2–inch furring blocks spaced 24 inches apart (24 inches on center). Galvanized or corrosion resistant screws are used, and they must be long enough to pass through the furring block and into the fascia board. Wood gutters of the best quality are made from the treated heartwood of cypress, redwood, and western red cedar. Square-cut butt joints are used to join wood gutter sections. The joints are fastened with dowels or splines, and sealed with white lead. A moisture-resistant paint or sealer can be applied to all surfaces to minimize water damage.

Designing a Roof Drainage System

A gutter and downspout system should be designed so that its capacity will meet the maximum anticipated rainfall for the area under normal weather conditions. When a gutter system becomes overloaded, it usually overflows between the downspouts or between a downspout and the end of a section of gutter. When a section of guttering has a downspout at one end, there will be a buildup of water in the central portion. This results from the fact that the gutter farthest from the downspout receives water runoff from the section of the roof directly above it, while the central portion of the guttering receives water from the roof above it as well as the water traveling along the gutter toward the downspout. At the downspout end of the gutter, the water discharges into the downspout and is carried away from the structure.

The maximum carrying capacity of a gutter system is commonly dictated by the size, shape, and cross-sectional area of the gutter. The size of the gutters is roughly determined by the size and spacing of the downspouts (fig. 16-7). If the downspouts are spaced 35 feet apart or less, the downspout should be the same area as the gutter. If the downspouts are spaced more than 35 feet apart, the width of the gutter should be increased proportionally. Maximum gutter size is limited, however, by such constraints as material strength, weight, volume of material required, and aesthetic appearance. To circumvent these constraints, architects have attempted to increase guttering capacity by mounting the gutters on a slope or by increasing the number of downspouts in the system. Sloping gutters have the disadvantage of looking unsightly when compared to the lines of the structure, and increasing the number of downspouts involves extra expense. A more recent attempt at increasing guttering capacity has been the redesign of the drop outlets. By using drop outlets that increase the flow of water down the downspouts, the capacity of the gutter system can be significantly increased without increasing the cross-sectional area of the guttering or adding more downspouts. All of these factors

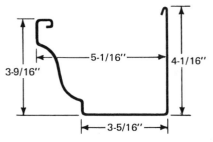

5" O. G. gutter

Fig. 16-7. Gutter dimensions. *(Courtesy Alcoa Building Products, Inc.)*

must be taken into consideration when designing a gutter and downspout system for a structure.

Gutters are sold in standard 10-foot lengths. If the roof length along the eaves measures 32 feet, 40 feet of gutter will have to be ordered with the excess 8 feet cut from one of the 10-foot sections. The order will be based on the total length of the roof that drains into the gutters. For example, a typical gable roof with no minor roofs, which measures 32 feet along the eaves, will require 64 feet of gutters (32 feet for each eave).

The steps to be followed when estimating and ordering the materials for a gutter system may be outlined as follows:

1. Draw a diagram of the existing or planned gutter system.
2. Measure and record on the diagram the length of each gutter run (straight gutter) along the roof eaves.
3. Divide each gutter run by ten to determine the number and location of joints or seams in the run, and the required number of standard 10-foot sections. Note that gutter lengths shorter than 10 feet must be cut from a 10-foot gutter section.
4. Record the number of drop outlets required for the gutter system, the number of downspout pipe straps, and the lengths of the downspouts. Allow two elbows for each drop outlet and one leader and splash block for each downspout.
5. Record the number of inside and outside mitered corners, right- and left-end gutter caps, slip joint connectors (one for each gutter joint), and fasteners required for the gutter system.

Use one slip connector for each cut section of gutter, three hanger straps for each 10-foot length of gutter, and two pipe bands for each 10-foot length of downspout. An elbow directing the water away from the wall should be used at each point a downspout empties onto the ground. The number of inside and outside miter sections will depend on the design of the structure. The same holds true for connections between gutters and downspouts. Most gutter systems require three elbows at each drop outlet (fig. 16-8).

Gutters should be pitched slightly downward toward the downspout. The water will flow better if this is done properly. The downward pitch of the gutters must be gradual and uniform. Look for low spots in a newly installed gutter system and correct them. Low spots will collect water, and the weight of this water may eventually loosen the joints between gutter sections.

Gutter drop outlet

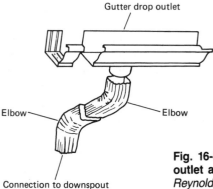

Elbow

Elbow

Connection to downspout

Fig. 16-8. Connection between drop outlet and downspout. *(Courtesy Reynolds Metal Co.)*

In general each downspout should not drain more than a 35-foot length gutter. On long roofs, downspouts should be placed at either end of the structure and at the center. Other downspouts should be added if necessary to meet the 35-foot rule.

The number of downspouts and their cross-sectional area will determine the water carrying capacity of a roof drainage system. For example, a 2 × 3–inch downspout will carry water from approximately 700 square feet of roof area, whereas a 3 × 4–inch downspout will drain approximately 1,200 square feet. These figures are based on *average* rainfall figures. In areas of the country where unusually heavy downpours occur, such as in Florida, additional downspouts should be added to the roof drainage system.

Expansion joints are used in a roof drainage system to relieve the stresses caused by the expansion and contraction of the metal. Unless properly relieved, this movement can loosen the fasteners securing the gutters and downspouts to the structure. In severe cases, it may even distort a metal system. Expansion joints should be used on all hip roof installations, on straight runs over 40 feet in length, and at any point in the roof drainage system subject to unusual stress or restricted movement.

Vinyl Gutters and Downspouts

The principal components and parts of a vinyl gutter and downspout system are illustrated in figure 16-9. Both formed and half-round gut-

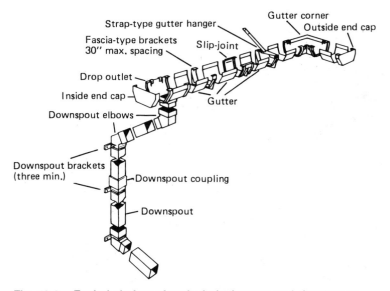

Fig. 16-9. Exploded view of typical vinyl gutter and downspout system. *(Courtesy Genova, Inc.)*

ters are available. As is the case with metal gutters, formed vinyl gutters are used with square or rectangular downspouts (see fig. 16-9), whereas half-round gutters are used with round downspouts (see fig. 16-5). Vinyl gutters and downspouts are not corrugated because the solid PVC plastic is considered strong enough to handle water runoff under most drainage conditions.

Installing Vinyl Gutters and Downspouts

Vinyl gutter and downspout systems can be installed with a minimum number of easily obtainable tools. These tools include a hacksaw with a fine-tooth blade for cutting the gutters and downspouts to the required length, a 10-foot measuring tape, a chalk line, a carpenter's level, a screwdriver, a soft-lead pencil, a drill and bits, a white cloth or paper towel, and a ladder. The gutter system manufacturer will supply the cement and cleaning solvent. If the surfaces are to be painted, they should be cleaned first with the manufacturer's cleaning solvent or acetone.

Fig. 16-10. Gutter brackets mounted on fascia. *(Courtesy Genova, Inc.)*

Vinyl gutters can be mounted on the fascia board with brackets or hung by strap hangers from the roof edge (figs. 16-10 and 16-11). Hanger straps are used on installations without suitable fascia boards or rafter tails for nailing. They must be bent to match the roof angle

Fig. 16-11. Strap hanger installation for vinyl gutters. *(Courtesy Genova, Inc.)*

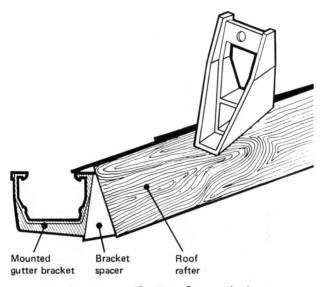

Fig. 16-12. Bracket spacer. *(Courtesy Genova, Inc.)*

and hold the gutter level. Vinyl gutter manufacturers also provide bracket spacers or adapters for mounting the gutters on nonvertical fascias or exposed rafter ends (fig. 16-12).

If there is an old gutter system in place, remove it and inspect the fascia boards and roof eaves for damage. Make any necessary repairs, and then clean and paint the surface.

The first step in installing the gutter system is to determine the location of the downspouts. If the gutters are to be mounted on the fascia boards with brackets, snap a level (or sloped) chalk line for each gutter run. The chalk line should be approximately $1\frac{1}{2}$ inches below the edge of the roof to allow proper water spillage into the gutter. If it is sloped, it must incline gently toward the downspout. Strap-type gutter hangers are aligned by means of a string stretched parallel to the roof eave.

Install the drop outlets, gutter elbows, and slip joints for each gutter run (fig. 16-13). The top edge of each fitting should be flush with the chalk line and screwed to the fascia board. If strap-type hangers are used to support the gutter sections, they should be nailed directly to the roof deck or over the shingle. The drop outlets and gutter elbows are then supported from both sides by gutter brackets that are

Fig. 16-13. Mounting drop outlet to fascia board with corrosion-resistant screws. *(Courtesy Genova, Inc.)*

screwed to the strap hangers. The top outside edge of the fittings should touch the alignment string stretched parallel to the eave.

Fascia-type brackets and strap-type gutter hangers should be evenly spaced 30 inches apart or less between the fittings. The maximum recommended spacing in areas where heavy snows are common is 24 inches. If the roof has insufficient shingle overhang to direct the flow of water from the roof into the gutter, install a metal drip edge under the shingles along the roof eave as in figure 16-14.

Lay the gutter sections in their brackets and snap them down in place (fig. 16-15). *Vinyl expands and contracts when temperatures change.* This expansion and contraction must be allowed for when installing a vinyl gutter system or sections may bend and pull loose from the roof. Short gutter sections can be cut to length with a fine-tooth saw. Gutters, downspouts, and fittings are joined with a silicone-lubricated, sliding rubber seal that allows for expansion and contraction at all joints (fig. 16-16).

Install the downspouts by working down from each drop outlet. Most installations require an offset to extend from the gutter at the roof edge back to the exterior wall of the structure. Brackets are used to fasten the downspouts to the wall. Connect elbows at the bottom of the downspouts to direct the water away from the walls.

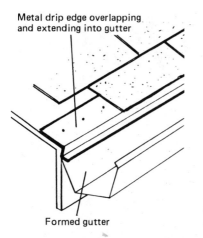

Metal drip edge overlapping and extending into gutter

Formed gutter

Fig. 16-14. Metal drip edge. *(Courtesy GSW Building Products Division)*

Vinyl gutters and downspout parts can be painted or left unpainted. If you wish to paint them, it is best to do so before assembly (fig. 16-17). Use only paints provided by the vinyl gutter manufacturer. These paints are specially formulated to bite into the PVC surface as though molded in. They will also serve as an excellent primer base for other paints. Do *not* paint the insides of the gutters.

Fig. 16-15. Installing gutter sections in gutter brackets. *(Courtesy Genova, Inc.)*

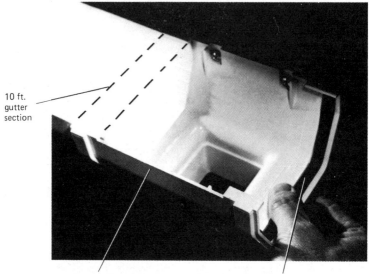

10 ft.
gutter
section

Drop outlet   Rubber sliding seal

Fig. 16-16.   Joint between gutter section and drop outlet. *(Courtesy Genova, Inc.)*

Fig. 16-17.   Painting vinyl parts. *(Courtesy Genova, Inc.)*

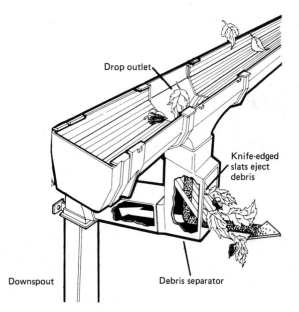

Drop outlet

Knife-edged
slats eject
debris

Downspout

Debris separator

Fig. 16-18. Drop outlet and debris separator. *(Courtesy Genova, Inc.)*

The gutters can be protected from debris with a gutter screen or by using specially designed drop outlets and separators. The Genova Raingo gutter system drop outlet shown in figure 16-18 is designed to eliminate the swirling action of the water at the juncture of the gutter and downspout. The steep slope of the gutter at the drop outlet opening to the downspout smooths out the water flow and creates a strong flushing action. As shown in figure 16-18, the debris separator is installed between the drop outlet and the downspout. It contains knife-edged slats that eject debris while allowing rain water to flow past the slats to the downspout without clogging.

Aluminum Gutters and Downspouts

Aluminum gutters and downspouts are light, durable, and easy to work with. The principal components and parts of an aluminum gutter

Fig. 16-19. Overlapping gutter sections. *(Courtesy Alcoa Building Products, Inc.)*

system are shown in figure 16-3. Most manufacturers provide slip connectors that make soldering unnecessary when joining sections. Gutter sections are also designed to overlap when joined (fig. 16-19). Neoprene expansion joints are used to join long runs, on hip roofs, or wherever movement is restricted (fig. 16-20). The neoprene section expands or contracts as required while at the same time providing a water-tight joint for the gutter sections. Because aluminum is a corrosion-resistant metal and will not rust, painting is not necessary. When paint is used, it is applied to match or contrast with the exterior siding and not for moisture or weather protection.

Some aluminum gutters and downspouts are manufactured with a baked-on enamel surface that is bonded to the metal. These bonded enamel surfaces are available in several different colors.

Both box-type (formed) and half-round aluminum gutters and downspouts are available. A box gutter will carry a smaller volume of water runoff per foot of roof area than the half-round type, but it is a structurally stronger and more rigid design.

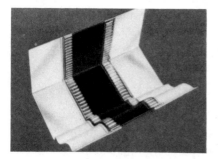

Fig. 16-20. Neoprene expansion joint. *(Courtesy Alcoa Building Products, Inc.)*

An aluminum mastic is used to seal the joints and provide a water-tight connection. To be on the safe side, the brand of aluminum mastic recommended by the gutter and downspout manufacturer should be used.

Aluminum gutters and downspouts are available in 10-foot lengths with either embossed or plain finishes, and they are generally sold at most local building supply outlets. Because aluminum gutters are lightweight, long runs can be assembled on the ground and placed in position without distorting the metal.

The tools used to install an aluminum gutter and downspout system are essentially the same as those used to install a vinyl system. Make sure the hacksaw blade is suitable for cutting aluminum. Aluminum gutters can be cut more easily by inserting a length of wood in the gutter and then cutting through both the wood and the metal. The wood insert prevents the metal from bending or wobbling as it is being cut. *Never* lean a ladder against an aluminum gutter. The metal is too soft to support the weight.

Installing Aluminum Gutters and Downspouts

Gutters and downspouts must be *tightly* assembled or they may leak and allow water to flow against the exterior wall surfaces. This will usually stain or damage the exterior wall and may cause leakage through the wall into the interior of the structure. This potentially expensive water damage can be avoided by correctly installing the gutters and downspouts and then inspecting them on a regular, semiannual basis.

Before assembling the gutters and downspouts, verify the dimensions by arranging the various components and parts of the drainage system along the sides of the structure. Inspect, clean, and make any necessary repairs before assembling the system.

In modern construction, formed gutters are generally assembled and attached to the roof after it has been shingled. The gutters are held in position by brackets that are screwed directly to the fascia or to a furring strip attached to the fascia. When formed gutters are attached to a furring strip, wrap-around hangers should be used for reinforcement. The hangers should be spaced 48 inches apart. Some of the methods used to mount aluminum gutters are shown in figures 16-21 to 16-24.

Downspouts should be fastened to the exterior wall by metal

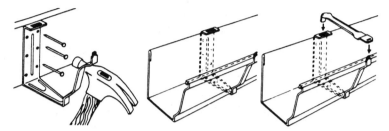

Fig. 16-21. Installing Snap-Lok bracket. *(Courtesy Howmet Aluminum Corp.)*

straps spaced approximately four to six feet apart. Use a strap that allows a space between the wall and downspout when it is fastened to the wall. The downspout must be installed in a vertical position. If the downspout is pulled toward the wall and out of its vertical position by the improper use of the strap, the downspout sections may eventually separate.

Downspout Drainage Connections

A downspout that empties the water directly onto the ground can cause soil erosion that will ruin the appearance of the lawn. In addition to killing any nearby flowers or shrubs, this can lead to the more serious problem of water seepage through the foundation walls.

The water flow from a downspout can be directed away from the structure by connecting the end of the downspout to a drain tile as shown in figure 16-25. The downspout is inserted into the bell end

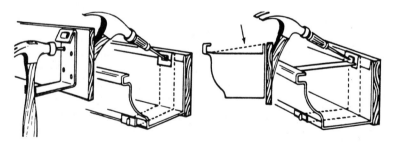

Fig. 16-22. Installing fascia-type gutter bracket. *(Courtesy Howmet Aluminum Corp.)*

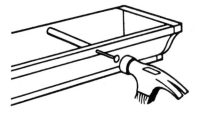

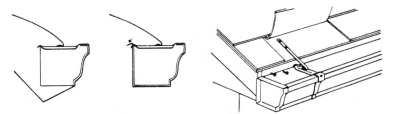

Fig. 16-23. Nailing spike through ferrule to fascia board. (Courtesy Howmet Aluminum Corp.)

Fig. 16-24. Wrap-around hanger strap installed. (Courtesy Howmet Aluminum Corp.)

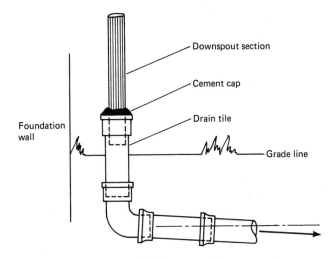

Fig. 16-25. Downspout and drain connection.

of the tile and the joint is sealed with cement. A piece of wire mesh or screen should be placed around the downspout inside the tile before the cement is applied. This will support the cement cap and prevent the cement from seeping down into the tile while it is still wet. The purpose of the cement cap is to plug the opening so that water will not back up and spill out of the tile during heavy water runoff from the roof. Capping the opening also keeps stones, dirt, and other debris out of the drain tile.

An 8-foot run of open-end drain tile installed below grade will carry the runoff a safe distance from the foundation walls. The joints in the open-end drain tiles do not have to be sealed. A large drain tile placed in a vertical position at the lower end of the run can be used to form a dry well. Place the drain tile on its end with the bell opening immediately below the opening of the last tile in the run.

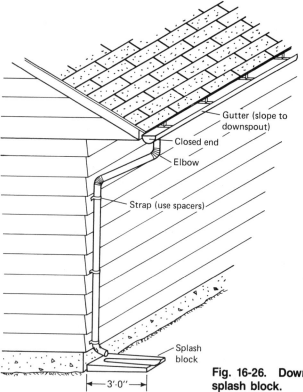

Gutter (slope to downspout)

Closed end

Elbow

Strap (use spacers)

Splash block

3'-0"

Fig. 16-26. Downspout and splash block.

Fill the tile with gravel and small stones. Then, cover the opening of the tile with a cast-concrete cap and fill in the hole with dirt up to grade level.

A splash block provides another means of directing water runoff away from the exterior walls, although it is not as effective as using drain tiles. A typical installation is shown in figure 16-26. The three-foot splash block is made of concrete with a channel that inclines downward and away from the downspout.

Repair and Maintenance

Make it a habit to regularly check all gutters and downspouts at least twice a year. The first inspection should take place in late fall *after* the leaves have fallen. The second inspection should be made in the early spring to check for damage caused by winter weather conditions. Other inspections are recommended after particularly heavy snowfalls or rains.

Clean out the gutters and downspouts in the fall after all the leaves have fallen from the trees. Accumulations of leaves in the gutters may plug the openings to downspouts and prevent adequate drainage.

Flush the gutters and downspouts with a garden hose and check them for leaks (fig. 16-27). Leaks should be repaired immediately because winter weather conditions will only make them worse.

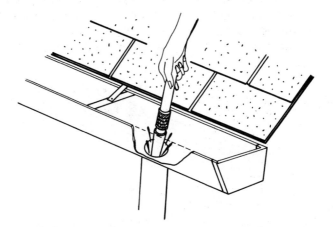

Fig. 16-27. Flushing gutters and downspouts with garden hose.

Replace damaged gutters or downspouts immediately. Check downspout straps for looseness or damage. Retighten the loose straps and replace the damaged ones. The same should be done with gutter hanger straps, brackets, and fasteners.

Check for sagging gutters. Sags in a gutter run must be eliminated or water will collect at these points and not flow toward the downspouts.

Clearing Plugged Downspouts

Leaves and other debris will plug the opening to the downspout unless a gutter screen is used. In cold weather, water will accumulate above the plugged area and freeze. The ice then expands and may cause the downspout to break open along its joint.

A plugged downspout can be cleared by the same kind of auger used to unplug stopped up drains. Once a passage has been opened, the downspout can be flushed with a garden hose. If the downspout is connected to a below-grade run of drain tile, it may be necessary to disconnect the downspout and use a drain auger to free the plug where the drain tile bends below grade.

Screening Gutters and Downspouts

The gutter outlet to the downspout can be protected by covering the opening with a strainer that permits the passage of water but does not allow entry of leaves and other debris that could plug the downspout. Strainers are available in several types and designs at local hardware stores and building supply outlets.

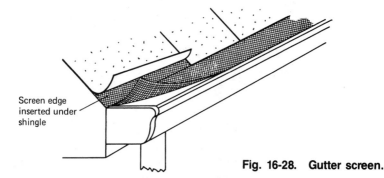

Screen edge inserted under shingle

Fig. 16-28. Gutter screen.

Screening the gutters is another effective method of preventing the accumulation of leaves and other debris. As shown in figure 16-28, one end of the screen should be inserted under the shingles and nailed to the roof deck.

Protection Against Water Damage

Aluminum and copper gutters will not rust. Galvanized sheet metal gutters are also normally rust-resistant as long as the protective galvanized coating remains intact. Check the coating on the gutters and downspouts for scratches or other damage. Damage to the galvanized coating can be caused by careless handling during installation, by the movement of a loose slip connector, gutter hanger, or downspout, or by the pulling apart of soldered joints. Clean the damaged area, dry it thoroughly, and cover it with a protective, corrosion-resistant coating such as roofing cement. These are asphalt-based coatings that are generally available through a local building supply outlet.

The outside surfaces of galvanized sheet metal gutters can be protected from rust and corrosion by coating them with a good corrosion-resistant paint. Always clean the surfaces thoroughly before applying the paint.

Wood gutters can be waterproofed by coating the inside surfaces with an asphalt-based roofing cement. Apply the coating and allow it time to dry thoroughly before installing the gutters.

Repairing Leaks

Leaks can develop from holes in gutters, broken gutter seams, or loose joints in gutter runs. Loose or broken gutter seams or gutter holes 1/4 inch in diameter or smaller can be sealed with plastic asphalt cement, butyl gutter and lap seal, or asphalt-saturated roofing tape. To insure good adhesion, use steel wool or a stiff wire brush to thoroughly clean the area over which the sealant will be applied. Remove loose dirt and other particles from the cleaning operation and apply the sealant. Extend the sealant coating out at least 4 inches in each direction from the hole.

Gutter holes larger than 1/4 inch in diameter must be patched and sealed. Patches may be cut from aluminum, galvanized steel, or canvas. Thoroughly clean the surface around the hole and brush away any dirt or other particles. If the gutters are made of galvanized steel, cover the cleaned surface area with a coat of primer and allow it time to dry. Apply a thick coat of plastic asphalt cement over the surface so that it extends at least 4 inches out from the hole in each direction, center the patch over the hole, and apply a second coat of cement.

Loose gutter joints should be sealed with a moisture-resistant caulking compound applied with a caulking gun. Squeeze the caulking compound into the joint and run a bead around the joint seam for additional protection.

Small holes or loose joints in downspouts can be repaired in the same manner as gutters. Large holes usually require replacement of the downspout because water pressure makes the retention of a patch very difficult.

Installing Heating Cables

Ice dams can form along the eaves and in the gutters in areas of the country where snow accumulates on the roof (fig. 16-29). These ice dams plug the gutters and downspout openings and cause water to back up under the roof covering. This backed-up water eventually leaks

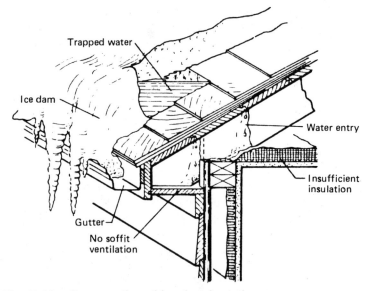

Fig. 16-29. Cross section of ice dam formation.

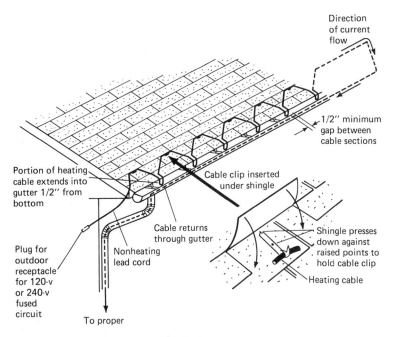

Direction
of current
flow

1/2" minimum
gap between
cable sections

Portion of heating
cable extends into
gutter 1/2" from
bottom

Cable clip inserted
under shingle

Cable returns
through gutter

Plug for
outdoor
receptacle
for 120-v
or 240-v
fused
circuit

Nonheating
lead cord

Shingle presses
down against
raised points to
hold cable clip

Heating cable

To proper

Fig. 16-30. Heating cable installation.

into the wall cavities or through the ceiling where it can cause extensive damage.

The direct cause of ice dam formation is the heat from inside the structure. As the warm air rises and reaches the roof, it melts the snow accumulated there. The water from the melted snow flows down to the roof eaves, where there is considerably less heat rising from the interior, refreezes, and forms an ice dam. Ice dams can also be formed when outside temperatures rise and fall in rapid succession.

Installing electric heating cables along the roof eaves is the easiest and most efficient method of preventing the formation of ice dams. These cables are available for use with either 115- or 230-volt power circuits, and can be purchased in lengths of up to 200 feet.

Electric heating cables are installed in the fall and removed in the winter. Special clips provided by the heating cable manufacturer are inserted under the roof covering. A hook at the exposed end supports the cable.

As shown in figure 16-30, the cable is mounted to form a zigzag pattern on the roof with short loops extending over the eave and into the gutter. When installing a heating cable, do not allow the bottom of the cable loop to touch the bottom of the gutter. Never allow any section of the heating cable to cross or overlap. Always space all sections of cable at least six inches apart.

For added protection, ground the gutter and downspout to a metal pipe. After grounding, extend the cable lead-in to an outdoor or indoor receptacle. The circuit for the heating cable must be properly fused.

APPENDIX A

Professional and Trade Associations

Many professional and trade associations have been organized to develop technical, educational, and promotional literature about roofing products. Useful information can be obtained by contacting these associations.

American Plywood Association (APA)
P.O. Box 11700
Tacoma, WA 98411

Asphalt Roofing Manufacturers Association (ARMA)
6288 Montrose Road
Rockville, MD 20832

California Redwood Association (CRA)
591 Redwood Hwy #3100
Mill Valley, CA 94941

National Roofing Foundation (NRF)
8600 Bryn Mawr Avenue
Chicago, IL 60631

National Tile Roofing Manufacturing Association (NTRMA)
3127 Los Feliz Blvd.
Los Angeles, CA 90039

Red Cedar Shingle & Handsplit Shake Bureau (RCSHSB)
515 116th Avenue, N.E., suite 275
Belleville, WA 98004

Roofing Industry Educational Institute (RIEI)
6851 S. Holly Circle, suite 250
Englewood, CO 80112

Western Red Cedar Lumber Association (WRCLA)
Yeon Building
Portland, OR 97204

Western Wood Products Association (WWPA)
Yeon Building
Portland, OR 97204

APPENDIX B

Roofing Manufacturers

Roofing manufacturers will often send promotional literature about their products, and provide the name, telephone number, and address of the nearest local supplier. The names of many of the major roofing manufacturers are listed in this appendix. Because their names are listed by product category, there is sometimes more than one listing for a manufacturer.

Asphalt Shingles, Roll Roofing, and Related Products

Bird & Son, Inc.
Washington Street
East Walpole, MA 02032

Celotex Corporation
P.O. Box 22602
Tampa, FL 33622

CertainTeed Products Corporation
P.O. Box 860
Valley Forge, PA 19482

Flintkote Company
777A Long Ridge Road
Stamford, CT 06902

GAF Corporation
140 West 51st Street
New York, NY 10020

Johns-Manville Corporation
Ken-Caryl Ranch
Denver, CO 80217

Logan-Long Company
6600 South Central Avenue
Chicago, IL 60638

Owens-Corning Fiberglas Corporation
Home Building Products Division
Fiberglas Tower
Toledo, OH 43659

Slate Roofing

Arvonia Buckingham Slate Company, Inc.
Rte 1, Box 5
Arvonia, VA 23004

Evergreen Slate Company, Inc.
68-T Potter Avenue
Granville, NY 12832

Vermont Structural Slate Company, Inc.
3 Prospect Street, Box 98
Fair Haven, VT 05743

Tile Roofing

California Life Tile
46111-T Industrial Drive
Fremont, CA 94538

Gladding, McBean & Company
P.O. Box 97
Lincoln, CA 95648

Ludowici-Celadon Division
CSC, Inc.
415-T West Golf Road
Arlington Heights, IL 60005

Monier Company
1091 North Batavia
Orange, CA 92666

Wood Shingles and Shakes

Foremost-McKesson Building Products, Inc.
985 South Sixth Street
San Jose, CA 94112

Koppers Company, Inc.
1420 Koppers Building
Pittsburgh, PA 15219

Shakertown Corporation
P.O. Box 400
Winlock, WV 98596

Index

DEL

e fully illustrated, up-to-date guides
nanuals mean a better job done for
cs, engineers, electricians, plumbers,
penters, and all skilled workers.

Contents

Guide to the 1984 Electrical Code®

Roland E. Palmquist
5½ x 8¼ Hardcover 664 pp. 225 illus.
ISBN: 0-672-23398-3 $13.95

sixth
edition

150 illus.

Authoritative guide to the National
Electrical Code® for all electricians,
contractors, inspectors, and homeowners: • terms and regulations
for wiring design and protection
• wiring methods and materials
• equipment for general use • special
occupancies • special equipment
and conditions • and communication
systems. Guide to the 1987 NEC®
will be available in mid-1987.

current
residen-
d illus-
tions
services
• under-
rcuit
uits
s and fit-
obile
ouse

Mathematics for Electricians and Electronics Technicians

Rex Miller
5½ x 8¼ Hardcover 312 pp. 115 illus.
ISBN: 0-8161-1700-4 $14.95

d by

335 illus.

Mathematical concepts, formulas,
and problem solving in electricity and
electronics: • resistors and resistance
• circuits • meters • alternating
current and inductance • alternating
current and capacitance • impedance
and phase angles • resonance in
circuits • special-purpose circuits.
Includes mathematical problems and
solutions.

on the
heir
etism
nd
nagnetic
nt
g calcu-
electric
and

Fractional Horsepower Electric Motors

Rex Miller and Mark Richard Miller
5½ x 8¼ Hardcover 436 pp. 285 illus.
ISBN: 0-672-23410-6 $15.95

Fully illustrated guide to small-to-
moderate-size electric motors in
home appliances and industrial
equipment: • terminology • repair
tools and supplies • small DC
and universal motors • split-phase,
capacitor-start, shaded pole, and
special motors • commutators and
brushes • shafts and bearings
• switches and relays • armatures
• stators • modification and replacement of motors.

Electric Motors

Edwin P. Anderson; revised by Rex Miller
5½ x 8¼ Hardcover 656 pp. 405 illus.
ISBN: 0-672-23376-2 $14.95

Complete guide to installation,
maintenance, and repair of all types
of electric motors: • AC generators
• synchronous motors • squirrel-cage
motors • wound rotor motors • DC
motors • fractional-horsepower
motors • magnetic contractors
• motor testing and maintenance
• motor calculations • meters • wiring
diagrams • armature windings
• DC armature rewinding procedure
• and stator and coil winding.

Home Appliance Servicing fourth edition

Edwin P. Anderson; revised by Rex Miller
5½ x 8¼ Hardcover 640 pp. 345 illus.
ISBN: 0-672-23379-7 $15.95

Step-by-step illustrated instruction
on all types of household appliances:
• irons • toasters • roasters and
broilers • electric coffee makers
• space heaters • water heaters • electric ranges and microwave ovens
• mixers and blenders • fans and blowers • vacuum cleaners and floor
polishers • washers and dryers • dishwashers and garbage disposals
• refrigerators • air conditioners and
dehumidifiers.

Television Service Manual

fifth edition
Robert G. Middleton; revised by Joseph G. Barrile
5½ x 8¼   Hardcover   512 pp.   395 illus.
ISBN: 0-672-23395-9 $15.95

Practical up-to-date guide to all aspects of television transmission and reception, for both black and white and color receivers: • step-by-step maintenance and repair • broadcasting • transmission • receivers • antennas and transmission lines • interference • RF tuners • the video channel • circuits • power supplies • alignment • test equipment.

Electrical Course for Apprentices and Journeymen

second edition
Roland E. Palmquist
5½ x 8¼   Hardcover   478 pp.   290 illus.
ISBN:0-672-23393-2 $14.95

Practical course on operational theory and applications for training and re-training in school or on the job: • electricity and matter • units and definitions • electrical symbols • magnets and magnetic fields • capacitors • resistance • electromagnetism • instruments and measurements • alternating currents • DC generators • circuits • transformers • motors • grounding and ground testing.

Questions and Answers for Electricians Examinations eighth edition

Roland E. Palmquist
5½ x 8¼   Hardcover   320 pp.   110 illus.
ISBN: 0-672-23399-1 $12.95

Based on the current National Electrical Code®, a review of exams for apprentice, journeyman, and master, with explanations of principles underlying each test subject: • Ohm's Law and other formulas • power and power factors • lighting • branch circuits and feeders • transformer principles and connections • wiring • batteries and rectification • voltage generation • motors • ground and ground testing.

Machine Shop and Mechanical Trades

Machinists Library

fourth edition   3 vols
Rex Miller
5½ x 8¼   Hardcover   1,352 pp.   1,120 illus.
ISBN: 0-672-23380-0 $44.85

Indispensable three-volume reference for machinists, tool and die makers, machine operators, metal workers, and those with home workshops.

Volume I, Basic Machine Shop
5½ x 8¼   Hardcover   392 pp.   375 illus.
ISBN: 0-672-23381-9 $14.95

• Blueprint reading • benchwork • layout and measurement • sheet-metal hand tools and machines • cutting tools • drills • reamers • taps • threading dies • milling machine cutters, arbors, collets, and adapters.

Volume II, Machine Shop
5½ x 8¼   Hardcover   528 pp.   445 illus
ISBN: 0-672-23382-7 $14.95

• Power saws • machine tool operations • drilling machines • boring • lathes • automatic screw machine • milling • metal spinning.

Volume III, Toolmakers Handy Book
5½ x 8¼   Hardcover   432 pp.   300 illus.
ISBN: 0-672-23383-5 $14.95

• Layout work • jigs and fixtures • gears and gear cutting • dies and diemaking • toolmaking operations • heat-treating furnaces • induction heating • furnace brazing • cold-treating processes.

Mathematics for Mechanical Technicians and Technologists

John D. Bies
5½ x 8¼   Hardcover   392 pp.   190 illus.
ISBN: 0-02-510620-1 $17.95

Practical sourcebook of concepts, formulas, and problem solving in industrial and mechanical technology: • basic and complex mechanics • strength of materials • fluidics • cams and gears • machine elements • machining operations • management controls • economics in machining • facility and human resources management.

Millwrights and Mechanics Guide

third edition
Carl A. Nelson
5½ x 8¼   Hardcover   1,040 pp.   880 illus.
ISBN: 0-672-23373-8 $24.95

Most comprehensive and authoritative guide available for millwrights and mechanics at all levels of work or supervision: • drawing and sketching • machinery and equipment installation • principles of mechanical power transmission • V-belt drives • flat belts • gears • chain drives • couplings • bearings • structural steel • screw threads • mechanical fasteners • pipe fittings and valves • carpentry • sheet-metal work • blacksmithing • rigging • electricity • welding • pumps • portable power tools • mensuration and mechanical calculations.

Welders Guide third edition

James E. Brumbaugh
5½ x 8 ¼   Hardcover   960 pp.   615 illus.
ISBN: 0-672-23374-6 $23.95

Practical, concise manual on theory, operation, and maintenance of all welding machines: • gas welding equipment, supplies, and process • arc welding equipment, supplies, and process • TIG and MIG welding • submerged-arc and other shielded-arc welding processes • resistance, thermit, and stud welding • solders and soldering • brazing and braze welding • welding plastics • safety and health measures • symbols and definitions • testing and inspecting welds. Terminology and definitions as standardized by American Welding Society.

Welder/Fitters Guide

John P. Stewart
8½ x 11   Paperback   160 pp.   195 illus.
ISBN: 0-672-23325-8 $7.95

Step-by-step instruction for welder/fitters during training or on the job: • basic assembly tools and aids • improving blueprint reading skills • marking and alignment techniques • using basic tools • simple work practices • guide to fabricating weldments • avoiding mistakes • exercises in blueprint reading • clamping devices • introduction to using hydraulic jacks • safety in weld fabrication plants • common welding shop terms.

Sheet Metal Work

John D. Bies
5½ x 8¼   Hardcover   456 pp.   215 illus.
ISBN: 0-8161-1706-3 $19.95

On-the-job sheet metal guide for manufacturing, construction, and home workshops: • mathematics for sheet metal work • principles of drafting • concepts of sheet metal drawing • sheet metal standards, specifications, and materials • safety practices • layout • shear cutting • holes • bending and folding • forming operations • notching and clipping • metal spinning • mechanical fastening • soldering and brazing • welding • surface preparation and finishes • production processes.

Power Plant Engineers Guide

third edition

Frank D. Graham; revised by Charlie Buffington

5½ x 8¼ Hardcover 960 pp. 530 illus.
ISBN: 0-672-23329-0 $27.50

All-inclusive question-and-answer guide to steam and diesel-power engines: • fuels • heat • combustion • types of boilers • shell or fire-tube boiler construction • strength of boiler materials • boiler calculations • boiler fixtures, fittings, and attachments • boiler feed pumps • condensers • cooling ponds and cooling towers • boiler installation, startup, operation, maintenance and repair • oil, gas, and waste-fuel burners • steam turbines • air compressors • plant safety.

Mechanical Trades Pocket Manual

second edition

Carl A. Nelson

4 × 6 Paperback 364 pp. 255 illus.
ISBN: 0-672-23378-9 $10.95

Comprehensive handbook of essentials, pocket-sized to fit in the tool box: • mechanical and isometric drawing • machinery installation and assembly • belts • drives • gears • couplings • screw threads • mechanical fasteners • packing and seals • bearings • portable power tools • welding • rigging • piping • automatic sprinkler systems • carpentry • stair layout • electricity • shop geometry and trigonometry.

Plumbing

Plumbers and Pipe Fitters Library

third edition 3 vols

Charles N. McConnell; revised by Tom Philbin

5½x8¼ Hardcover 952 pp. 560 illus.
ISBN: 0-672-23384-3 $34.95

Comprehensive three-volume set with up-to-date information for master plumbers, journeymen, apprentices, engineers, and those in building trades.

Volume 1, Materials, Tools, Roughing-In
5½ x 8¼ Hardcover 304 pp. 240 illus.
ISBN: 0-672-23385-1 $12.95

• Materials • tools • pipe fitting • pipe joints • blueprints • fixtures • valves and faucets.

Volume 2, Welding, Heating, Air Conditioning
5½ x 8¼ Hardcover 384 pp. 220 illus.
ISBN: 0-672-23386-x $13.95

• Brazing and welding • planning a heating system • steam heating systems • hot water heating systems • boiler fittings • fuel-oil tank installation • gas piping • air conditioning.

Volume 3, Water Supply, Drainage, Calculations
5½ x 8¼ Hardcover 100 pp. 100 illus.
ISBN: 0-672-23387-8 $12.95

• Drainage and venting • sewage disposal • soldering • lead work • mathematics and physics for plumbers and pipe fitters.

Home Plumbing Handbook

third edition

Charles N. McConnell

8½ x 11 Paperback 200 pp. 100 illus.
ISBN: 0-672-23413-0 $13.95

Clear, concise, up-to-date fully illustrated guide to home plumbing installation and repair: • repairing and replacing faucets • repairing toilet tanks • repairing a trip-lever bath drain • dealing with stopped-up drains • working with copper tubing • measuring and cutting pipe • PVC and CPVC pipe and fittings • installing a garbage disposals • replacing dishwashers • repairing and replacing water heaters • installing or resetting toilets • caulking around plumbing fixtures and tile • water conditioning • working with cast-iron soil pipe • septic tanks and disposal fields • private water systems.

The Plumbers Handbook

seventh edition

Joseph P. Almond, Sr.

4 × 6 Paperback 352 pp. 170 illus.
ISBN: 0-672-23419-x $10.95

Comprehensive, handy guide for plumbers, pipe fitters, and apprentices that fits in the tool box or pocket: • plumbing tools • how to read blueprints • heating systems • water supply • fixtures, valves, and fittings • working drawings • roughing and repair • outside sewage lift station • pipes and pipelines • vents, drain lines, and septic systems • lead work • silver brazing and soft soldering • plumbing systems • abbreviations, definitions, symbols, and formulas.

Questions and Answers for Plumbers Examinations

second edition

Jules Oravetz

5½ x 8¼ Paperback 256 pp. 145 illus.
ISBN: 0-8161-1703-9 $9.95

Practical, fully illustrated study guide to licensing exams for apprentice, journeyman, or master plumber: • definitions, specifications, and regulations set by National Bureau of Standards and by various state codes

• basic plumbing installation • drawings and typical plumbing system layout • mathematics • materials and fittings • joints and connections • traps, cleanouts, and backwater valves • fixtures • drainage, vents, and vent piping • water supply and distribution • plastic pipe and fittings • steam and hot water heating.

HVAC

Air Conditioning: Home and Commercial

second edition

Edwin P. Anderson; revised by Rex Miller

5½ x 8¼ Hardcover 528 pp. 180 illus.
ISBN: 0-672-23397-5 $15.95

Complete guide to construction, installation, operation, maintenance, and repair of home, commercial, and industrial air conditioning systems, with troubleshooting charts: • heat leakage • ventilation requirements • room air conditioners • refrigerants • compressors • condensing equipment • evaporators • water-cooling systems • central air conditioning • automobile air conditioning • motors and motor control.

Heating, Ventilating and Air Conditioning Library

second edition 3 vols

James E. Brumbaugh

5½ x 8¼ Hardcover 1,840 pp. 1,275 illus.
ISBN: 0-672-23388-6 $47.95

Authoritative three-volume reference for those who install, operate, maintain, and repair HVAC equipment commercially, industrially, or at home. Each volume fully illustrated with photographs, drawings, tables and charts.

Volume I, Heating Fundamentals, Furnaces, Boilers, Boiler Conversions
5½ x 8¼ Hardcover 656 pp. 405 illus.
ISBN: 0-672-23389-4 $16.95

• Insulation principles • heating calculations • fuels • warm-air, hot water, steam, and electrical heating systems • gas-fired, oil-fired, coal-fired, and electric-fired furnaces • boilers and boiler fittings • boiler and furnace conversion.

Volume II, Oil, Gas and Coal Burners, Controls, Ducts, Piping, Valves
5½ x 8¼ Hardcover 592 pp. 455 illus.
ISBN: 0-672-23390-8 $15.95

• Coal firing methods • thermostats and humidistats • gas and oil controls and other automatic controls •

ducts and duct systems • pipes, pipe fittings, and piping details • valves and valve installation • steam and hot-water line controls.

Volume III, Radiant Heating, Water Heaters, Ventilation, Air Conditioning, Heat Pumps, Air Cleaners

5 1/2 x 8 1/4 Hardcover 592 pp. 415 illus.
ISBN: 0-672-23391-6 $17.95

• Radiators, convectors, and unit heaters • fireplaces, stoves, and chimneys • ventilation principles • fan selection and operation • air conditioning equipment • humidifiers and dehumidifiers • air cleaners and filters.

Oil Burners fourth edition

Edwin M. Field
5 1/2 x 8 1/4 Hardcover 360 pp. 170 illus.
ISBN: 0-672-23394-0 $15.95

Up-to-date sourcebook on the construction, installation, operation, testing, servicing, and repair of all types of oil burners, both industrial and domestic: • general electrical hookup and wiring diagrams of automatic control systems • ignition system • high-voltage transportation • operational sequence of limit controls, thermostats, and various relays • combustion chambers • drafts • chimneys • drive couplings • fans or blowers • burner nozzles • fuel pumps.

Refrigeration: Home and Commercial second edition

Edwin P. Anderson; revised by Rex Miller
5 1/2 x 8 1/4 Hardcover 768 pp. 285 illus.
ISBN: 0-672-23396-7 $17.95

Practical, comprehensive reference for technicians, plant engineers, and homeowners on the installation, operation, servicing, and repair of everything from single refrigeration units to commercial and industrial systems: • refrigerants • compressors • thermoelectric cooling • service equipment and tools • cabinet maintenance and repairs • compressor lubrication systems • brine systems • supermarket and grocery refrigeration • locker plants • fans and blowers • piping • heat leakage • refrigeration-load calculations.

Pneumatics and Hydraulics

Hydraulics for Off-the-Road Equipment second edition

Harry L. Stewart; revised by Tom Philbin
5 1/2 x 8 1/4 Hardcover 256 pp. 175 illus.
ISBN: 0-8161-1701-2 $13.95

Complete reference manual for those who own and operate heavy equipment and for engineers, designers, installation and maintenance technicians, and shop mechanics: • hydraulic pumps, accumulators, and motors • force components • hydraulic control components • filters and filtration, lines and fittings, and fluids • hydrostatic transmissions • maintenance • troubleshooting.

Pneumatics and Hydraulics fourth edition

Harry L. Stewart; revised by Tom Philbin
5 1/2 x 8 1/4 Hardcover 512 pp. 315 illus.
ISBN: 0-672-23412-2 $19.95

Practical guide to the principles and applications of fluid power for engineers, designers, process planners, tool men, shop foremen, and mechanics: • pressure, work and power • general features of machines • hydraulic and pneumatic symbols • pressure boosters • air compressors and accessories • hydraulic power devices • hydraulic fluids • piping • air filters, pressure regulators, and lubricators • flow and pressure controls • pneumatic motors and tools • rotary hydraulic motors and hydraulic transmissions • pneumatic circuits • hydraulic circuits • servo systems.

Pumps fourth edition

Harry L. Stewart; revised by Tom Philbin
5 1/2 x 8 1/4 Hardcover 508 pp. 360 illus.
ISBN: 0-672-23400-9 $15.95

Comprehensive guide for operators, engineers, maintenance workers, inspectors, superintendents, and mechanics on principles and day-to-day operations of pumps: • centrifugal, rotary, reciprocating, and special service pumps • hydraulic accumulators • power transmission • hydraulic power tools • hydraulic cylinders • control valves • hydraulic fluids • fluid lines and fittings.

Carpentry and Construction

Carpenters and Builders Library

fifth edition 4 vols
John E. Ball; revised by Tom Philbin
5 1/2 x 8 1/4 Hardcover 1,224 pp. 1,010 illus.
ISBN: 0-672-23369-x $43.95
Also available in a new boxed set at no extra cost:
ISBN: 0-02-506450-9 $43.95

These profusely illustrated volumes, available in a handsome boxed edition, have set the professional standard for carpenters, joiners, and woodworkers.

Volume 1, Tools, Steel Square, Joinery

5 1/2 x 8 1/4 Hardcover 384 pp. 345 illus.
ISBN: 0-672-23365-9 $10.95

• Woods • nails • screws • bolts • the workbench • tools • using the steel square • joints and joinery • cabinetmaking joints • wood patternmaking • and kitchen cabinet construction.

Volume 2, Builders Math, Plans, Specifications

5 1/2 x 8 1/4 Hardcover 304 pp. 205 illus.
ISBN: 0-672-23366-5 $10.95

• Surveying • strength of timbers • practical drawing • architectural drawing • barn construction • small house construction • and home workshop layout.

Volume 3, Layouts, Foundations, Framing

5 1/2 x 8 1/4 Hardcover 272 pp. 215 illus.
ISBN: 0-672-23367-3 $10.95

• Foundations • concrete forms • concrete block construction • framing, girders and sills • skylights • porches and patios • chimneys, fireplaces, and stoves • insulation • solar energy and paneling.

Volume 4, Millwork, Power Tools, Painting

5 1/2 x 8 1/4 Hardcover 344 pp. 245 illus.
ISBN: 0-672-23368-1 $10.95

• Roofing, miter work • doors • windows, sheathing and siding • stairs • flooring • table saws, band saws, and jigsaws • wood lathes • sanders and combination tools • portable power tools • painting.

Complete Building Construction

second edition
John Phelps; revised by Tom Philbin
5 1/2 x 8 1/4 Hardcover 744 pp. 645 illus.
ISBN: 0-672-23377-0 $19.95

Comprehensive guide to constructing a frame or brick building from the footings to the ridge: • laying out building and excavation lines • making concrete forms and pouring fittings and foundation • making concrete slabs, walks, and driveways • laying concrete block, brick, and tile • building chimneys and fireplaces • framing, siding, and roofing • insulating • finishing the inside • building stairs • installing windows • hanging doors.

Complete Roofing Handbook
James E. Brumbaugh
5¹⁄₂ x 8¹⁄₄   Hardcover   536 pp.   510 illus.
ISBN: 0-02-517850-4 $29.95

Authoritative text and highly detailed drawings and photographs,on all aspects of roofing: • types of roofs • roofing and reroofing • roof and attic insulation and ventilation • skylights and roof openings • dormer construction • roof flashing details • shingles • roll roofing • built-up roofing • roofing with wood shingles and shakes • slate and tile roofing • installing gutters and downspouts • listings of professional and trade associations and roofing manufacturers.

Complete Siding Handbook
James E. Brumbaugh
5¹⁄₂ x 8¹⁄₄   Hardcover   512 pp.   450 illus.
ISBN: 0-02-517880-6 $23.95

Companion to *Complete Roofing Handbook*, with step-by-step instructions and drawings on every aspect of siding: • sidewalls and siding • wall preparation • wood board siding • plywood panel and lap siding • hardboard panel and lap siding • wood shingle and shake siding • aluminum and steel siding • vinyl siding • exterior paints and stains • refinishing of siding, gutter and downspout systems • listings of professional and trade associations and siding manufacturers.

Masons and Builders Library
second edition   2 vols
Louis M. Dezettel; revised by Tom Philbin
5¹⁄₂ x 8¹⁄₄   Hardcover   688 pp.   500 illus.
ISBN: 0-672-23401-7 $27.95

Two-volume set on practical instruction in all aspects of materials and methods of bricklaying and masonry: • brick • mortar • tools • bonding • corners, openings, and arches • chimneys and fireplaces • structural clay tile and glass block • brick walks, floors, and terraces • repair and maintenance • plasterboard and plaster • stone and rock masonry • reading blueprints.

Volume 1, Concrete, Block, Tile, Terrazzo
5¹⁄₂ x 8¹⁄₄   Hardcover   304 pp.   190 illus.
ISBN: 0-672-23402-5 $13.95

Volume 2, Bricklaying, Plastering, Rock Masonry, Clay Tile
5¹⁄₂ x 8¹⁄₄   Hardcover   384 pp.   310 illus.
ISBN: 0-672-23403-3 $12.95

Woodworking

Woodworking and Cabinetmaking
F. Richard Boller
5¹⁄₂ x 8¹⁄₄   Hardcover   360 pp.   455 illus.
ISBN: 0-02-512800-0 $18.95

Compact one-volume guide to the essentials of all aspects of woodworking: • properties of softwoods, hardwoods, plywood, and composition wood • design, function, appearance, and structure • project planning • hand tools • machines • portable electric tools • construction • the home workshop • and the projects themselves — stereo cabinet, speaker cabinets, bookcase, desk, platform bed, kitchen cabinets, bathroom vanity.

Wood Furniture: Finishing, Refinishing, Repairing second edition
James E. Brumbaugh
5¹⁄₂ x 8¹⁄₄   Hardcover   352 pp.   185 illus.
ISBN: 0-672-23409-2 $12.95

Complete, fully illustrated guide to repairing furniture and to finishing and refinishing wood surfaces for professional woodworkers and do-it-yourselfers: • tools and supplies • types of wood • veneering • inlaying • repairing, restoring, and stripping • wood preparation • staining • shellac, varnish, lacquer, paint and enamel, and oil and wax finishes • antiquing • gilding and bronzing • decorating furniture.

Maintenance and Repair

Building Maintenance second edition
Jules Oravetz
5¹⁄₂ x 8¹⁄₄   Hardcover   384 pp.   210 illus.
ISBN: 0-672-23278-2 $9.95

Complete information on professional maintenance procedures used in office, educational, and commercial buildings: • painting and decorating • plumbing and pipe fitting

• concrete and masonry • carpentry • roofing • glazing and caulking • sheet metal • electricity • air conditioning and refrigeration • insect and rodent control • heating • maintenance management • custodial practices.

Gardening, Landscaping and Grounds Maintenance
third edition
Jules Oravetz
5¹⁄₂ x 8¹⁄₄   Hardcover   424 pp.   340 illus.
ISBN: 0-672-23417-3 $15.95

Practical information for those who maintain lawns, gardens, and industrial, municipal, and estate grounds: • flowers, vegetables, berries, and house plants • greenhouses • lawns • hedges and vines • flowering shrubs and trees • shade, fruit and nut trees • evergreens • bird sanctuaries • fences • insect and rodent control • weed and brush control • roads, walks, and pavements • drainage • maintenance equipment • golf course planning and maintenance.

Home Maintenance and Repair: Walls, Ceilings and Floors
Gary D. Branson
8¹⁄₂ x 11   Paperback   80 pp.   80 illus.
ISBN: 0-672-23281-2 $6.95

Do-it-yourselfer's step-by-step guide to interior remodeling with professional results: • general maintenance • wallboard installation and repair • wallboard taping • plaster repair • texture paints • wallpaper techniques • paneling • sound control • ceiling tile • bath tile • energy conservation.

Painting and Decorating
Rex Miller and Glenn E. Baker
5¹⁄₂ x 8¹⁄₄   Hardcover   464 pp.   325 illus.
ISBN: 0-672-23405-x $18.95

Practical guide for painters, decorators, and homeowners to the most up-to-date materials and techniques: • job planning • tools and equipment needed • finishing materials • surface preparation • applying paint and stains · decorating with coverings • repairs and maintenance • color and decorating principles.

Tree Care second edition
John M. Haller
8½ x 11 Paperback 224 pp. 305 illus.
ISBN: 0-02-062870-6 $16.95

New edition of a standard in the field, for growers, nursery owners, foresters, landscapers, and homeowners: • planting • pruning • fertilizing • bracing and cabling • wound repair • grafting • spraying • disease and insect management • coping with environmental damage • removal • structure and physiology • recreational use.

Upholstering
updated
James E. Brumbaugh
5½ x 8¼ Hardcover 400 pp. 380 illus.
ISBN: 0-672-23372-x $14.95

Essentials of upholstering for professional, apprentice, and hobbyist: • furniture styles • tools and equipment stripping • frame construction and repairs • finishing and refinishing wood surfaces • webbing • springs • burlap, stuffing, and muslin • pattern layout • cushions • foam padding • covers • channels and tufts • padded seats and slip seats • fabrics • plastics • furniture care.

Automotive and Engines

Diesel Engine Manual fourth edition
Perry O. Black; revised by William E. Scahill
5½ x 8¼ Hardcover 512 pp. 255 illus.
ISBN: 0-672-23371-1 $15.95

Detailed guide for mechanics, students, and others to all aspects of typical two- and four-cycle engines: • operating principles • fuel oil • diesel injection pumps • basic Mercedes diesels • diesel engine cylinders • lubrication • cooling systems • horsepower • engine-room procedures • diesel engine installation • automotive diesel engine • marine diesel engine • diesel electrical power plant • diesel engine service.

Gas Engine Manual third edition
Edwin P. Anderson; revised by Charles G. Facklam
5½ x 8¼ Hardcover 424 pp. 225 illus.
ISBN: 0-8161-1707-1 $12.95

Indispensable sourcebook for those who operate, maintain, and repair gas engines of all types and sizes: • fundamentals and classifications of engines · engine parts • pistons • crankshafts • valves • lubrication, cooling, fuel, ignition, emission

control and electrical systems • engine tune-up • servicing of pistons and piston rings, cylinder blocks, connecting rods and crankshafts, valves and valve gears, carburetors, and electrical systems.

Small Gasoline Engines
Rex Miller and Mark Richard Miller
5 ½ x 8¼ Hardcover 640 pp. 525 illus.
ISBN: 0-672-23414-9 $16.95

Practical information for those who repair, maintain, and overhaul two- and four-cycle engines – with emphasis on one-cylinder motors – including lawn mowers, edgers, grass sweepers, snowblowers, emergency electrical generators, outboard motors, and other equipment up to ten horsepower: • carburetors, emission controls, and ignition systems • starting systems • hand tools • safety • power generation • engine operations • lubrication systems • power drivers • preventive maintenance • step-by-step overhauling procedures • troubleshooting • testing and inspection • cylinder block servicing.

Truck Guide Library 3 vols
James E. Brumbaugh
5½ x 8¼ Hardcover 2,144 pp. 1,715 illus.
ISBN: 0-672-23392-4 $45.95

Three-volume comprehensive and profusely illustrated reference on truck operation and maintenance.

Volume 1, Engines
5½ x 8¼ Hardcover 416 pp. 290 illus.
ISBN: 0-672-23356-8 $16.95

• Basic components · engine operating principles • troubleshooting • cylinder blocks • connecting rods, pistons, and rings • crankshafts, main bearings, and flywheels • camshafts and valve trains • engine valves.

Volume 2, Engine Auxiliary Systems
5½ x 8¼ Hardcover 704 pp. 520 illus.
ISBN: 0-672-23357-6 $16.95

• Battery and electrical systems • spark plugs • ignition systems, charging and starting systems • lubricating, cooling, and fuel systems • carburetors and governors • diesel systems • exhaust and emission-control systems.

Volume 3, Transmissions, Steering, and Brakes
5½ x 8¼ Hardcover 1,024 pp. 905 illus.
ISBN: 0-672-23406-8 $16.95

• Clutches • manual, auxiliary, and automatic transmissions • frame and suspension systems • differentials and axles, manual and power steering • front-end alignment • hydraulic, power, and air brakes • wheels and tires • trailers.

Drafting

Answers on Blueprint Reading
fourth edition
Roland E. Palmquist; revised by Thomas J. Morrisey
5½ x 8¼ Hardcover 320 pp. 275 illus.
ISBN: 0-8161-1704-7 $12.95

Complete question-and-answer instruction manual on blueprints of machines and tools, electrical systems, and architecture: • drafting scale • drafting instruments • conventional lines and representations • pictorial drawings • geometry of drafting • orthographic and working drawings • surfaces • detail drawing • sketching • map and topographical drawings • graphic symbols • architectural drawings • electrical blueprints • computer-aided design and drafting. Also included is an appendix of measurements • metric conversions • screw threads and tap drill sizes • number and letter sizes of drills with decimal equivalents • double depth of threads • tapers and angles.

Hobbies

Complete Course in Stained Glass
Pepe Mendez
8½ x 11 Paperback 80 pp. 50 illus.
ISBN: 0-672-23287-1 $8.95

Guide to the tools, materials, and techniques of the art of stained glass, with ten fully illustrated lessons: • how to cut glass • cartoon and pattern drawing • assembling and cementing • making lamps using various techniques • electrical components for completing lamps • sources of materials • glossary of terminology and techniques of stained glasswork.

Macmillan Practical Arts Library
Books for and by the Craftsman

World Woods in Color
W.A. Lincoln
7 × 10 Hardcover 300 pages
300 photos
ISBN: 0-02-572350-2 $38.41

Large full-color photographs show the natural grain and features of nearly 300 woods: • commercial and botanical names • physical characteristics, mechanical properties, seasoning, working properties, durability, and uses • the height, diameter, bark, and places of distribution of each tree • indexing of botanical, trade, commercial, local, and family names • a full bibliography of publications on timber study and identification.

The Woodturner's Art: Fundamentals and Projects
Ron Roszkiewicz
8 × 10 Hardcover 256 pages 300 illus.
ISBN: 0-02-605250-4 $28.80

A master woodturner shows how to design and create increasingly difficult projects step-by-step in this book suitable for the beginner and the more advanced student: • spindle and faceplate turning • tools • techniques • classic turnings from various historical periods • more than 30 types of projects including boxes, furniture, vases, and candlesticks • making duplicates • projects using combinations of techniques and more than one kind of wood. Author has also written *The Woodturner's Companion.*

The Woodworker's Bible
Alf Martensson
8 × 10 Paperback 288 pages 900 illus.
ISBN: 0-02-011940-2 $13.95

For the craftsperson familiar with basic carpentry skills, a guide to creating professional-quality furniture, cabinetry, and objects d'art in the home workshop: • techniques and expert advice on fine craftsmanship whether tooled by hand or machine • joint-making • assembling to ensure fit • finishes. Author, who lives in London and runs a workshop called Woodstock, has also written *The Book of Furnituremaking.*

Cabinetmaking and Millwork
John L. Feirer
7⅛ × 9½ Hardcover 992 pages
2,350 illus. (32 pp. in color)
ISBN: 0-02-537350-1 $47.50

The classic on cabinetmaking that covers in detail all of the materials, tools, machines, and processes used in building cabinets and interiors, the production of furniture, and other work of the finish carpenter and millwright: • fixed installations such as paneling, built-ins, and cabinets • movable wood products such as furniture and fixtures • which woods to use, and why and how to use them in the interiors of homes and commercial buildings • metrics and plastics in furniture construction.

Cabinetmaking: The Professional Approach
Alan Peters
8½ × 11 Hardcover 208 pages 175 illus.
(8 pp. color)
ISBN: 0-02-596200-0 $28.00

A unique guide to all aspects of professional furniture making, from an English master craftsman: • the Cotswold School and the birth of the furniture movement • setting up a professional shop • equipment • finance and business efficiency • furniture design • working to commission • batch production, training, and techniques • plans for nine projects.

Carpentry and Building Construction
John L. Feirer and Gilbert R. Hutchings
7½ × 9½ hardcover 1,120 pages
2,000 photos (8 pp. in color)
ISBN: 0-02-537360-9 $50.00

A classic by Feirer on each detail of modern construction: • the various machines, tools, and equipment from which the builder can choose • laying of a foundation • building frames for each part of a building • details of interior and exterior work • painting and finishing • reading plans • chimneys and fireplaces • ventilation • assembling prefabricated houses.